Finite Element Analysis
with Personal Computers

MECHANICAL ENGINEERING

A Series of Textbooks and Reference Books

EDITORS

L. L. FAULKNER

*Columbus Division
Battelle Memorial Institute*

and

*Department of Mechanical Engineering
The Ohio State University
Columbus, Ohio*

S. B. MENKES

*Department of Mechanical Engineering
The City College of the
City University of New York
New York, New York*

Additional Volumes in Preparation

Mechanical Engineering Software

Spring Design with an IBM PC, *by Al Dietrich*

Mechanical Design Failure Analysis: With Failure Analysis System Software for the IBM PC, *by David G. Ullman*

Finite Element Analysis with Personal Computers

Edward R. Champion, Jr.
J. Michael Ensminger

Consultants
Atlanta, Georgia

MARCEL DEKKER, INC.   New York and Basel

Library of Congress Cataloging-in-Publication Data

Champion, Edward R.
  Finite element analysis with personal computers / Edward R.
Champion, Jr., J. Michael Ensminger.
      p.   cm. -- (Mechanical engineering ; 64)
    Bibliography: p.
    Includes index.
    ISBN 0-8247-7981-9
    1. Finite element method--Data processing.   2. Microcomputers--
--Programming. I. Ensminger, J. Michael. II. Title. III. Series:
Mechanical engineering (Marcel Dekker, Inc.) ; 64.
TA347.F5C46 1988
620'.0015'15353--dc19   88-15574
                                                              CIP

MARCEL DEKKER, INC.
270 Madison Avenue, New York, New York   10016

Current printing (last digit):
10 9 8 7 6 5 4 3 2 1

PRINTED IN THE UNITED STATES OF AMERICA

To Ginger, Caroline and Trey
and to the memory of my father,
Edward Ray Champion, Sr. — Thank you!

To Darlene and Stephenie

Preface

This book is intended to be a basic guide for those practicing engineers who have a need for the use of finite element analysis (FEA) in order to solve typical and not so typical problems. No previous exposure to FEA is assumed, but a basic understanding of differential equations is desirable. Moreover, this book is applicable to students in the science and engineering fields who wish to obtain a concise introduction to the subject of applied FEA and how to learn to apply this analysis technique to their particular situation.

In keeping with this theme, the reader will find the text directed toward helping the user solve general classes of engineering problems with FEA on personal computers. In general, the types of problems that may be solved on personal computers are primarily limited by the amount of memory and speed of execution of the system. Work problems that previously involved cumbersome or intractable solution can now be analyzed if the engineer owns or has access to a personal computer and the proper software.

There are advantages and disadvantages in using the personal computer in FEA. Certainly a motivating factor giving rise to the popularity of the PC/FEA combination is the relatively low cost of this type of analysis. However, there are restrictions on problem size and, in addition, long execution times are often encountered. These problems are addressed in the main text of the book. While this book is not an exhaustive treatise on the subject, we have tried to cover the major areas of applications: structural, fluids, heat transfer, and advanced topics. The section on advanced topics covers areas where FEA is being applied to such diverse topics as bioengineering and solid-state physics.

Chapter 1 addresses the history of FEA and why FEA is becoming a necessary tool for the solution of a wide variety of problems encountered in the professional engineer's career. The reader will find sections dealing with how one uses this book relative to the experience level of the user. We recognize that readers of this book have varying levels of experience in FEA. Those wishing a more detailed introduction to the mathematical theory behind applied FEA should consult the references at the end of Chapter 1.

Chapter 2 explains the hardware requirements necessary for analysis and the peripherals recommended for the enhancement of the modeling process. Discussions in this chapter range from the basic PC unit to the varied input and output devices available to the end user. The aim of Chapter 2 is to provide a basis for defining a system that will supply the necessary knowledge to "build" an efficient system of hardware and software for FEA.

Chapter 3 highlights the fundamentals of starting an analysis using the finite element technique. We discuss how to select the proper mesh size for a particular element and the mesh size/solution accuracy question, how to select the proper element, what basic elements are available in commercial packages, and what kinds of problems can be solved. We also discuss commonly used elements for various classes of problems. In addition, we include sections that discuss, in general, the steps in applying a typical commercially available FEA program to a problem.

In Chapter 4 a detailed presentation for a classical structural problem is discussed. The problem solution is presented from the classical standpoint and a detailed FEA evaluation of the problem is also given. This chapter provides an insight in tying together the theoretical aspects and practical aspects of FEA. In most cases, we will provide both the theoretical and FEA solutions for all applied problems.

Chapters 5, 6, 7, and 8 address specific topics of interest. Chapter 5 is concerned with the application of FEA to structural problems. We include discussions on the types of problems generally encountered and the type of FEA software that is applicable. For the various types of problems encountered, we discuss the types of elements normally used, mesh size, the setting up of a problem, the degrees of freedom that can be expected in light of problem size limitations, and an interpretation of the results. Sample structural problems are given that highlight the discussion areas addressed above. We try to impart to the reader the important parameters that should be addressed when considering the use or purchase of a particular modeling package. The last section describes commercially available FEA programs for structural analysis. We directly compare models available to us and note this comparison in the concluding sections of each chapter. Given a sample problem, the reader will find tables comparing features such as the number of elements available, problem size limitations, machine requirements, machines the programs will run on, advantages and disadvantages and hints on using a particular model.

Chapters 6 and 7 address the thermal/fluid areas of FEA. These areas of application are not as fully developed as FEA structural models. However, some models do include thermal/fluid elements. The format of both chapters follows Chapter 5. Again one of the strong areas of each chapter is the review of the commercially available programs and tables comparing each program feature by feature.

Chapter 8 covers the concluding remarks, suggestions for possible advanced applications, and reviews of the models used in the book. These are topics that have their roots in Chapters 5, 6, and 7 and are now being explored through the use of FEA. This chapter is intended to stimulate the user to apply existing FEA models to new and diverse problems.

Finally, Appendix A lists the necessary information from each FEA vendor with respect to programs and prices. Appendix B covers separate pre- and postprocessor programs available.

In developing and writing this book, many people assisted us in obtaining the finite element models discussed herein. We wish to thank the following individuals and the staff at each organization for their particular help and discussions:

Mr. Peter Brooks (Brooks Scientific, Inc.)—PCTRAN Plus
Mr. Ken Tashiro (Celestial Software, Inc.)—IMAGES 3D
Mr. Perry Grant (MacNeal-Schwendler Corp.)—MSC/Pal2
Dr. M. Lashkari (Structural Research and Analysis Corp.)—COSMOS/M
Ms. Monica Kozar (Swanson Analysis Systems, Inc.)—ANSYS-PC/LINEAR
Mr. Michael Brodgly (Intercept Software, Inc.)—LIBRA
Dr. James T. Gordon (Number Cruncher Microsystems, Inc.)—SAP86
Mr. Ellis Morgan (Structural Analysis Corp.)—MICROTAB
Mr. Charles Paulsen (Algor Interactive Engineering, Inc.)—SUPERSAP

We also would like to thank the publisher Marcel Dekker, Inc., and in particular, Dr. Eileen Gardiner, Ms. Elizabeth Fox, and Mr. John Bottomley. Finally, but no less importantly, we are grateful to Mrs. Julie Goethe for reviewing the manuscript, Mr. Jimmy Ward for assisting in the drawings, and Dr. Douglas Boylan. Finally, we thank our families for their patience and encouragement.

Edward R. Champion, Jr.
J. Michael Ensminger

Contents

Contents

Finite Element Analysis
with Personal Computers

1
Introduction

1.1 A BRIEF HISTORY OF FINITE ELEMENT ANALYSIS (FEA)

What is the finite element method? The finite element method is a numerical analysis procedure to obtain approximate solutions to problems posed in every field of engineering. Often problems encountered in everyday practical applications do not lend themselves to closed-form solutions. This is readily apparent if, for example, one wishes to analyze the flow of a fluid through a duct of varying cross section subjected to nonuniform boundary conditions. We can write the governing equations with the appropriate initial and boundary conditions. However, it is apparent that no simple analytic solution exists. This is found not only in the fluid mechanics area, but also in the area of structural mechanics. From the inability to solve many complex structural problems arose the foundations of the finite element method as we know it today.

The label "finite element method" appeared in 1960 in a paper by Clough (1) concerning plane elasticity problems. However, the basis for the finite element method goes back further to the early 1940s in the applied mathematics literature. We will not cover that period of development for there are sufficient reviews available concerning this area. During the mid- to late 1950s, the mathematical literature on the finite element method increased significantly. Books and monographs (2-4) were written to explain the mathematical foundations of the method during the early 1970s. Not only were the mathematicians exploring the finite element method basics, but also the physicists were developing ideas and concepts along similar lines. One such example is the development of the hypercircle

1

method, a concept in function space, that allows a geometric interpretation of minimum principles when associated with the classical theory of elasticity.

During the mid- to late 1950s and early 1960s a series of papers were published covering linear structural analysis and methods to effect efficient solutions to these problems. Turner, Clough, Martin, and Topp (5) derived solutions to plane stress problems via triangular elements with properties determined from elasticity theory. With the development of the digital computer, these solution techniques allowed the analysis of more complex problems.

Concepts of the method began to be drawn into focus by the publishing of many papers and articles (6-10). The method was recognized as a form of the Ritz method, and the method was recognized as a technique to solve elastic continuum problems. In 1965, Zienkiewicz and Cheung (11) illustrated the applicability of the finite element method to any field problem that could be formulated by variational means. Most of the applications of the finite element method during the early- to mid 1970s were in the structural analysis area. This is the reason why most FEA programs today are heavily oriented toward the structural analysis area.

As the structural analysis are of FEA was refined, other areas of applications of FEA were nurtured during the late 1970s and early 1980s (6-12). Attention was and is being given to the application of FEA to the thermal/fluids areas. For example, many papers in the mid- to late 1970s applied the finite element method to the Navier-Stokes equations. Papers by Girault and Raviart (12) and Teman (13) have lead to further development of this method in fluids applications. The reader should refer to the works of Fortin (14), Griffith (15), Thomasset (16), Heywood and Rannacher (17), and Cullen (18) for basic understanding of the fluids modeling effort. In addition, Chung (19) gives an excellent text on FEA in fluid dynamics. If the user wishes a fundamental text concerning FEA in general, we recommend either the text by Gallagher (20) or that by Segerlind (21). Several mainframe programs are available to solve problems in these areas as well as in the structural areas.

1.2 APPLICATIONS OF FEA FOR PRACTICING ENGINEERS

Every area of engineering can use the power of the finite element method of analysis. From the history of the analytic technique, the most obvious application is structural analysis. The fields of civil and aerospace engineering rely heavily on FEA methods to analyze various types of structures ranging from buildings to spacecraft design. The analysis involves the determination of static deflections and stresses and the determination of natural frequencies and modes of vibration. In addition to these types of problems, the ability exists to analyze the stability of structures.

Other FEA applications lie in the thermal/fluids areas. Many models contain thermal and fluid elements which will allow the determination of temperature distributions in structures and such items as velocity, pressure, and concentration distributions in fluid flows.

There are three basic types of problems, from a mathematical standpoint, that the practicing engineer faces. These are time independent (equilibrium) problems, eigenvalue problems, and time dependent (propagation) problems.

When engineers are faced with problems that do not lend themselves to easily tractable solutions, they have to make judgments as to the best solution approach in light of their background and familiarity with the tools available. The available tools include, but are not limited to, computer resources and programs to utilize the computer resources.

Unfortunately, not every engineer has at his disposal a large mainframe computer, the finite element code, and the time to solve whatever problems arise. Consequently, the engineer is forced to make approximations based on similarity to classical, closed-form soluble problems or the engineer may rely on physical prototyping. Physical prototyping is generally expensive and increases the cost of the design process. These approximations may or may not result in an adequate design. An overdesigned (or overengineered) product may be noncompetitive from a cost standpoint and/or may actually perform less well than intended. However, this design process is often tempered by experience.

With the proliferation of personal computers and with the availability of software to aid the design process, it is no longer necessary to rely heavily on intuition and settle for approximate solutions to real-world problems. FEA programs are available to aid the design engineer in developing a product to withstand its environment.

1.3 HOW TO GET THE MOST FROM THIS BOOK

If the engineer is not as experienced as he wishes to be, then this book can aid in the development of a better foundation of understanding in the basics of analysis. The application of FEA to a wide variety of problems that may be encountered in day-to-day engineering activities can be readily handled.

The following section addresses how one uses this book as an aid and guide to FEA with personal computers. Since the basic objective of this text is to familiarize the novice as well as the experienced user with what tools are available and necessary to analyze and solve problems with finite element methods, we suggest the following two approaches in the use of this book.

1.3.1 The Novice or Casual User

If you are a part of the novice category, the first question to be answered is "Do I have an application or problem that requires the use of finite element techniques?". We recommend that you review Chapter 3. Section 3.1 addresses the criteria for selecting the finite element approach to solving a problem. Perhaps there may be alternative solution techniques you may have overlooked. However, if you are convinced FEA is the direction you wish to pursue, then continue in Chapter 3 and develop a familiarity with the basic concepts and terms of FEA.

The next step is to evaluate the equipment requirements in order to perform an analysis. We have attempted in Chapter 2 to cover the minimum system requirements for most programs. In the sections of Chapter 2, you will find discussions of many kinds of peripheral equipment available. This equipment and combinations of equipment enhance the actual modeling process for a given problem. In most cases, the right combination of equipment makes the analysis an enjoyable task rather than a tedious task.

Following the selection of the equipment, you must define the type of problem to be analyzed. We suggest that you review Chapters 5, 6, 7, or 8 so that you can familiarize yourself with the ins and outs of setting up and running a particular type of problem. For a more in-depth look at FEA, read Chapter 4. In Chapter 4 we set up and run a simple finite element problem. This example lends some insight into the operation of commercially available programs.

The next step in selecting a program is often a difficult choice. In Chapter 8 we have listed several programs that we are familiar with and have run on various computers. The programs listed in no way carry our or the publisher's stamp of approval. However, we feel that they do represent the present state-of-the-art programs available for finite element analysis with personal computers.

In summary, you must consider the extent of your modeling efforts, both short term and long term. It is best to purchase a system that will provide a comfortable environment when performing an analysis. In addition, the FEA program should fit your needs. Some programs allow purchase of modules depending on the type of analysis and depth of analysis you need. Again, one purpose of this book is to assist in your selection of the correct hardware/software combination for your particular situation.

1.3.2 The Experienced User

If you are familiar with computers and the use of computers in the analysis of scientific and engineering problems, this book provides an introduction to FEA or a refresher to the subject. In addition, we have attempted to provide up-to-date information on the types of new products available to aid in your analysis. Therefore, a review of Chapter 2 would be in order.

Chapter 3 may be bypassed if you regularly use FEA. If not, this chapter provides an excellent refresher on the analysis technique. Chapter 4 addresses in depth a sample finite element application.

Chapters 5, 6, 7, and 8 provide useful information in specific areas and areas that show how to extend the application to the fields of medicine to solid-state physics.

1.4 SUMMARY

The programs used in and referenced in this book have been developed and verified. The user must learn how to apply a particular program or model to his analysis problem. Note that it is not necessary to learn all of the capabilities of a particular program or model. The goal is to learn how to apply that part of a program that will, in an efficient and simple manner, solve the problems at hand.

Since finite element analysis methods are widely used and have proven to be a necessary analytical tool in many areas of science and engineering, the information obtained from this book will be a stepping stone to further development of the user, not as just a user, but perhaps as a program developer.

With many finite element programs available for the personal computer, it is difficult to select the "right" program. The user could review the program demos to help him select the correct program for him. Manuals would also be helpful. However, the demo programs will feature a very professional graphics presentation and they do not let the user interface with the program to a great degree. In evaluating a program, the user must examine the following items:

Program Accuracy

In order to validate the program accuracy, the program developers must include numerous test problems that have a classical or analytical solution. The accuracy of each type of element and capability must be checked out as well as the application of the program to both small and large problems.

Problem Size

Many advertisements for programs promise the solution of thousands of static and dynamic degrees of freedom. The user should verify that the program(s) will actually handle a large problem. Requests should be made of the program developers to provide the documentation.

Capabilities

The user should review current requirements and future requirements as to the type of problems he may encounter. If large problems are to be solved, the program should contain solution options such as guyan reduction, substructuring, or cyclic symmetry. Perhaps buckling capabilities or composite material analysis should be included.

State-of-the-Art Techniques

As with any technique, advances are being and will be made. The user should determine if the code contains the latest in element formulation, numerical techniques, and the development of new finite element techniques.

Mesh Generation

A time-consuming task is the problem formulation and mesh generation. All programs reviewed in this book contain some form of mesh generation. This may work well for simple rectangular plates, cylinders, or conical shells. In order to check the mesh generation capabilities try generating meshes for intersecting pipes or plates with elliptical holes.

Program Execution

It is imperative that the user check the execution times of benchmark programs. Speed of execution is of importance, especially in the execution of large problems.

Graphics

In order to evaluate correctly and in a timely manner the program must be capable of displaying the undeformed and deformed structure, stress contours, and animation of the results of the dynamic analysis.

Manuals

In order to use the program for a long period of time, the program should come with thorough and complete documentation. This includes program use, verification problems, and a theoretical background for the program.

Support

This is a most critical item, especially for the novice user. The user must determine whether the organization has the background to support the program.

Mainframe Capability

For some large problems the PC-based finite element program must be capable of interfacing to some of the more well-known mainframe programs such as NASTRAN or ANSYS.

Cost

The cost of the programs vary and this cost does not seem to depend on the program capabilities or the efficiency of the program. Some programs have limited capability and are expensive. Other programs contain all the capabilities the user may need and the cost of these programs is quite reasonable.

2

The PC: Hardware Requirements for FEA

With the advent of the personal computer (PC), scientists and engineers have had available an invaluable resource with which to perform a wide variety of computing tasks. Tasks ranging from simple word processing, spread sheets, and report generation to sophisticated data analysis, machine control, and computer aided engineering (CAE). Convenience is the prime motivation behind the ever-increasing numbers of PCs. In general the PC can address a wide variety of computing needs at a fraction of the cost of its larger cousins, the much larger mainframes. While it is still true the PC may never compete with the mainframe, newer and faster PCs are reaching the market with steady resolve.

This chapter deals with the general hardware requirements needed to address finite element analysis (FEA) with the personal computer. The chapter begins with two important performance items one may address when determining hardware requirements for a PC which will be used to perform FEA, the memory and disk subsystems. Next, we shall discuss devices and methods with which to input and output data for subsequent analysis. In addition, we shall look at memory size, input/output devices, and PC performance as a function of processor speed and the particular microprocessor chip used.

2.1 GENERAL HARDWARE REQUIREMENTS

The basic hardware requirements for FEA are not particularly special. Almost any "off-the-shelf" PCs with not much more than the main chassis assembly and a keyboard can provide a basis for reliable FEA. But, to derive the maximum benefit

7

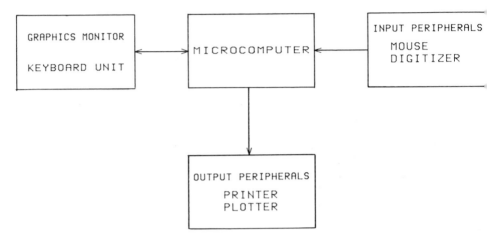

Figure 2.1   Typical PC-based FEA hardware configuration.

from a PC-based FEA system, a minimum hardware configuration is usually specified by the FEA software vendor. The following paragraphs in this section address some of the options an FEA program may require in order to perform its function efficiently. Figure 2.1 presents a pictorial example of a typical PC-based FEA hardware configuration.

2.1.1 The Main PC Unit

The most popular PC (1987) is the ubiquitous **IBM PC** and its compatibles. The IBM PC product line consists of the PC, PC/XT, PC/AT, and more recently the IBM PS/2 series. Each of these computers is able to execute the same operating system, PC DOS, but each offers additional performance over the previous unit. The original IBM PC offered basic computing capability utilizing the 8088 CPU chip. It had the ability to utilize the 8087 math co-processor, but lacked a high capacity hard disk drive. The next unit, IBM's PC/XT added the hard disk and provided additional capacity for optional plug in cards. The PC/AT was a much needed architectural change for the PC line. This computer offered additional speed, increased memory capacity, and increased hard disk storage. Moreover, the PC/AT uses the 80286 CPU instead of the 8088 thereby providing a true 16 bit data path. The 16 bit data path increases the system throughput which increases performance.

IBM's success spawned a large and aggressive compatible market. In order to compete, the PC compatibles generally offer enhanced features such as higher clock speeds and larger standard memory sizes. When selecting a PC compatible

Table 2.1 List of Compatibles

Name	Model	Compatibility
Compac	Desk Pro	XT
Compac	Desk pro 286	AT
Sanyo	Model 880	XT
Sanyo	Model 555	PC
Zenith	150	XT
PC's Limited	286	AT

one should verify that it will indeed provide the necessary base for your FEA analysis. One should make sure that the graphics system is compatible. This feature is probably the most important when choosing a compatible machine. In general though, PC compatibility is no longer the issue it used to be. Table 2.1 lists several of the most popular PC compatible machines.

2.1.2 Memory

The amount of random access memory (RAM) available will, in general, determine the overall performance of an FEA calculation. This is based on the fact that a disk access takes longer than an access to RAM. Therefore many manufacturers of FEA programs design their programs to perform all calculations within the main memory of the PC. This feature allows the PC to calculate at its maximum clock speed. Unfortunately most FEA programs themselves require a large portion of the available memory in a typical PC system. This leaves little room for the data. Therefore one can expect to perform only problems of limited size using a PC with only 640K of main memory.

One technique to enhance the performance using additional RAM is to use extended-memory cards. By creating a special partition in this extended memory one may simulate a disk in RAM. The simulation of a disk in RAM enhances the access time by four to five orders of magnitude. Clearly this is a significant increase in system performance. Nearly all manufacturers of these extension memories provide the software required to perform this function. In addition this provides the ability to process larger FEA problems. Still, one may be faced with an FEA problem which requires more memory resource than is available on a PC. The solution to this dilemma is addressed in the following section.

2.1.3 Disks

The amount of RAM is a key to the timely execution of an FEA model. But the FEA models themselves are quite large and usually have many support programs.

Given the restriction of 360K bytes of data per floppy disk drive, one quickly concludes that a hard disk is an important system asset. This is true even in light of the fact that producers of FEA models will generally provide a means for using two floppy disk drives, but at the expense of convenience and performance.

There are two key attributes one should examine when selecting a disk drive. These are the average seek time of the disk and the disk data transfer rate to main memory. The data transfer rate can be similar in both floppy and hard disk drives, but hard disk systems are typically faster. The transfer rate is anywhere from 250K bytes per second to 2M bytes per second. Seek time is probably the most important attribute. This is where hard disks have a definite advantage. Typical floppy disk access times are about 300 ms compared to hard disk access times of 100 ms or less. This is a three to one increase in performance.

2.2 DATA INPUT DEVICES

Our discussions of the various input and output devices will address mostly those aspects which pertain to the input and output of graphical data. Textual and numeric data would generally be input or output for the purpose of transferring data to an auxiliary program. As an example, one may wish to send a numeric data file to a mainframe computer after using a PC to build the model. (The data was probably generated using a graphics input device.) The major thrust behind using any imput device is to provide a fast way of entering data. To this end it is clear that those devices which provide a graphical (mouse, light pen, or digitizer) rather than alphanumeric (keyboard) input are more desirable.

2.2.1 Keyboards

From the standpoint of graphical data input, keyboards are probably the most cumbersome method of entering graphical data. This fact is obvious since anyone who has ever typed reams of data on a keypunch machine or at a CRT knows how tiresome this is. Still, the keyboard as a method of graphical data entry has a place even with the existence of simpler methods. As an example, making small changes to a graphics data file may be easier via keyboard entry rather than executing a special program to perform the same task. One of the keyboard's drawbacks is the speed with which one may enter data. Data entry through a keyboard is the least costly method since it is an integral part of the PC. Keyboard data input is primarily used in the issuance of commands and the entry of a few system variables to the FEA program.

2.2.2 Mouse

The entry of graphical data via a mouse is a definite step in the right direction. The mouse adds the convenience of speed and ease of use as well as being low cost. The mouse generally operates in one of two ways. The first method uses what is called the inverted trackball. This method has the advantage of being able to operate on almost any surface. The second method uses optics. With this method two pieces are needed, the mouse and a tablet which has been etched with grid pattern on its surface. This technique of implementing a mouse is most useful when a limited amount of area is available since the mouse works only on the grid surface and these are usually about 6 inches on a side in size. The primary drawback to a mouse entry device is that slippage may occur (even in the case of the optical type) which makes it unsuitable for application where accurate repositioning of the input device may be needed.

2.2.3 Digitizers

Digitizers are probably the most efficient method of data entry. Most digitizers work in one of two ways, using ultrasonics or using magnetics. A more exotic method of digitizing uses laser light to gather graphical data. The typical digitizer works in two dimensions but there exist three-dimensional digitizers. A digitizer consist of a "tablet" and a locating device. The locating device is either a hand-held cursor (sometimes called a "puck") or a pencil-like "stylus." Most digitizer tablets today operate using ultrasonics. The stylus is placed at any arbitrary location on the tablet; this position is then determined through the use of two microphones. The implementation of three-dimensional digitizers is a simple extension of this principal. By adding another microphone in the Z direction we have accomplished our task.

The spatial resolution of the digitizer is typically 0.001 inches. The digitizing surface area ranges from 6 by 6 inches to 4 by 6 feet.

2.3 DATA OUTPUT DEVICES

Data output devices are used to present or transmit (to other programs) the results of an FEA analysis. As with input devices we are, for the most part, interested in graphical output devices. The following section will describe various graphic display methods and attempt to provide the basic advantages and disadvantages of each method.

Before we begin, two quantities used to describe graphical output devices should be addressed. These are spatial resolution and spectral resolution. Spatial resolution addresses the "fineness" of detail we perceive when we view an object. Spectral resolution addresses the information content of a discrete unit (usually called a pixel) of information contained within the object of interest.

2.3.1 Graphics Boards

The graphics display board within the PC is used to generate a video signal, either analog or RGB, for display on the PC's system monitor. One of the most important and often overlooked aspects of building an FEA model is the resolution of the graphics display board. A standard color graphics adapter (CGA) provides reasonable graphics capability but at 320 by 200 pixel resolution a user will find the displayed image somewhat lacking in quality. The displayed image of the CGA will produce jagged edges for diagonal lines and limit the size of the displayed model. To overcome this, manufacturers of graphics boards have produced the enhanced color graphics (EGA) board. This board provides up to 640 by 350 pixel resolution. Some manufacturers have provided even more resolution (up to 1280 by 1024) to even further enhance displayed graphics. One of the advantages of higher resolution is the ability to draw larger models since more picture elements are available within the displayed area of the CRT. Another feature of many of the add in graphics boards is the ability to display more than 16 colors (or gray shades). The number of colors or gray shades are not as important when defining a model because only one intensity is needed to draw the model structure. But, the number of colors (or gray shades) does add to the quality of the video output and can be important when viewing stress output files. While generally not a problem, the refresh rate of the graphics board should be considered when selecting a display board. Refresh rate is the rate at which the graphics board update the video information contained in video RAM to the CRT display. In general, the faster the refresh rate the more solid and flicker free the image appears. A refresh rate of 30 Hz will in most cases appear to flicker while a 60-Hz refresh rate is perceived as a steady image.

Many FEA software vendors supply driver programs which support many of the popular high resolution graphics boards. Table 2.2 lists several vendors of graphics boards.

Table 2.2 Graphics Board Suppliers

Adage

Cambridge Computer Graphics

Imagraph Corp.

Matrox

Number Nine Computer Corporation

PGX, Inc.

Sigma Designs

Vectrix Corporation

Vermont Microsystems, Inc.

Verticom, Inc.

2.3.2 CRT Displays

The CRT is the least costly since it exists as part of the basic requirements of the FEA/PC system. The resolution of the CRT display can cover a wide range. Although the standard IBM color adapter provides limited temporal and spatial resolution, one may find several display system which provide up to 1280 by 1024 pixels and 4096 colors out of a pallet of millions.

CRT displays have other performance criterion which provide a measure of the viewing quality one perceives. The persistence of the phosphor used in the manufacture of a CRT determines the decay rate of the image displayed. If the phosphors persistence is too short, fading of the image will occur and depending on the refresh rate, flicker may appear. Flicker is an undesirable attribute and is reduced in one of two ways: first by increasing the refresh rate and second by increasing the persistence of the phosphor in the CRT.

2.3.3 Printers

The printer may be used to provide alphanumeric data as well as graphical data (if the printer is provided with a graphics option). Although a printer's spatial resolution is limited, it does provide an inexpensive method of producing hardcopy results of an FEA analysis. Printer resolutions range from 60 dots per inch to 240 dots per inch. Several newer printers also offer multicolor capability.

2.3.4 Plotters

By far the most desirable method for the display of graphical results is the plotter. A plotter can range in cost from several hundred dollars to several thousand dollars. Plotters offer greater resolution (generally a function of cost) and speed (also a function of cost). Plotters have the advantage of drawing straight lines at any angle.

One interesting fact is that printers can now offer a greater range of colors than plotters. This is primarily due to the fact that plotters require the use of "pens," each pen having only one color.

2.4 PERFORMANCE

2.4.1 Clock Speed

Raw CPU clock speed is a measure of performance when comparing apples to apples. If we limit our discussion to the IBM PC, PC/XT and their compatibles, we may make this comparison. If we discuss the IBM AT and its compatibles, we must take other aspects into consideration. Clock speeds range from 4.77 MHz to over 8 MHz depending on which PC one chooses.

2.4.2   Add-in Co-Processors

Several manufacturers of PC compatible products provide add-in hardware which provides capabilities approaching medium-sized minis. These hardware add-ins are typically called array processors and for the most part do indeed function in the same manner as their bigger brothers. The drawback to these devices is that they are generally not supported by the FEA models currently on the market. Therefore, adaptation to one of these models may or may not be possible.

2.5   SYSTEM ENHANCEMENTS

2.5.1   The 8087/80287 Math Co-Processor

The 8087 or 80287 math co-processor chip is typically specified as a required piece of hardware by most FEA software vendors. This is basically due to the complexity of FEA models. The co-processor chip can increase the speed of some calculations by two or three orders of magnitude.

2.5.2   Software Additions

Hardware determines the baseline performance of a PC system but additional convenience may be had with some software packages. CAD software can provide a faster method of model data entry that most FEA programs. One simply draws the model using a mouse or digitizer, then submits the resulting file to a conversion program generally provided by the FEA software vendor.

3
Fundamentals of FEA

3.1 BASIC CONCEPTS

The finite element method is an analytical tool for the performing of stress and vibration analysis, and thermal and fluid flow analysis of systems and structures. A typical analysis using the finite element technique requires the following information:

1. Nodal point spatial locations (geometry)
2. Elements connecting the nodal points
3. Mass properties
4. Boundary conditions or restraints
5. Loading or forcing function details
6. Analysis options

This information only directs the user toward the final solution for the particular problem in question. The user must, in an intelligent manner, interpret the results and apply the results to the actual system behavior. This is a goal of the application of finite element analysis to practical scientific and engineering problems.

Certainly, there are other goals beyond the system behavior. If the user is addressing, for example, a structure of some nature, a goal is to develop an understanding of the structural integrity of the system. If an optimum design of a structure is wanted, the user can use the finite element method to examine how a structure (system) responds to design modifications. Furthermore, from a cost

15

standpoint, the use of finite element modeling to assess or simulate system testing or test results presents a goal of product development cost reduction.

Can finite element analysis really be done on personal computers? In reviewing the literature, there appear to be three opinions concerning the use of personal computers and finite element analysis as a unit.

The first opinion views the combination as merely a tool with which to learn basic finite element techniques and to solve "small" problems.

The second opinion is that full-featured linear finite element analysis is possible on personal computers.

Finally, the third opinion is that a full-featured finite element analysis (linear and nonlinear) is possible on enhanced versions of PC/ATs and compatibles.

The following sections examine in detail these aspects of finite element modeling and the details in the various components of a finite element model.

3.1.1 The Physical Problem

To start a finite element analysis, one must first survey the item or situation to be modeled and all boundary conditions and restraints. A careful problem review may lead to an application of the precise element; otherwise, one may be tempted to use a too complex element for the problem under consideration. The use of the inappropriate element can lead to long computational times and solutions that are not accurate. Furthermore, the answer, although correct, may have been arrived at with fewer and less complicated elements.

The use of FEA relies on the user's understanding of the underlying theory of the types of elements and their applications and on the experience of the user in the field in which he is working. It is suggested that the users of this text may wish to build small verification models particular to their field of expertise. This type of building and learning affords the user the confidence level to pursue more complicated models.

3.1.2 Criteria for Using FEA

When is it necessary to use FEA and when should one use some other method? A new initiate to FEA may wish to solve every problem that may arise using this technique. The more one uses and explores the capabilities of a FEA program, the better able he is to adapt the techniques to other problems. However, one must not lose sight of the fundamentals behind the analysis of a problem using FEA. We encourage the solving of simpler problems, on occasion, by conventional methods.

One should determine whether FEA is the quickest and most cost effective solution to product design and development. If this is not the case, then FEA should not be done. Inexpensive hardware prototyping and the availability of reliable testing generally preclude the use of FEA.

FEA can be used as the necessary tool for the design and development when any of the following conditions (although not an exhaustive list) apply:

1. The product is not a fabricated product.
2. Tool modifications may be required.
3. The production item would not perform as a fabricated part.
4. The design process calls for material optimization.
5. Prohibitive model construction costs are involved.
6. Prohibitive (expensive and/or dangerous) test requirements are anticipated.
7. Legal aspects of product design are important.

It is very important to use FEA as early in the design stage of a project as possible. An insight into possible problems will provide a more reliable product.

Table 3.1 Engineering Units for Commonly Encountered Variables

Length	L	inch	meter
Mass	M	lb-sec^2/inch	kilogram
Time	T	second	second
Area	L^2	inch^2	m^2
Volume	L^3	inch^3	m^3
Velocity	LT^{-1}	inch/sec	m/sec
Acceleration	LT^{-2}	inch/sec^2	m/sec^2
Rotation	–	radian	radian
Rotational velocity	T^{-1}	rad/sec	rad/sec
Rotational acceleration	T^{-2}	rad/sec^2	rad/sec^2
Frequency	T^{-1}	Hertz	Hertz
Force	MLT^{-2}	pound	newton
Weight	MLT^{-2}	pound	newton
Mass density	ML^{-3}	lb-sec^2/inch^4	kg/m^3
Young's modulus	$ML^{-1}T^{-2}$	lb/inch^2	N/m^2
Poisson's ratio	–	–	–
Shear modulus	$ML^{-1}T^{-2}$	lb/inch^2	N/m^2
Moment	ML^2T^{-2}	inch-lb	N⁻m
Area moment of inertia	L^4	inch^4	m^4
Torsional moment of inertia	L^4	inch^4/rad	m^4/rad
Mass moment of inertia	ML^2	inch-lb-sec^2	kg-m^2
Stress	$ML^{-1}T^{-2}$	lb/inch^2	N/m^2
Strain	–	–	–

3.1.3 Using the Correct Engineering Units

FEA programs do not contain a fixed set of units within which to work. The user selects the units that are best suited to the program. The proper selection of units and the adhering to the selected set of units is important in achieving the correct solution to a problem. Table 3.1 lists the various quantities and the units particular to the quantities in terms of mass (M), length (L), and time (T). Common English and metric units are given.

3.2 DISCRETIZATION

Discretization of the physical problem into subdivisions or regions is the first step in any analysis. If a structure or system is discretized, the meaning is that the system is represented by a discrete grid or node points connected by elements. The finite element method is a discrete representation of a continuous physical system. We do not believe that a rigid theoretical basis to perform the subdividing of a region into elements exists. Engineering judgment is the key at this stage of the analysis. Decisions must be made regarding the number and the size and shape of the subdivisions. This solution process is balanced against the selection of the proper elements that produce accurate and useful results in a reasonable amount of computational time and computational efficiency.

The common "rule of thumb" in discussions about discretization is that the greater number of nodes and elements the more accurate the solution. In theory, this assumption is correct. We suggest to use as few nodes as possible that follows the purpose of the analysis. If a detailed stress analysis is a requirement, the nodal density in the regions of large stress gradients is increased. For example, these areas occur in regions of applied forces. If only deflections are required, perhaps even fewer node points are required.

3.2.1 Selecting the Proper Elements

For the wide variety of problems that one may face, there is no one element that can satisfy all requirements. There are specific elements for one-dimensional, two-dimensional, and three-dimensional problems. There are elements for pure stress, stress plus bending, thermal elements, fluid elements, and so on. There are triangular plate elements, quadrilateral plate elements, eight-node solid brick elements, and so on. The purpose of this section is to not provide an exhaustive look at the different types of elements, but to look at the more common elements generally used in the solving of typical problems.

A finite element analysis is better developed if the user has a concept of the expected behavior of the system being modeled and element behavior when applied to that system. This expected behavior of the actual system comes from prior knowledge of the system or similar systems and from calculations performed on simplified models. There are no exact formulas or rules to aid in this development process and/or selection of elements.

3.2.2 Element Types

The following section describes the more commonly used elements for structural, thermal, and fluid analyses. Individual programs offer these and other advanced elements. The user should determine his requirements and needs accordingly.

Commonly Used Structural Elements

Three-Dimensional Two-Node Truss The three-dimensional two-node truss element is perhaps one of the simplest elements available. This element has a cross-sectional area and is shown in Figure 3.1 as a line segment. There are two nodes, one at each end. Each node has three degrees of freedom and is not capable of carrying bending loads. The stress is assumed to be constant over the entire element. This element is used in problems involving two-force (truss) members.

Three-Dimensional Two-Node Beam The three-dimensional beam element, unlike the two-node truss element, allows bending. The user must specify the "moments of inertia" in both the local x and y directions as well as the shear areas in the local directions. The degrees of freedom are translation in the local x, y, and z directions as well as rotation in each of the local directions. The beam must not have a zero length or area. The moments of inertia, however, may be zero. The beam can have any cross-sectional area for which the moments of inertia can be computed. This element is used where a typical beam may be used in real-world problems. For example, this element should be used in the analysis of three-dimensional frames such as bridges and powerline towers. This element is illustrated in Figure 3.2.

Three-Dimensional Two-Node Spring The three-dimensional two-node spring element is allowed translation in each of the local directions. As shown in Figure

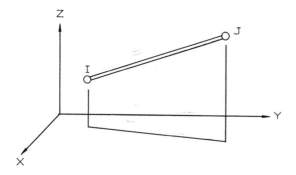

Figure 3.1 Three-dimensional two-node truss element.

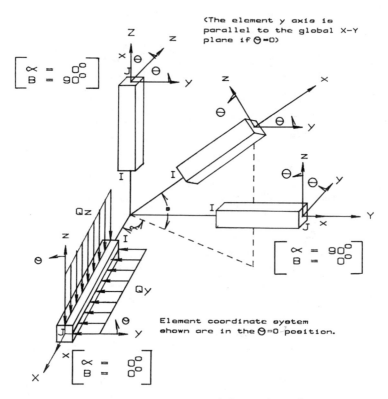

Figure 3.2   Three-dimensional two-node beam element.

3.3, the only quantity that must be known is the stiffness of the element. This element can be quite useful in cases where there is no clear-cut "fixity" of a particular node. As with other elements, some trial and error may be required and a substantial amount of engineering judgment may have to be brought into play to successfully integrate this element into an analysis.

Three-Dimensional Four-Node Plate   The three-dimensional four-node plate (and variations of this element) are very useful in the modeling of structures where bending (out-of-plane) and/or membrane (in-plane) stresses play equally important roles in the behavior of that particular structure. A careful analysis of the structure and the reaction of the structure to the applied loads should be given much consideration.

Each node has six degrees of freedom, translation in each of the local axes and rotation in each of the local axes. The user must supply the correct plate thickness. The reader should refer to the appropriate sections of this book that are concerned with solution accuracy. This element is illustrated in Figure 3.4.

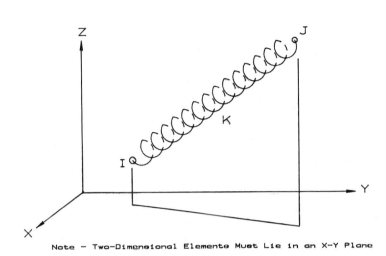

Note — Two-Dimensional Elements Must Lie in an X-Y Plane

Figure 3.3   Three-dimensional two-node spring element.

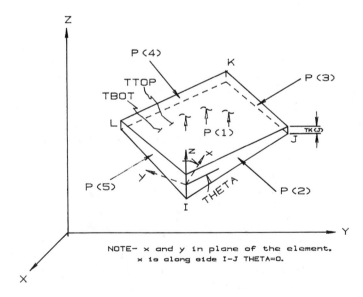

NOTE— x and y in plane of the element.
x is along side I-J THETA=0.

Figure 3.4   Three-dimensional four-node plate.

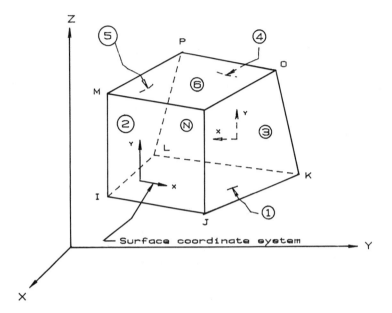

Figure 3.5 Three-dimensional eight-node solid brick.

Three-Dimensional Eight-Node Solid Brick The three-dimensional eight-node solid brick element has three translational degrees of freedom per node. No rotational degrees of freedom are allowed. This element as shown in Figure 3.5 is most useful in the modeling of items such as bolts, washers, and heavy metal casings or practically any solid structure. This is a linear element in that the gradients are not constant but are a linear function of one of the coordinate directions.

Two-Dimensional Four-Node Solid The two-dimensional four-node solid is also referred to as the isoparametric quadrilateral element. This is perhaps one of the more common elements used in two-dimensional stress problems and natural frequency analysis problems for solid structures. The element is considered thin in that the stress magnitude in the third direction is considered constant over the element thickness. In addition to being used as a biaxial plane element for stress, the element can also be used as a plain strain element. There are two translational degrees of freedom and no rotational degrees of freedom. Figure 3.6 shows this element.

Two-Dimensional Two-Node Truss The two-dimensional two-node truss is similar to the three-dimensional two-node truss in that the element is not capable

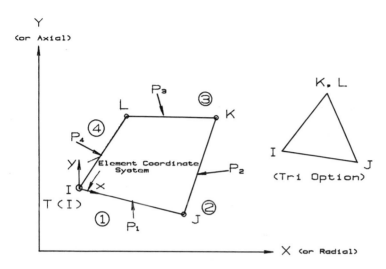

Figure 3.6   Three-dimensional four-node solid.

of carrying bending loads. This element is an uniaxial tension-compression element. It assumes a straight bar with uniform properties from end to end. The length must be nonzero and the only real constant that must be provided is the cross-sectional area. Only two degrees of freedom are allowed per node. There is only translation in two coordinate directions and no rotational degrees of freedom are allowed. This element is illustrated in Figure 3.7.

Two-Dimensional Two-Node Beam   The two-dimensional two-node beam is similar to the three-dimensional beam except that the degrees of freedom are limited to two translational (x and y) and one rotational degree of freedom about the z-axis. The beam element can have any cross-sectional shape for which the moment of inertia can be computed. The element is defined by the two end (nodal) points, the cross-sectional area, the area moment of inertia, the height, and the material properties. As with the two-dimensional two-node truss, the element must not have zero length or zero area. This beam element is used for those problems requiring such application. This element is shown in Figure 3.8.

Two-Dimensional Axisymmetric Solid   The two-dimensional axisymmetric solid is similar to the two-dimensional four-node solid with only translational degrees of freedom. This element is particularly suited to solids of revolution that are subjected to axisymmetric loading. Problems of this nature possess two independent components of displacement and therefore can be analyzed as a two-dimensional problem. Figure 3.9 illustrates this element.

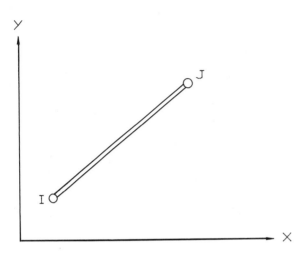

Figure 3.7 Two-dimensional two-node truss element.

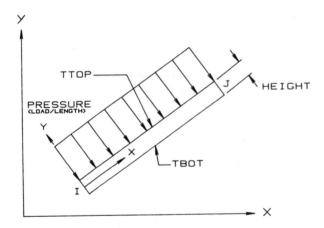

Figure 3.8 Two-dimensional two-node beam element.

Nodal Mass Elements One-, two-, and three-dimensional nodal mass elements are used in natural frequency analyses to lump masses at specified nodes. The particular mass element used is dictated by the degrees of freedom that a specific node has. This is illustrated in Figure 3.10.

Commonly Used Fluid/Thermal Elements

One-Dimensional Two-Node Thermal Truss The one-dimensional two-node thermal truss has only a one-dimensional conduction capability with one degree

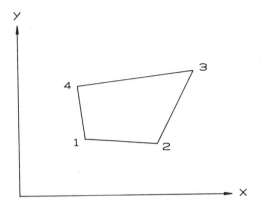

Figure 3.9 Two-dimensional axisymmetric solid.

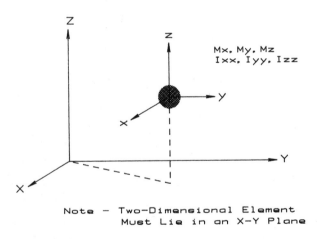

Figure 3.10 One-, two-, and three-dimensional mass elements.

of freedom per node. The degree of freedom is temperature. Convection and radiation are allowed at each node. Figure 3.11 illustrates the element. This element cannot have zero length or zero cross-sectional area.

Two-Dimensional Elements   Figure 3.12 illustrates the two-dimensional isoparametric thermal solid. This element can be used as a biaxial plane element or as an axisymmetric ring element with a two-dimensional thermal conduction capability. The element has four nodes. Each node has a single degree of freedom (temperature). The element is applicable to a two-dimensional steady state analysis.

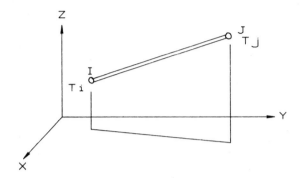

Figure 3.11   One-dimensional two-node truss element.

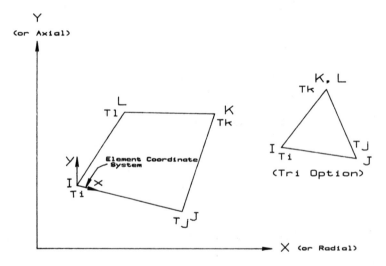

Figure 3.12   Two-dimensional isoparametric thermal solid element.

If the model containing the isoparametric temperature element is to be analyzed structurally, the element should be replaced by an equivalent structural element.

The element must not have a negative or zero area. As with all quadrilateral elements the node numbering must be counterclockwise (in most FEA programs) relative to the local coordinate system.

The two-dimensional fluid element shown in Figure 3.13 is a modification of the isoparametric solid element. This fluid element is well suited for modeling

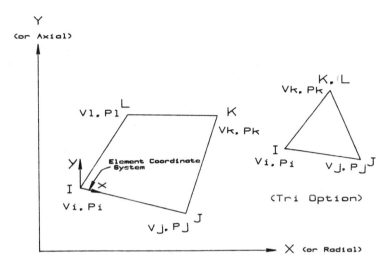

Figure 3.13 Two-dimensional fluid element.

hydrostatic pressures and fluid/solid interactions. The element is defined by four-nodal points having two degrees of freedom in the local x and y directions. The area of the element must be positive and the numbering of the nodes must be counterclockwise in the coordinate system. The amount of flow permitted is limited to that which will not cause significant distortions in the element. When used for a static application, the free surface must be input as flat.

Three-Dimensional Element Currently, there are no PC-based three-dimensional fluid elements available. Three-dimensional elements are available for thermal analysis. They are generally isoparametric elements described earlier. The user can determine temperature profiles in general solid structures. These elements allow orthotropic materials.

3.2.3 Elements, Nodes, and Degrees of Freedom

General Discussion

Many commercially available programs advertise and emphasize the number of noes available to the user for modeling purposes. While this information is necessary, the feature that the user must focus on is a combination of the number of nodes, the type of element chosen for the analysis, and the degrees of freedom associated with a node of a particular element. However, the degrees of freedom associated with the model should be sufficient and contain only those degrees of freedom necessary to characterize the actual model. The reason behind using only those degrees of freedom necessary to characterize a physical problem is that in using a personal computer, the user is limited to the maximum amount of storage available in the particular personal computer configuration to solve a problem. To allow a large problem to be run on a personal computer, certain algorithms are used to reduce the storage requirements.

The basic method of increasing the allowable problem size is a technique of reducing the storage requirements for the stiffness matrix. In general, the full stiffness matrix is not stored and is reduced during the solution process. In essence, the technique stores only the nonzero terms of the matrix. These nonzero terms are contained within a "band" as illustrated in Figure 3.14. The bandwidth of a structure is proportional to the maximum nodal difference among the connecting nodes for all of the elements in the model. In other words, the bandwidth can be calculated from the following equation:

BW = (high node number – low node number +1) X degs of freedom

The maximum value of this group of numbers defines the bandwidth of the structure.

All programs contain a node renumbering algorithm that maps the user defined node numbers to equivalent system numbers. The system node numbers allow the program to run more efficiently during the analysis. The user only sees the results at the nodes he defines.

By renumbering the nodes and by using the banded solution technique, the solution time is significantly reduced as is the amount of disk storage required.

In the development of a finite element model, the most important item to focus on is the bandwidth.

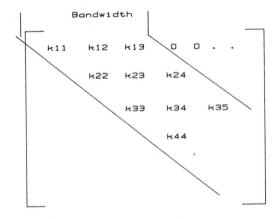

Figure 3.14 Nonzero terms are contained within a band.

Mesh Size

One of the more frequently asked questions concerns the generation and selection of the proper mesh size in order to analyze a problem. Each problem is different and there are no definite rules to develop the proper mesh size. Engineering judgment and intuition are called for to "know" where the regions of excessive stress or strain will be located. The problem geometry will dictate the areas where prominent changes in geometry occur such that a finer mesh will be required in that particular area. Also, the personal computer configuration does influence the mesh size via the available storage capacity of the unit.

First Runs with Selected Mesh

The user must review the output of his modeling efforts. This is especially true for the first solution obtained from a particular program. There are items to look for that will determine if the mesh size was too coarse or too fine for the particular application. Look for disproportionate stress level changes from node to node or plate to plate and large adjacent node displacement differences. If a sign change occurs, this is an indication that the elements are too large to produce accurate results. If the mesh density can be doubled with a change of one-percent in the results, either model is satisfactory. Although mesh density studies are time-consuming, this technique may prove to be worthy where many analyses are to be performed on similar components. The user must use caution and not trust the finite element program explicitly. Paying attention to the details will result in the finite element program answers being accurate.

Solution Accuracy

A distinct concern of every engineer who uses the finite element method in solving problems is the accuracy of the solution. Again, the fundamental concept of finite element modeling is that any continuous quantity can be approximated by a discrete model composed of a set of piecewise continuous functions defined over a finite number of subdomains. The geometry of the problem under consideration is defined in the finite element approach by dividing the structure into subdivisions or elements. The elements are connected by nodes. As the number of nodes is increased (and in turn the number of elements is increased), a closer approximation to the real-world physical problem is achieved. This, however, leads to very lengthy calculation times on the personal computer. Defining the model accurately is the user's primary concern. Nodes should be defined at locations where changes of geometry or loading occur. Changes in geometry relate to thickness, material, and/or curvature. The accuracy in the locations of loads and restraints and the proper definition of the restraints are critical in obtaining a reasonable solution. Modeling of the geometry is perhaps the easiest area in which to improve the accuracy of the solution.

Every model is different. To verify the solution, a load case should be used with which an approximate answer can be obtained. One may wish to use measured data if a physical model is available. This will allow verification of the model's accuracy quickly. In addition, the verified model can be used to examine "what if" options with confidence in the results.

Refining the Mesh in Critical Areas

A straightforward check for accuracy of the model is to reduce the element size by 50-percent and compare the results. If large changes in stresses or deflections occur, or if a sign change occurs, the original element size was too large to give accurate results. The following discussion focuses on the selection of two commonly used plate elements and their influence on solution accuracy. A complete dissertation on all elements and their influence on the accuracy of the solution is beyond the scope of this text.

To study the plate element and the accuracy of a solution derived from the use of the plate element, two basic elements are selected: (1) the isoparametric quad element, and (2) the triangular element.

In any selection of the type of element to use, the engineer must carefully consider how the structure will react to the loads imposed on the structure.

There are two basic types of load carrying mechanisms for plates. The first is membrane, and the second is bending. If the engineer feels that both types of forces are expected, the better element to use is the one that allows both load carrying mechanisms. Experience has shown that better stress results are obtained with the quad element than with the triangular element. Triangular elements are best used where dictated by geometry. The triangular elements are often used for modeling transition regions between fine and coarse grids, for modeling irregular structures, for modeling warped surfaces, and so on. When using triangular elements in a rectangular array of nodal points, the engineer can obtain the best results from an element pattern having alternating diagonal directions as shown in Figure 3.15. Triangular two-dimensional solid elements are less accurate than equivalent-sized quadrilateral elements and should not be used in highly stressed

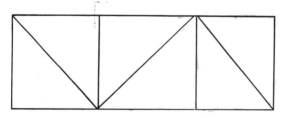

Figure 3.15 Element pattern having alternating diagonal directions.

areas unless absolutely necessary. Quadrilateral plate elements should lie in an exact flat plane. Warped areas should be modeled with triangular elements. Warped quadrilateral elements will cause a loss of equilibrium since the element resisting stiffness is based upon the element plane defined by the first three nodes. If the fourth node does not lie in the element plane, a moment imbalance results. If the model is one that has, for example, a thick shell, then one finds the thick-plate variations of the elements the better ones to use for an analysis. Note that a plate is considered to be thick if the overall length to thickness ratio is less than four. For applications involving cylindrical coordinates, a thick plate should be used when the radius to thickness ratio is less than four.

Practical Suggestions, Limitations and Interfacing to Mainframes

There are several additional items to consider in the modeling of a physical problem beyond the type of element to select. These are aspect ratio, flexibility of the physical structure, boundary conditions, thickness, and applied forces.

The aspect ratio of an element is the ratio of the length of a defined base to the height of the triangle. For the quad element, either regular or skewed, the ratio is the length of one side to an adjacent side. This concept is shown in Figure 3.16. Triangular elements should have an aspect ratio between one and two. If the isoparametric quad element is used, the aspect ratio should be below two. Deflections vary greatly with theory if the aspect ratio is greater than two. Note that some special quad elements such as the Crisfield quad can have an aspect ratio up to ten without a loss of accuracy if membrane stresses are not significant.

The flexibility of a structure influences the number of elements to be used. For example, a very flexible structure to model, such as a flexible hose, would require many more elements than a physically equivalent metal hose or concrete duct.

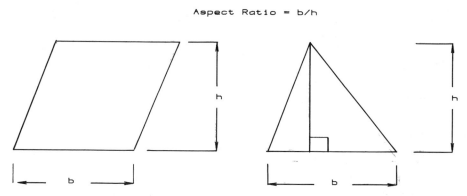

Figure 3.16   Aspect ratio of an element is the ratio of the length of a defined base to the height of the triangle.

The engineer must pay more attention to the aspect ratio of the elements used. Experience can lead to the correct choice of appropriate constraints based on the model geometry and materials.

Boundary conditions become important in evaluating the results. Structures with boundaries having a degree of "fixity" between "simply supported" and "rigid" require the use of experienced engineering judgment in the setting of the boundary conditions. Boundary conditions can be a source of singularities in a model. A singularity exists in a problem whenever an indeterminant or undefined solution is possible. Singularities can be caused by the following boundary related conditions:

1. *Unconstrained structure.* All structures should be constrained to prevent rigid body motions.
2. *Unconstrained joints.* Singularities can exist at a particular degree of freedom due to the element arrangements.

Thickness is a concern only for extremely thick shells. If the thick-plate element is used, then the aspect ratio becomes an important parameter because of the shear deflection.

It is often difficult to define the types of forces, magnitude, location, and distribution. Next to properly defining the boundary conditions, selecting the forces to be applied and the eventual quantification of those forces may be one of the most difficult aspects of finite element modeling. Applied forces include, but are not limited to, static point, static distributed, transient, harmonic vibratory, and thermally induced loadings. That the forces are applied correctly from the standpoint of magnitude, distribution, direction, and frequency and phasing if necessary, is imperative.

There are occasions when it is impossible to accurately model a given problem with the limitations of the personal computer. Many FEA programs allow interface to mainframe computers that have versions of codes such as ANSYS or NASTRAN.

The available finite element programs reviewed in this book allow the user to create files compatible with these programs. Transfer of the node and element files often occurs via communication software and a modem. Therefore, the PC-based finite element software is used as a preprocessor for the mainframe codes. This reduces the user's connect time to the mainframe during the model building process.

3.3 PREPROCESSING

Preprocessing is the initial part of any modeling process. This includes the areas discussed above for the selection of nodes, elements, loads, and constraints. Most FEA programs have a form of preprocessing, although that particular feature of the model may not be called out as such in a particular FEA program.

The preprocessing or model building is an important aspect of any finite element program. Several programs are available that allow the user to independently develop the model. By using a software-based translator, the user is able to mesh the model with most of the programs reviewed in this book. One such program is mTAB (formerly MICROTAB), from Structural Analysis, Inc.

The mTAB system is an interactive microcomputer application package used to generate two- and three-dimensional finite element models. The model data includes nodes, element definition, material properties, boundary conditions, and loads of varying types. These can be processed by finite element programs. After a finite element model is built, the model data is stored in a neutral data file that is translated to input data for the various finite element programs that handle a batch file input. mTAB will interface with mainframe finite element codes such as ANSYS and NASTRAN. The programs reviewed in this text that have mTAB translators are SAP86, ANSYS, and MSC/Pal 2.

mTAB is designed in modules with each module having a command menu screen. The modules are listed below:

1. Global set up
2. Node development
3. Element development
4. Loads, restraints, and boundary conditions
5. Offline plotting
6. Writing SAP data and saving the neutral file

mTAB has a variety of geometry generating options. An important feature is the ability to develop complex surface intersections. The program supports a variety of graphic output devices.

3.3.1 Assembling the Model

At this point in the analysis of a particular problem, the user must generate the required input to model the physical problem. Nodes are selected to represent points on the boundaries, the surface(s), and any positions interior to the physical problem.

Other constants must be defined in order for the model to run. These include, but are not limited to, the material properties (modulus of elasticity, Poisson's ratio, material density, material conductivity, material coefficient of thermal expansion, etc.), the cross-sectional areas, the moment of inertia(s) (I_{xx}, I_{yy}, I_{zz}), the plate thickness (if applicable), and so on.

3.3.2 Defining Constraints (Restraints)

The definition of the proper restraints for a problem is, in most cases, not an easy or straightforward task. Often, this part of the analysis is not clearly defined and the user is required to rely on his best engineering judgment. If his assumptions are not correct, then the model may act as a rigid body. This topic is addressed in Section 3.2.3 under the heading "Practical Suggestions, Limitations and Interfacing to Mainframes." The FEA programs do not model the behavior of rigid bodies.

A restraint is applied to a node. This restraint restricts movement in a given direction. This will in turn require an output from the program of a reaction force at that node. Another term to become familiar with is "member end release." A member end release is the removal of a force-carrying ability on the member and it allows movement.

What Restraints Correspond To

Each unrestrained node has six degrees of freedom: X, Y, and Z translation and rotation about the X, Y, and Z axes. The entire structure also has six rigid body degrees of freedom. Since only a few structures have rigid body motion, boundary conditions must be applied in order to restrain rigid body motion.

Two-dimensional planar structures must have the out-of-plane degrees of freedom zeroed. A planar structure defined in the XY plane would have the X and Y rotations zeroed and the Z translation zeroed. In addition, at least three additional degrees of freedom in the XY plane must also be zeroed in order to constrain rigid body motion in the XY plane.

In a two-dimensional problem and considering the XY plane, a clamped boundary condition would have the X and Y translations equal to zero and the Z rotation set equal to zero. A knife edge support would have the X and Y translations set equal to zero and the Z rotation would not be set equal to zero. If modeling a "roller" type boundary condition, the user would force the Y translation and Z rotation to be set equal to zero and allow the X translation to be free.

Three-dimensional structures have six rigid body degrees of freedom. At least six degrees of freedom need to be restrained to prevent rigid body motion. This can be accomplished by zeroing translations and rotations as dictated by the problem.

3.3.3   Defining Loads

Nodal Loading

Each FEA program has the ability to apply forces at nodal locations in a model in order to simulate the actual structure loading. These loads (Fx, Fy, Fz, Mx, My, Mz, temperature, convective boundaries, heat flows, and/or heat fluxes) are input at a given node in and about the respective global coordinate directions. The loads at these nodes can also be specified as applied nodal displacements.

Plate Loading

In some applications, the loading of a model may be determined by internal and/ or external pressures. A typical example would be the modeling of a soft drink can shaken vigorously. A model of this system requires the input of the increased pressure internal to the soft drink can and the outside ambient pressure. Some FEA models allow the input of plate pressures. The plate pressure input is converted to an equivalent nodal loading by taking into account the shape functions and area of the plate associated with a particular node. If the user has a program that does not contain this option of specifying this particular type of loading, a "first cut" at this type of analysis would be to assume that all nodes are loaded equally.

3.3.4 Defining Nodal Weights

FEA Program-Generated Weights

In any structure, weight is involved. FEA programs will automatically assign a weight to each node. In general this weight is the total structural weight divided by the number of nodes if the model of the structure has uniform material properties. For those models constructed with elements of different materials, the appropriate nodal weight will be assigned by the FEA program. The user should be aware that some programs do not consider the weight of the structure in the static stress analysis portions of the program. This is illustrated in the concrete dam example in Chapter 5.

User-Generated Weights

There may be occasions when the user must apply additional nodel weights to compensate for real world situations that will have a noticeable effect on the solution. This option allows a closer approximation to proper simulation without the need of additional modeling such as the use of more nodes and elements. Most FEA programs allow the user to specify these additional weights, either instead of or in addition to, the FEA program generated weights.

3.4 EXECUTING THE MODEL

With all parts of the model defined (nodes, elements, restraints, and loadings), the analysis phase of the model is ready to begin. Stresses, deflections, temperatures, pressures, velocities, and vibration modes can be determined. The solution of the equations is accomplished in one of two ways. The first is to use a "wavefront" solution. In general, node numbers are arbitrary and certain nodes are ignored in the analysis. In bandwidth-based programs, nodes must be numbered and sequenced. This operation is usually done with certain mathematical algorithms. In the analysis phase, there are certain occurrences basic to any FEA model that must happen to effect the solution to any particular problem. The three following sections address the sequence of events necessary to execute a static stress model, a vibration model, and a thermal model.

The first step for any of the three types of analyses is to perform a check of the preprocessing activities. All FEA programs include routines to check the geometry of the model. The areas checked include nodes (check for duplication), elements (check for type, proper connectivity, all specified constants), and material properties (check for modulus of elasticity, density, Poisson's ration, conductivity, thermal expansion coefficient, etc.).

The next step is to renumber the nodes for the internal system of equations. Renumbering is beneficial from the standpoint of reducing used disk space and run time for the type of analysis to be performed. As discussed earlier, reducing the bandwidth of the stiffness matrix is imperative. This is an important point that cannot be overstated. As the user becomes familiar with the FEA program he is using, he can construct better models that reflect user-inspired small bandwidths.

After renumbering the nodes, the next step is to calculate the element stiffness matrices (or element conductivities if performing a heat transfer analysis) and assemble these into the global stiffness file. All FEA programs generally use the same technique to accomplish this task. The difference lies in the algorithm used to perform the stiffness assembly task.

3.4.1 Static Stress Analysis

After the model check, node renumbering, and element stiffness assembly, the next step is to solve the matrix equations for the nodal displacements. For the static stress analysis the form of the equation to solve is:

$$[K] \ [U] = [F]$$

where K = stiffness matrix, U = nodal displacements, and F = nodal forces.

The solution to this set of equations is accomplished by the Gauss elimination technique or some derivative of the technique.

The final step in this process is to calculate the stress associated with the deflected shape. The details of this calculation are beyond the scope of this book. It is sufficient to say that the user should refer to a mechanics of materials text to review or develop an understanding of the stress displacement relationship. Computer implementation from that point is straightforward.

Stresses are calculated at the element centroids and are averaged for postprocessing at the nodes.

In a static analysis, there are certain points to remember in the development of a consistent and error free model.

Nodes

Increasing the number of nodes used for the model increases the accuracy of the model; however, solution and debug time increase. Node point spacing is not required to be uniform. We recommend using a fine mesh in regions of high stress or where the stresses are changing rapidly. A coarse mesh should be used in areas of low stress or where the stresses are nearly constant. Transition in mesh density should be smooth.

Loads

Loads can be applied to each node point in a model if desired. At times, concentrated loads should be applied across a minimum of two nodes for continuous structures. This allows the element to be loaded rather than one point of the element.

3.4.2 Thermal Analysis

A thermal analysis is quite similar to the static stress analysis discussed in Section 3.5.1. The stiffness matrix is replaced by a conductivity matrix with the solution technique similar to that for the static case. The theory behind the conductivity matrix is that the conductivity of the item must be discretized as the stiffness in the static stress analysis. The output from this analysis is all nodal temperatures.

The FEA heat transfer calculations that output the nodal temperatures are very useful by themselves. However, most users are interested in the ability of a FEA program to calculate thermal stresses. Thermally induced stresses and displacements are caused by the following conditions:

1. Differential expansion of dissimilar materials in the same structure or system
2. Uneven or transient heat flux applied to the system
3. A rigid constraint

The use of FEA for thermal analysis will generally fall into one of three categories: steady state heat transfer, transient heat transfer, and thermally induced displacements and stresses. Steady state heat transfer equations for conduction are basically the Laplace equation for each node point. The basic equation, in matrix form, is given by:

$$[C] \ [T] = [Q]$$

where C is the global conductivity matrix, T is the nodal temperatures, and Q is the nodal heat flux vector. The heat flux vector is resolved into nodal loadings. If convection boundary conditions are specified, the set of matrix equations is given by:

$$[C] \ [T] = [Q] + [h(T)] \ [T]$$

where $h(T)$ is the set of convection coefficients.

Transient heat conduction is expressed by the following equation:

$$K \ d^2T/dx^2 + pc/k \ dT/dt = 0$$

In the above equation, p is the mass density, c is the material specific heat, and K is the material conductivity. Programs that provide transient thermal capabilities contain specific time-marching algorithms in order to implement a time dependent FEA solution formulation.

For the thermally induced displacements and stresses, nodal temperatures are included in the static displacement and stress solution as a set of nodal loadings as follows:

$$[k] \ [d] = [F] + [FT(T)]$$

where k is the stiffness matrix, d is the displacement vector, F are the nodal loadings other than thermally induced, and $[FT(T)]$ is the nodal loading vector due to temperature considerations.

3.4.3   Vibration (Dynamic) Analysis

A vibration analysis is more complicated than the static stress case. Element mass matrices must be calculated. For natural frequency analysis, the mass of the object must be discretized, as is the stiffness of the element. The elements use a lumped mass technique such that the total mass of each element is distributed among its nodes with a method similar to the resolution of the system stiffness to the stiffness between the nodal points.

The dynamic response of a system is a function of three parameters:

1. System stiffness and mass properties
2. Certain dynamic forces driven by amplitude, frequency, and relative phase angle
3. Damping due to material damping, viscous damping, and Coulomb damping.

Most of the commercially available programs reviewed in this book do not allow the user to individually specify the individual damping components. Some programs do allow the user to specify an overall damping number or coefficient. COSMOS/M is an example of such a program.

After assembling the stiffness and mass matrices, the set of equations is solved for the eigenvalues and eigenvectors. The natural frequencies come from the eigenvalues and the deflected mode shapes relate to the eigenvectors. The solution of the set of equations requires an interative solution. Each FEA program effects a solution in a different manner. Due to the number of calculations and storage requirements for a dynamic analysis, some programs rely on condensation techniques to decrease the size of the global problem and reduce the dynamic degrees of freedom sufficient to characterize the dynamic response of the structure. A popular form of dynamic condensation is Guyan reduction. Using this procedure requires the reduction of the original stiffness and mass matrices to a specified number of dynamic degrees of freedom. Stiffness terms are reduced independently of the mass terms. The mass terms are redistributed according to the reduced stiffness matrix.

Another method to solve for the natural frequencies is called subspace iteration. This method does not require the problem size to be reduced. The procedure begins by solving for the lowest modes first. Upon completion of this step, the mode shapes as well as the system stresses are calculated.

In equation form:

$$[K] * [D] = [M] * [D] * [W]^2$$

where D = displacement matrix and W = diagonal matrix containing the eigenvalues.

3.5 POSTPROCESSING

The postprocessing of the data generated from an analysis is an important phase of any problem. The postprocessors associated with the FEA programs reviewed in this text are quite good and offer much insight into the program generated results.

In addition to organization of the results in an orderly fashion, the FEA postprocessors offer graphical output to the monitor, printer, and/or plotter.

In the case of the static stress and thermal calculations, most programs offer the capability of plotting the stress/temperature contours as well as full element color shading of stress/thermal gradients on the undeflected as well as the deflected model.

Output from a vibration analysis is usually given as the deflected shape with the accelerations and stresses associated with the frequency analysis. A particularly useful form of the output is the animation of the calculated mode shapes. Animation helps the user visualize exactly what is happening to the item modeled and allows the user to assess the reality of the modeling process.

3.6 DESIGN OPTIMIZATION

After the completion of a structural analysis, a question often asked relates to whether or not a part is overdesigned from either a weight or cost standpoint. Optimized designs are those designs which meet all of the strength requirements while minimizing such factors as cost and weight.

In order to achieve the optimum structural design, the following four criteria must be satisfied:

1. The design satisfies all necessary standard engineering practices.
2. All stresses are below the allowable stresses.
3. All vibration, natural frequency, and deflections (both static and dynamic) are within specifications.
4. Minimize the structural weight and/or cost.

Structural optimization is an iterative process. This process involves the repeated analysis of a structural system while changing the design variable values based on the previous analysis result. The user may select any design variables.

Both upper and lower bounds must be selected by the user. The bounds may be physical constraints or behavioral constraints. Physical constraints could include minimum or maximum allowable cross-sectional areas. Behavioral constraints place a limit on stress or perhaps one of the fundamental structural resonant points.

The goal of the optimization process is to produce a design where each element is at or near a fully stressed state for at least one of the specified load cases. However, the fully stressed state does not necessarily represent the minimum weight goal. In turn, the minimum weight goal could lead to a structure that is unstable from a buckling or resonant frequency standpoint.

4

A Simple Model

4.1 PROBLEM DEFINITION

The purpose of this chapter is to examine the application of FEA to a simple problem. The simple problem is a three-dimensional truss and is illustrated in Figure 4.1. All relevant data is given in the figure. The analytical solution (22) is given in Section 4.2 and the finite element solution is given in Section 4.3. The user will see each step in the setup, the execution, and the postprocessing. This detailed presentation will allow the user of this book to become more familiar with the process involved in the development and use of a commercially available finite element model.

4.2 ANALYTICAL SOLUTION

To solve the problem illustrated in Figure 4.1 for the stresses in each member it is necessary to determine the tension in each of the truss members. Point 4 is chosen as a free body and is subjected to four forces. The unknowns are the tensions in the members: (41), (42), and (43). The coordinate system and the unit vectors i, j, and k are shown in Figure 4.2.

It is necessary to determine the components and magnitudes of the vectors (41), (42), and (53). Let λ be the unit vector along (41). Therefore:

$\lambda(41) = 4i + 6j - 2k$ $[41] = 7.4833$

         $= (41)/7.4833 = .5345i + .8018j - .2673k$

$\lambda(42) = -4j + 6j - 2k$ $[42] = 7.4833$

The tension in each cable is written as follows:

$$= (42)/7.4833 = -.5345i + .8018j - .2673k$$
$$\lambda(43) = -4i + 6j + 4k \qquad\qquad [43] = 8.2462$$
$$= (43)/8.2462 = -.4851i + .7276j + .4851k$$
$$T41 = .5345\,T41i + .8018\,T41j - .2673\,T41k$$

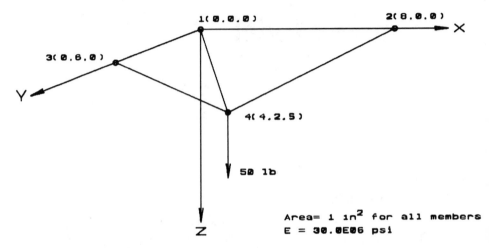

Figure 4.1   Three-dimensional truss problem.

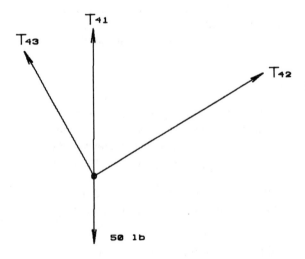

Figure 4.2   Coordinate system for three-dimensional truss problem.

$$T42 = -.5345\ T42i + .8018\ T42j - .2673\ T42k$$
$$T43 = -.4851\ T43i + .7276\ T43j + .4851\ T43k$$

Since point 4 is in equilibrium, the sum of all the forces is zero.

$$F = 0:\ T41 + T42 + T43 + W = 0$$

Substituting the above expressions for Tmn and setting the coefficients of i, j, and k equal to zero results in three equations for the three unknowns. The solution to the set of equations can be arrived at by putting the equations into matrix form and using any standard technique such as the Guassian elimination technique:

$$[Amn] \times [Tmn] = [Fm]$$

The solution is

$$T41 = 10.391\ lb$$
$$T42 = 31.180\ lb$$
$$T43 = 22.906\ lb$$

The stress in each truss member is determined by dividing the tension in each member by the cross-sectional area. For this example problem, the cross-sectional area is 1 in.^{2}

4.3 FINITE ELEMENT SOLUTION: THREE-DIMENSIONAL TRUSS

The FEA solution to the above problem is quite straightforward. All steps used to arrive at a solution are given and can be run on any of the FEA models discussed in Chapter 5.

FINITE ELEMENT SOLUTION: THREE-DIMENSIONAL TRUSS

```
PCTRAN

C)echo off


P C T R A N    P l u s
(C) Copyright 1985, 1986    Brooks Scientific, Inc.
Version 3.2


ENTER JOB NAME: FOUR1


Portions of the PCTRAN menu system are copyrighted
by Alpha Software Corporation.
```

FINITE ELEMENT SOLUTION: THREE-DIMENSIONAL TRUSS

```
PCTRAN Plus 3.2 PREPROCESSOR
(C) Copyright 1985, 1986  Brooks Scientific, Inc.

    ... Please wait ...

FOUR1       is a new job

* N,1,

Node:      1 =    .0000E+00    .0000E+00    .0000E+00
* N,2,8,

Node:      2 =   8.0000E+00    .0000E+00    .0000E+00
* N,3,0,6

Node:      3 =    .0000E+00   6.0000E+00    .0000E+00
* N,4,4,2,6

Node:      4 =   4.0000E+00   2.0000E+00   6.0000E+00
* ETYPE,1,1

Element type  1 is STIFF    1    3-D ROD
IEOPT = 0          2 Nodes;  3 DOF per node.
* E,1,1,4

Element     1 =   1  1  1    1    4    0    0    0    0    0    0
* E,2,2,4

Element     2 =   1  1  1    2    4    0    0    0    0    0    0
* E,3,3,4

Element     3 =   1  1  1    3    4    0    0    0    0    0    0
* R,1,1

Real set   1. value(s) =     1.0000E+00    .0000E+00    .0000E+00    .0000E+00
   .0000E+00    .0000E+00
* EX,1,30.0E6

Elasticity for material number   1 is   3.0000E+07
* DENS,1,0.28

Density for material number 1 is   2.8000E-01
* NUXY,1,.3

Poissons ratio for material number   1 is   3.0000E-01
* D,1,UX,,3

Define displacement from node      1 with label UX   and value   .0000E+00
to node      3 in steps of      1
* D,1,UY,,3

Define displacement from node      1 with label UY   and value   .0000E+00
to node      3 in steps of      1
* D,1,UZ,,3

Define displacement from node      1 with label UZ   and value   .0000E+00
to node      3 in steps of      1
* F,4,FZ,-500
```

```
Load case   1
Define force from node    4 with label FZ   and value  -5.0000E+01
to node     4 in steps of     1
* CHECK

Model checking routine . . .

    *** No errors detected ***
* EXIT


  Statistics for job FOUR1
  **********************************

Analysis type: 0 STATIC STRESS
Active coordinate system:   0    .0000E+00   .0000E+00   .0000E+00
Number of load cases: 1
Max node number:     4
Number of system nodes:    0
Max element number:    3
Number of applied displacements:     9
Half bandwidth:     0
Input file echo: ON
Reorder load cases in ascending order
Load Case, # of Loads
    1       1
Reorder displacements in ascending order
Copying "PCTX" files back to job FOUR1
Exit preprocessor .....

C>STATICS

C>echo off

PCTRAN Plus 3.2 NODE RENUMBERING
(C) Copyright 1985, 1986  Brooks Scientific, Inc.

MAXN:         4
MAXE:         3
NSN:          4

NODE RENUMBERING [Y/N/(E to exit)]? Y
Enter number of starting nodes (30 max) NSTART =
Enter the  0 starting nodes [N1,N2, ... etc.]:

Nodes for element    1 renumbered
Nodes for element    2 renumbered
Nodes for element    3 renumbered

NUMBER OF NODES        =    4
NUMBER OF SYSTEM NODES =    4
HALF BANDWIDTH         =    9
DEGREES OF FREEDOM     =   12

NODE RENUMBERING [Y/N/(E to exit)]? E

PCTRAN Plus 3.2 STIFFNESS ASSEMBLY
(C) Copyright 1985, 1986  Brooks Scientific, Inc.

Max node number:     4
```

FINITE ELEMENT SOLUTION: THREE-DIMENSIONAL TRUSS

```
Max element number:    3
Half bandwidth:      9
Total DOF:     12

Initializing stiffness file ...

Assembling element      1
Assembling element      2
Assembling element      3
EXIT STIFF . . .

PCTRAN Plus 3.2 BANDED EQUATION SOLVER
(C) Copyright 1985, 1986  Brooks Scientific, Inc.

NUMBER OF NODES:                 4
NUMBER OF ELEMENTS:              3
NUMBER OF LOAD CASES:            1
# OF LOADS IN CASE  1:     1
DEGREES OF FREEDOM:             12
NUMBER OF DISPLACEMENTS:         9
ACTIVE DEGREES OF FREEDOM:       3
HALF BANDWIDTH:                  9


LOADING GLOBAL STIFFNESS MATRIX ....


DECOMPOSITION ...

BACKWARD SUBSTITUTION

REACTION FORCE CALCULATION


PCTRAN Plus 3.2 STRESS CALCULATION
(C) Copyright 1985, 1986  Brooks Scientific, Inc.

READ POST CODES FROM FILE "PCTRAN.CSD"
MAX NODE NUMBER:      4
MAX ELEMENT NUMBER:      3
HALF BANDWIDTH:      9
TOTAL DOF:     12
NUMBER OF SYSTEM NODES:      4
STRESS CALCULATION: ELEMENT      1
LOAD CASE  1, ROD      1: STRESS = -1.0393E+01
LOAD CASE  1, ROD      2: STRESS = -3.1180E+01
LOAD CASE  1, ROD      3: STRESS = -2.2906E+01
EXIT STRESS ....

C>POST1

C>echo off

PCTRAN Plus 3.2 POSTPROCESSOR
(C) Copyright 1985, 1986  Brooks Scientific, Inc.

DEVICE DRIVER IS /PCTIBMG.DEV/       , MODE =   0
       ... Please wait ...

FOUR1      IS AN OLD JOB
```

```
CALCULATE MIN,MAX POST DATA

  Statistics for FOUR1
************************

Analysis type:  0 STATIC STRESS
Active coordinate system:   0   .0000E+00   .0000E+00   .0000E+00
Current load case:  1
Max node number:     4
Max element number:      3
Number of displacemenets:      9
Half bandwidth:      9

Screen echo: OFF  , Input file echo: ON
Reading POST codes from file "PCTRAN.CSD"
* DEFL,OVR

Plot label DEFL is: OVR
* PLOT

* PRDISP,1,4

                   DISPLACEMENTS: LOAD CASE  1

 NODE       UX          UY          UZ         ROTX        ROTY        ROTZ

   1     .0000E+00   .0000E+00   .0000E+00   .0000E+00   .0000E+00   .0000E+00
   2     .0000E+00   .0000E+00   .0000E+00   .0000E+00   .0000E+00   .0000E+00
   3     .0000E+00   .0000E+00   .0000E+00   .0000E+00   .0000E+00   .0000E+00
   4    4.8503E-06  5.4199E-06 -8.2737E-06   .0000E+00   .0000E+00   .0000E+00

                        ** MAX DEFLECTION **
 NODE:       4           4           4           0           0           0
          4.8503E-06  5.4199E-06 -8.2737E-06   .0000E+00   .0000E+00   .0000E+00
  *
```

4.3.1 Nodes

After entering the program and getting into the preprocessing portion of the program, the first step is to input the nodes and the node coordinates. For this particular problem, the nodes and node coordinates are given below.

Node	x	y	z
1	0.0	0.0	0.0
2	8.0	0.0	0.0
3	0.0	6.0	0.0
4	4.0	2.0	6.0

All node coordinates are in feet.

4.3.2 Material Properties

The only material properties needed for a three-dimensional truss element static analysis is the modulus of elasticity, density, and Poisson's ratio. Assume that the truss element is steel. Therefore, E is equal to 30.0E06 lb/in.^{2}, the density is equal to 0.28 lb/in.^{3} and Poisson's ratio is 0.3.

4.3.3 Real Constant(s)

Beam and truss elements require the input of various physical data such as cross-sectional area and various moments of inertia. The three-dimensional truss requires only the cross-sectional area. The value is 1.0 in.^{2}

4.3.4 Truss Connectivity

The following list is the element connectivity. Included are truss length, from node/to node, and the direction cosines (as determined from the problem definition) for each beam.

Beam	Nodes			Direction cosines		
No.	from	to	Length	x	y	z
1	1	4	7.483	-0.15	0.96	-0.22
2	2	4	7.483	0.15	0.96	-0.22
3	3	4	8.246	0.27	0.87	0.40

4.3.5 Restraints

In any problem, the determination of the proper restraint to apply at each node is very important. In this problem, the use of a three-dimensional truss element dictates that any rotations about the x, y, or z axes is prohibited. Only translations in each coordinate direction are allowed. The movements in each coordinate direction for nodes 1-3 will be restrained; that is, displacements in the x, y, and z directions for nodes 1, 2, and 3 will be equal to zero.

4.3.6 Loads

The only load for this problem is one of 50 pounds applied at node 4 in the z direction.

4.3.7 Solution

After inputing all of the above information, the next step is to check the input for consistency. Most programs have subroutines to check if all information required to perform an analysis has been imputed to the program. Following this step is the requirement to "renumber the nodes." This is internal to the program and allows an improvement in the solution efficiency.

The solution to the problem is in most programs carried out in an automatic or semiautomatic fashion. We ran this simple program on several of the FEA programs reviewed for this text. The FEA output from each program matched the classical solution at least to the third decimal place.

This technique of using a simple example is often quite useful for checking the accuracy of the model proposed and is useful for understanding the basics and principles of the application of finite element analysis to common textbook problems. This in no way suggests not studying or understanding the principles involved. FEA is simply an analytical tool that allows solutions to complex problems.

4.4 ADDITIONAL VERIFICATION PROBLEMS

Many typical engineering problems can be found in any number of texts from the various engineering fields. During the development of this text, we have relied on problems from several texts that have allowed us to experiment with the various FEA programs reviewed in this book. In addition to the textbooks, one program has aided in the FEA program verification. This program is MSC/CASE (Computer Aided Solutions for Engineers) by the MacNeal-Schwendler Corporation.

MSC/CASE provides analytical solutions to equations needed in structural and mechanical engineering analysis and design. The equations are those encountered in engineering texts on strength of materials and theory of elasticity.

The solutions available from the MSC/CASE program represent one of the most comprehensive collections available. The user has the ability to quickly and efficiently solve many problems from beams to thin shells with any number of boundary conditions. The solutions are based on the already developed analytical solutions thus freeing the user from lengthy and sometimes error-prone calculations.

5
Structural Models

5.1 INTRODUCTION

This chapter discusses the application of several commercially available finite element programs to four illustrative problems. These problems were selected for the following reasons:

1. The problems involve practical design considerations.
2. The problems illustrate the range of applications.
3. The problems show in detail how the models are developed.

Certainly, the examples should suggest further and more intricate applications.

The problems to be used as examples include a stress analysis of a concrete dam, the natural frequency calculation of a hyperbolic natural draft cooling tower, the analysis of the stress in a plate with a hole, and the torsion of a box beam. Some problems are based on actual experience while others are demonstration problems from the various finite element program packages.

5.2 STRESS ANALYSIS OF A CONCRETE DAM

This problem addresses the state of stress in a simplified dam cross section when subjected to its own weight and by the pressure of the water exerted on one face of the dam. The structure is shown in Figure 5.1. The structure is a simplified concrete gravity dam that rests upon bedrock. The dimensions are 500 feet high, 500 feet wide at the base, and 300 feet wide at the top of the dam. It is required to find the maximum tensile stress throughout the cross section.

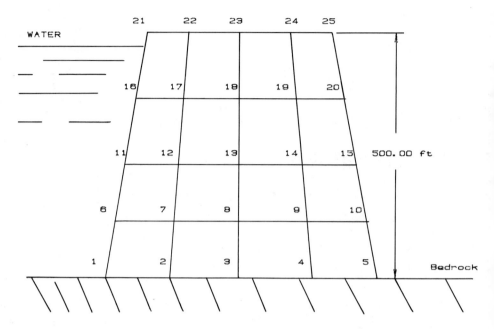

Figure 5.1 Concrete Dam.

The analysis is performed in two dimensions since the third dimension is assumed to be very long. The element used to model the center section of the dam can be either a two-dimensional plane strain element (out-of-plane strain assumed to be zero) or a two-dimensional plane stress element (out-of-plane stress assumed to be zero).

5.2.1 Input Data

This section contains all input data needed to solve the problem. Table 5.1 contains the coordinate data for each node point. In general, it is not necessary to input every node point since all programs used in this book contain provisions to generate node points based on an initial and final node and by using a node generate or node fill command. Since there are no dramatic shape changes (i.e., sharp corners $\leqslant 90°$ inside) and a uniform loading, the stress gradient within each element is not expected to be high. Therefore only 25 node points and 16 elements were used in this model definition.

Elements are defined with the material properties. Finite element programs vary somewhat in their data input. The user must be careful in the definition of the material properties. A wrong value in any of the material properties translates directly to an invalid problem solution. As an example, consider the material

Table 1 Input Geometry for Concrete Dam Problem

Node	x	y	z
1	0	0	0
2	125	0	0
3	250	0	0
4	375	0	0
5	500	0	0
6	25	125	0
7	137.5	125	0
8	250	125	0
9	362.5	125	0
10	475	125	0
11	50	250	0
12	150	250	0
13	250	250	0
14	350	250	0
15	450	250	0
16	75	375	0
17	162.5	375	0
18	250	375	0
19	337.5	375	0
20	425	375	0
21	100	500	0
22	175	500	0
23	250	500	0
24	325	500	0
25	400	500	0

Table 5.2 Material Properties for Concrete Dam Problem

E (modulus of elasticity)	5.76×10^8 lb/ft^2
	4.00×10^6 lb/ft^2
D (density)	4.66 slugs/ft^3
	150.05 lbm/ft^3
	0.0864 lbm/in.^{3}
Nu (Poisson's ratio)	0.2

Table 5.3   Elements for the Concrete Dam Problem

Element	i	j	k	l
1	1	2	7	6
2	2	3	8	7
3	3	4	9	8
4	4	5	10	9
5	6	7	12	11
6	7	8	13	12
7	8	9	14	13
8	9	10	15	14
9	11	12	17	16
10	12	13	18	17
11	13	14	19	18
12	14	15	20	19
13	16	17	22	21
14	17	18	23	22
15	18	19	24	23
16	19	20	25	24

density. A program may require this variable to be input as the mass density or as the weight density. It is important to be consistent with units when inputing data. Table 5.2 displays the material properties required for this problem.

The element used in this problem is the four-node isoparametric solid. The size of each element is consistent with the basics outlined in Chapter 3, and the section entitled "Practical Suggestions, Limitations and Interfacing to Mainframes" in particular, that is concerned with the element aspect ratio. Since the problem is node numbered consecutively left to right and bottom to top it is possible to define the first element and generate the remainder of the elements. Table 5.3 lists the element numbers and the nodes connected by the specific element.

The restraints are assumed to be along the line defined by nodes 1, 2, 3, 4, and 5. That is, nodes 1 through 5 are assumed to be firmly attached to the bedrock so that displacements are zero. The nodes could be allowed to translate by defining the spring-damper characteristics of the bedrock. This is beyond the scope of this problem. Deflections in both the x and y directions are set equal to zero. These are given in Table 5.4.

Table 5.4 Restraints for the Concrete Dam Problem

Node	Displacement	
	Ux	Uy
1	0.0	0.0
2	0.0	0.0
3	0.0	0.0
4	0.0	0.0
5	0.0	0.0

Some programs allow a pressure to be applied to a face of an element. In other programs, the user must resolve the pressure into appropriate forces in the three coordinate directions and estimate the probable distribution of forces at each node point of a particular element. In the problem under investigation, the assumption is made that a linearly varying water pressure is applied to the upstream face of the dam and the pressure is constant on each element face. The pressure is calculated from the equation:

$$P = pgh$$

where P is the pressure to be specified on each element face, p is the density of the water, g is the acceleration of gravity, and h is the depth below the water surface of the center of the element face.

To complete the model input data, the weight of each element must be accounted for to accurately model the stresses in the dam. The user must verify that the particular model being used takes into account the weight of each element. The weight of each element is accounted for in most finite element models by the specification of the acceleration of gravity and the direction in which the gravity vector acts.

5.2.2 Solution

Two solutions are given for the concrete dam problem. The first solution illustrates the use of a program called ANSYS-PC/Linear. ANSYS-PC/Linear contains the program-generated pressure input, and the accounting of the weight of the structure in the stress calculations. The second solution illustrates the use of the Program PCTRAN Plus. This version of PCTRAN Plus requires the user to input the pressure on the element faces and the weight of the structure.

A review of the results of both analyses indicates that the calculated stress levels are similar as is the stress distribution. Differences are accounted for in the element pressure input and element weight loading.

SIMPLIFIED CONCRETE DAM

```
8087 ON

C>VIRTL
ANSYS Large Virtual Memory Driver
(C)Copyright Swanson Analysis Systems,Inc. 1986

C>MEDIUM
ANSYS Medium Resolution Graphics Driver
(C)Copyright Swanson Analysis Systems,Inc. 1986

C>PREP

ANSYS-PC/LINEAR PREP MODULE
Copyright(C)  1971 1978 1982 1985 1986
Swanson Analysis Systems, Inc.
As an Unpublished Work.
PROPRIETARY DATA - Unauthorized Use, Distribution
 or Duplication is Prohibited.
All Rights Reserved.


   *** ANSYS REV 4.2 A3    ENSMINGER/DEMO    CP=     0.33 ***
   FOR SUPPORT CALL CUSTOMER SUPPORT   PHONE (412) 746-3304    TWX 510-690-8655


   TITLE
**ANSYS VERSION FOR DEMONSTRATION PURPOSES ONLY**

      ***** ANSYS ANALYSIS DEFINITION (PREP7) *****

ENTER   RESUME   TO RESUME EXISTING MODEL
ENTER   INFO     FOR PREP7 INFORMATION
ENTER   FINISH   TO LEAVE PREP7
THE VIRTUAL MEMORY SIZE IS 4194304 32 BIT WORDS

IMMEDIATE MODE IS AVAILABLE ON THIS GRAPHIC DEVICE
 ENTER   /IMMED,YES  TO TURN ON  IMMEDIATE MODE
 ENTER   /IMMED,NO   TO TURN OFF IMMEDIATE MODE
 Immediate Mode Requires User Graphic Scaling
   Remember to Set /USER and Define /VIEW, /FOCUS and /DIST
PREP7 -INP=/TITLE, CONCRETE DAM


NEW TITLE=  CONCRETE DAM


PREP7 -INP=ET,1,42,0,0,2


ELEMENT TYPE  1 USES STIF 42
  KEYOPT(1-9)=  0   0   2    0   0   0     0   0   0 INOTPR= 0
  NUMBER OF NODES=   4

STRESS SOLID, 2-D

CURRENT NODAL DOF SET IS  UX     UY
  TWO-DIMENSIONAL STRUCTURE
PREP7 -INP=EX,1,5.76E8
```

```
MATERIAL    1          COEFFICIENTS OF EX   VS. TEMP EQUATION
  CO =  0.5760000E+09

PROPERTY TABLE EX     MAT=   1  NUM. POINTS=  2
     TEMPERATURE      DATA          TEMPERATURE       DATA
   0.00000E+00 0.57600E+09    2300.0      0.57600E+09
PREP7 -INP=DENS,1,4.66


MATERIAL    1          COEFFICIENTS OF DENS VS. TEMP EQUATION
  CO =   4.660000

PROPERTY TABLE DENS  MAT=   1  NUM. POINTS=  2
     TEMPERATURE      DATA          TEMPERATURE       DATA
   0.00000E+00  4.6600        2300.0       4.6600
PREP7 -INP=NUXY,1,.2


MATERIAL    1          COEFFICIENTS OF NUXY VS. TEMP EQUATION
  CO =  0.2000000

PROPERTY TABLE NUXY   MAT=   1  NUM. POINTS=  2
     TEMPERATURE      DATA          TEMPERATURE       DATA
   0.00000E+00 0.20000        2300.0      0.20000
PREP7 -INP=N,1


NODE    1  KCS= 0  X,Y,Z= 0.00000E+00 0.00000E+00 0.00000E+00
PREP7 -INP=N,5,500


NODE    5  KCS= 0  X,Y,Z=  500.00      0.00000E+00 0.00000E+00
PREP7 -INP=FILL


FILL     3 POINTS BETWEEN NODE    1 AND NODE    5
  START WITH NODE     2 AND INCREMENT BY     1
PREP7 -INP=N,21,100,500


NODE   21  KCS= 0  X,Y,Z=  100.00       500.00      0.00000E+00
PREP7 -INP=N,25,400,500


NODE   25  KCS= 0  X,Y,Z=  400.00       500.00      0.00000E+00
PREP7 -INP=FILL


FILL     3 POINTS BETWEEN NODE   21 AND NODE   25
  START WITH NODE    22 AND INCREMENT BY     1
PREP7 -INP=FILL,1,21,3,,,5,1


FILL     3 POINTS BETWEEN NODE    1 AND NODE   21
  START WITH NODE     6 AND INCREMENT BY     5
  REPEAT THIS FILL OPERATION     5 TIMES WITH INCREMENT OF     1
PREP7 -INP=E,1,2,7,6

ELEMENT    1      1     2     7     6
PREP7 -INP=EGEN,4,1,-1
```

SIMPLIFIED CONCRETE DAM

```
GENERATE     4 TOTAL SETS OF ELEMENTS WITH NODE INCREMENT OF      1
   SET IS SELECTED ELEMENTS IN RANGE      1 TO     1 IN STEPS OF      1
NUMBER OF ELEMENTS=     4
PREP7 -INP=EGEN,4,5,-4


GENERATE     4 TOTAL SETS OF ELEMENTS WITH NODE INCREMENT OF      5
   SET IS SELECTED ELEMENTS IN RANGE      1 TO     4 IN STEPS OF      1
NUMBER OF ELEMENTS=    16
PREP7 -INP=/NUM,-1


NUMBER KEY SET TO -1 -1=NONE   0=COLOR  1=NUMBER
PREP7 -INP=D,1,ALL,,,5,1


SPECIFIED DISP ALL  FOR SELECTED NODES IN RANGE      1 TO     5 BY     1
   VALUES= 0.00000E+00 0.00000E+00  ADDITIONAL DOFS=
PREP7 -INP=P,1,6,27300


SPECIFIED PRESSURE FROM NODE      1 TO NODE      1 BY     1
   NODES=     1     6     0     0 PRESSURE=   27300.0000

NUMBER OF ELEMENT PRESSURES=     1
PREP7 -INP=P,6,11,19500


SPECIFIED PRESSURE FROM NODE      6 TO NODE      6 BY     1
   NODES=     6    11     0     0 PRESSURE=   19500.0000

NUMBER OF ELEMENT PRESSURES=     2
PREP7 -INP=P,11,16,11700


SPECIFIED PRESSURE FROM NODE     11 TO NODE     11 BY     1
   NODES=    11    16     0     0 PRESSURE=   11700.0000

NUMBER OF ELEMENT PRESSURES=     3
PREP7 -INP=P,16,21,3900


SPECIFIED PRESSURE FROM NODE     16 TO NODE     16 BY     1
   NODES=    16    21     0     0 PRESSURE=   3900.00000

NUMBER OF ELEMENT PRESSURES=     4
PREP7 -INP=ACEL,,32.2


ACEL=   0.00000E+00   32.200      0.00000E+00
PREP7 -INP=SFWRIT


*** NOTE ***
NPRINT IS ZERO OR GREATER THAN NITTER. SOLUTION PRINTOUT
WILL BE SUPPRESSED

*** NOTE ***
```

```
DATA CHECKED - NO ERRORS FOUND

SOLUTION DATA (BINARY) WRITTEN ON FILE3

ENTER  FINISH  TO LEAVE PREP7

ENTER  ANSYS  TO RUN THIS ANALYSIS
PARAMETERS FOR THE EXECUTION ARE IN  CONTROL.ANS
PREP7 -INP=FINISH


ALL CURRENT PREP7 DATA WRITTEN TO FILE16
 FOR POSSIBLE RESUME FROM THIS POINT
Execution terminated : 0


C>POSTS

ANSYS-PC/LINEAR SOLUTION MODULE
Copyright(C)  1971 1978 1982 1985 1986
Swanson Analysis Systems, Inc.
As an Unpublished Work.
PROPRIETARY DATA - Unauthorized Use, Distribution
 or Duplication is Prohibited.
All Rights Reserved.
```

SIMPLIFIED CONCRETE DAM

***** NOTICE ***** THIS IS THE ANSYS-PC/LINEAR FINITE
ELEMENT PROGRAM. NEITHER SWANSON ANALYSIS SYSTEMS, INC.
NOR THE DISTRIBUTOR SUPPLYING THIS PROGRAM ASSUME ANY
RESPONSIBILITY FOR THE VALIDITY, ACCURACY, OR APPLICABILITY
OF ANY RESULTS OBTAINED FROM THE ANSYS SYSTEM. USERS
MUST VERIFY THEIR OWN RESULTS.

THE VIRTUAL MEMORY SIZE IS 4194304 32 BIT WORDS

   *** ANSYS REV 4.2 A3   ENSMINGER/DEMO   CP=   5.32 ***
   FOR SUPPORT CALL CUSTOMER SUPPORT   PHONE (412) 746-3304   TWX 510-690-8655

CONCRETE DAM
ANSYS VERSION FOR DEMONSTRATION PURPOSES ONLY

     ***** CENTROID, MASS, AND MASS MOMENTS OF INERTIA *****

CALCULATIONS ASSUME ELEMENT MASS AT ELEMENT CENTROID

TOTAL MASS = 0.93200E+06

	CENTROID	MOM. OF INERTIA ABOUT ORIGIN		MOM. OF INERTIA ABOUT CENTROID	
XC =	250.00	IXX =	0.6735E+11	IXX =	0.1785E+11
YC =	230.47	IYY =	0.7058E+11	IYY =	0.1233E+11
ZC =	0.00000E+00	IZZ =	0.1379E+12	IZZ =	0.3018E+11
		IXY =	-0.5370E+11	IXY =	0.9201E-06
		IYZ =	0.0000E+00	IYZ =	0.0000E+00
		IZX =	0.0000E+00	IZX =	0.0000E+00

*** MASS SUMMARY BY ELEMENT TYPE ***

TYPE   MASS
  1   932000.

RANGE OF ELEMENT MAXIMUM STIFFNESS IN GLOBAL COORDINATES
MAXIMUM= 0.395579E+09 AT ELEMENT   16.
MINIMUM= 0.289220E+09 AT ELEMENT   3.

 *** ELEMENT STIFFNESS FORMULATION TIMES
TYPE   NUMBER   STIF   TOTAL CP   AVE CP

  1   16   42   6.33   0.396
TIME AT END OF ELEMENT STIFFNESS FORMULATION CP=   71.510
MAXIMUM WAVE FRONT ALLOWED= 200.
EQUATION SOLUTION   ELEM=   1 L.S.=   1 ITER=   1 CP=   79.25
MAXIMUM IN-CORE WAVE FRONT=   14.
  MATRIX SOLUTION TIMES
 READ IN ELEMENT STIFFNESSES   CP=   0.770
 NODAL COORD. TRANSFORMATION   CP=   0.000
 MATRIX TRIANGULARIZATION   CP=   1.700
TIME AT END OF MATRIX TRIANGULARIZATION CP=   94.080

```
TIME AT START OF BACK SUBSTITUTION    CP=      98.640

*** ELEM. STRESS CALC. TIMES
TYPE  NUMBER  STIF  TOTAL CP    AVE CP

 1      16     42    2.59    0.162

*** NODAL FORCE CALC. TIMES
TYPE  NUMBER  STIF  TOTAL CP    AVE CP

 1      16     42    0.38    0.024


*** PROBLEM STATISTICS
NO. OF ACTIVE DEGREES OF FREEDOM =      40
R.M.S. WAVEFRONT SIZE =      11.2
TOTAL CP TIME=       121.270
Execution terminated : 0


C>POST

ANSYS-PC/LINEAR POST MODULE
Copyright(C)  1971 1978 1982 1985 1986
Swanson Analysis Systems, Inc.
As an Unpublished Work.
PROPRIETARY DATA - Unauthorized Use, Distribution
 or Duplication is Prohibited.
All Rights Reserved.


*** ANSYS REV 4.2 A3    ENSMINGER/DEMO   CP=       0.33 ***
  FOR SUPPORT CALL CUSTOMER SUPPORT  PHONE (412) 746-3304    TWX 510-690-8655


**ANSYS VERSION FOR DEMONSTRATION PURPOSES ONLY**

        ***** ANSYS RESULTS INTERPRETATION (POST1) *****
THE VIRTUAL MEMORY SIZE IS 4194304 32 BIT WORDS

ENTER  INFO   FOR POST1 DOCUMENTATION
ENTER  FINISH  TO LEAVE POST1
ENTER  /IMMED,YES  FOR MENU SYSTEM
POST1 -INP=SET,1,1


USE LOAD STEP    1  ITERATION     1  SECTION    1  FOR LOAD CASE    1


GEOMETRY STORED FOR    25  NODES    16  ELEMENTS
  TITLE=  CONCRETE DAM


DISPLACEMENT STORED FOR    25 NODES

NODAL STRESSES AND TEMPS. STORED FOR   16 ELEMENTS

NODAL FORCES STORED FOR   16 ELEMENTS
```

SIMPLIFIED CONCRETE DAM

REACTIONS STORED FOR   10 REACTIONS

FOR LOAD STEP=   1 ITERATION=   1 SECTION=   1
TIME= 0.000000E+00   LOAD CASE= 1
TITLE= CONCRETE DAM

POST1 -INP=PRDISP

PRINT NODAL DISPLACEMENTS

            ***** POST1 NODAL DISPLACEMENT LISTING *****

LOAD STEP   1 ITERATION=   1 SECTION= 1
TIME=   0.00000E+00   LOAD CASE= 1

THE FOLLOWING X,Y,Z DISPLACEMENTS ARE IN NODAL COORDINATES

NODE   UX   UY
   1   0.0000E+00 0.0000E+00
   2   0.0000E+00 0.0000E+00
   3   0.0000E+00 0.0000E+00
   4   0.0000E+00 0.0000E+00
   5   0.0000E+00 0.0000E+00
   6   0.9231E-02-0.5068E-02
   7   0.7025E-02-0.1038E-01
   8   0.6885E-02-0.1184E-01
   9   0.7698E-02-0.1272E-01
  10   0.9815E-02-0.1395E-01
  11   0.1542E-01-0.1258E-01
  12   0.1434E-01-0.1737E-01
MORE (YES,NO OR CONTINUOUS)=YES

            ***** POST1 NODAL DISPLACEMENT LISTING *****

LOAD STEP   1 ITERATION=   1 SECTION= 1
TIME=   0.00000E+00   LOAD CASE= 1

THE FOLLOWING X,Y,Z DISPLACEMENTS ARE IN NODAL COORDINATES

NODE   UX   UY
  13   0.1381E-01-0.2026E-01
  14   0.1425E-01-0.2216E-01
  15   0.1526E-01-0.2348E-01
  16   0.2022E-01-0.1859E-01
  17   0.1973E-01-0.2220E-01
  18   0.1943E-01-0.2518E-01
  19   0.1940E-01-0.2747E-01
  20   0.1977E-01-0.2922E-01
  21   0.2442E-01-0.2144E-01
  22   0.2419E-01-0.2421E-01
  23   0.2380E-01-0.2679E-01
  24   0.2350E-01-0.2897E-01
MORE (YES,NO OR CONTINUOUS)=YES

            ***** POST1 NODAL DISPLACEMENT LISTING *****

```
LOAD STEP      1  ITERATION=      1  SECTION=   1
TIME=    0.00000E+00      LOAD CASE=   1

THE FOLLOWING X,Y,Z DISPLACEMENTS ARE IN NODAL COORDINATES

NODE      UX           UY
  25  0.2346E-01-0.3088E-01

MAXIMUMS
NODE      21          25
VALUE   0.2442E-01-0.3088E-01
POST1 -INP=PRNSTR


PRINT  PRIN  NODAL STRESSES PER NODE

          ***** POST1 NODAL STRESS LISTING *****

LOAD STEP      1  ITERATION=      1  SECTION=   1
TIME=    0.00000E+00      LOAD CASE=   1

NODE      SIG1          SIG2          SIG3          SI          SIGE
  1   -2946.87      -6000.15      -27053.9      24107.0      22734.6
  2   -9277.37      -12355.1      -52498.2      43220.8      41787.4
  3   -10951.1      -14246.7      -60282.4      49331.3      47778.3
  4   -11387.2      -15207.4      -64650.0      53262.8      51463.0
  5   -10974.5      -16814.2      -73096.6      62122.1      59417.9
  6   -10105.6      -16416.9      -34111.1      24005.5      21887.6
  7   -11674.3      -15116.9      -45930.3      34255.9      32888.8
  8   -8138.66      -12561.0      -54666.6      46527.9      44490.2
  9   -3082.17      -12471.7      -59276.1      56194.0      52171.2
 10  -0.461823      -12569.5      -62847.1      62846.6      57612.4
 11   -8949.75      -14082.1      -30666.7      21716.9      19673.0
 12   -8389.78      -11655.8      -30293.1      21903.3      20583.9
 13   -6674.39      -8973.76      -36196.1      29521.7      28454.9
MORE (YES,NO OR CONTINUOUS)=YES


          ***** POST1 NODAL STRESS LISTING *****

LOAD STEP      1  ITERATION=      1  SECTION=   1
TIME=    0.00000E+00      LOAD CASE=   1

NODE      SIG1          SIG2          SIG3          SI          SIGE
 14   -3064.75      -8578.22      -39826.4      36761.6      34354.6
 15   -1705.69      -8520.64      -40897.5      39191.8      36284.4
 16   -5279.25      -7583.29      -18813.0      13533.7      12587.0
 17   -4695.50      -6864.65      -16612.9      11917.4      11106.5
 18   -4443.94      -5111.55      -18112.1      13668.2      13376.2
 19   -2629.02      -4512.74      -19934.7      17305.6      16471.5
 20   -1403.92      -4654.32      -21867.7      20463.7      19071.0
 21   -3044.25      -4146.76      -11074.5      8030.23      7539.68
 22   -2846.53      -4836.55      -9396.08      6549.56      5840.80
 23   -2914.17      -4875.39      -9695.48      6781.30      6054.62
 24   -2707.75      -3370.56      -10666.2      7958.42      7661.77
 25   -2251.84      -2887.73      -12186.8      9934.96      9632.77

MAXIMUMS
NODE       7            5             5            10            5
VALUE  -11674.3      -16814.2      -73096.6      62846.6      59417.9
```

SIMPLIFIED CONCRETE DAM

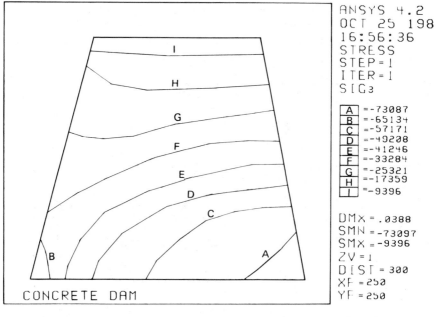

```
ANSYS 4.2
OCT 25 198
16:56:36
STRESS
STEP=1
ITER=1
SIG3

A =-73087
B =-65134
C =-57171
D =-49208
E =-41246
F =-33284
G =-25321
H =-17359
I =-9396

DMX=.0388
SMN=-73097
SMX=-9396
ZV=1
DIST=300
XF=250
YF=250
```

CONCRETE DAM

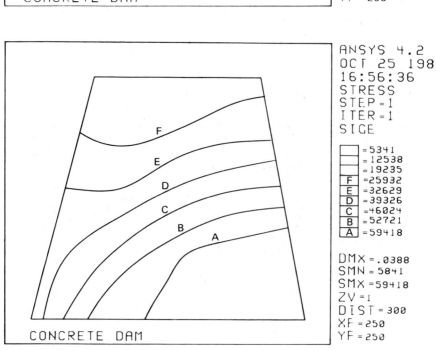

```
ANSYS 4.2
OCT 25 198
16:56:36
STRESS
STEP=1
ITER=1
SIGE

=5341
=12538
=19235
F =25932
E =32629
D =39326
C =46024
B =52721
A =59418

DMX=.0388
SMN=5841
SMX=59418
ZV=1
DIST=300
XF=250
YF=250
```

CONCRETE DAM

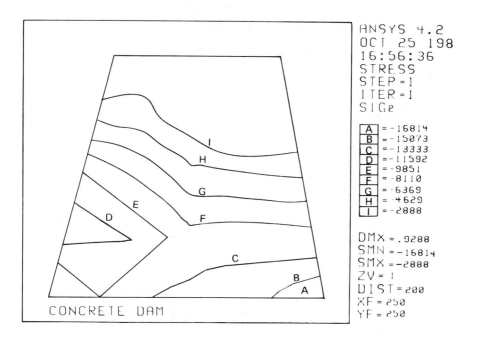

```
ANSYS 4.2
OCT 25 198
16:56:36
STRESS
STEP=1
ITER=1
SIG2

A = -16814
B = -15073
C = -13333
D = -11592
E = -9851
F = -8110
G = -6369
H = -4629
I = -2888

DMX = .9288
SMN = -16814
SMX = -2888
ZV = 1
DIST = 200
XF = 250
YF = 250
```

CONCRETE DAM

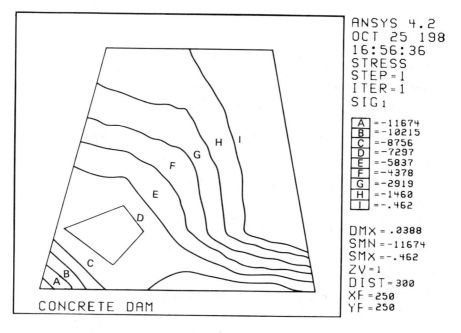

```
ANSYS 4.2
OCT 25 198
16:56:36
STRESS
STEP=1
ITER=1
SIG1

A = -11674
B = -10215
C = -8756
D = -7297
E = -5837
F = -4378
G = -2919
H = -1460
I = -.462

DMX = .0388
SMN = -11674
SMX = -.462
ZV = 1
DIST = 300
XF = 250
YF = 250
```

CONCRETE DAM

CONCRETE DAM SOLUTION: PCTRAN PLUS

```
?PCTRAN

C>echo off


P C T R A N    P l u s
(C) Copyright 1985, 1986    Brooks Scientific, Inc.
Version 3.2


ENTER JOB NAME: CONCDAM




Portions of the PCTRAN menu system are copyrighted
by Alpha Software Corporation.

C>PREP1

C>echo off

PCTRAN Plus 3.2 PREPROCESSOR
(C) Copyright 1985, 1986  Brooks Scientific, Inc.

      ... Please wait ...

CONCDAM     is a new job

* N,1,

Node:     1 =    .0000E+00    .0000E+00    .0000E+00
* NGEN,4,1,1,1,1,125

Generate    4 sets of nodes with nodal increment        1
from node pattern      1 to      1 in steps of    1
DX =    125.0000 DY =       .0000 DZ =        .0000
* N,6,25,125,

Node:      6 =   2.5000E+01  1.2500E+02    .0000E+00
* NGEN,4,1,6,6,1,112.5

Generate    4 sets of nodes with nodal increment        1
from node pattern      6 to      6 in steps of    1
DX =    112.5000 DY =       .0000 DZ =        .0000
* N,11,50,250,

Node:     11 =   5.0000E+01  2.5000E+02    .0000E+00
* NGEN,4,1,11,11,1,100

Generate    4 sets of nodes with nodal increment        1
from node pattern     11 to     11 in steps of    1
DX =    100.0000 DY =       .0000 DZ =        .0000
* N,16,75,375,

Node:     16 =   7.5000E+01  3.7500E+02    .0000E+00
* NGEN,4,1,16,16,1,87.5
```

```
Generate    4 sets of nodes with nodal increment       1
from node pattern     16 to     16 in steps of    1
DX =      87.5000 DY =        .0000 DZ =        .0000
* N,21,100,500

Node:     21 =    1.0000E+02   5.0000E+02     .0000E+00
* NGEN,4,1,21,21,1,75

Generate    4 sets of nodes with nodal increment       1
from node pattern     21 to     21 in steps of    1
DX =      75.0000 DY =        .0000 DZ =        .0000
* ETYPE,1,6

Element type   1 is STIFF    6      2-D SOLID (MEMBRANE)
IEOPT =   0        4 Nodes;   2 DOF per node.
* E,1,1,2,7,6

Element       1 =   1  1  1      1     2     7     6    0    0    0    0
* EGEN,3,1,1,1,1

Generate    3 sets of elements with nodal increment       1
from the element pattern      1 to      1 in steps of    1
* PLOT

* EGEN,3,5,1,4,1

Generate    3 sets of elements with nodal increment       5
from the element pattern      1 to      4 in steps of    1
* PLOT

* R,1,12

Real set    1. value(s) =      1.2000E+01     .0000E+00     .0000E+00     .0000E+00
     .0000E+00      .0000E+00
* EX,1,5.76E8

Elasticity for material number    1 is    5.7600E+08
* DENS,1,.0864

Density for material number   1 is    8.6400E-02
* NUXY,1,.2

Poissons ratio for material number    1 is    2.0000E-01
* D,1,UX,0.0,5,1

Define displacement from node       1 with label UX    and value     .0000E+00
to node      5 in steps of    1
* D,1,UY,0.0,5,1

Define displacement from node       1 with label UY    and value     .0000E+00
to node      5 in steps of    1
* F,16,FY,772106885200,4

Load case    1
Define force from node      16 with label FY    and value  -7.2108E+05
to node     20 in steps of    4
* F,17,FY,-1.4422E6,19,1

Load case    1
```

CONCRETE DAM SOLUTION: PCTRAN PLUS

```
Define force from node     17 with label FY    and value   -1.4422E+06
to node      19 in steps of      1
* F,11,FY,-8.3205E5,15,4

Load case    1
Define force from node     11 with label FY    and value   -8.3205E+05
to node      15 in steps of      4
* F,12,FY,-1.6641E6,14,1

Load case    1
Define force from node     12 with label FY    and value   -1.6641E+06
to node      14 in steps of      1
* F,6,FY,-9.4295E5,10,4

Load case    1
Define force from node      6 with label FY    and value   -9.4295E+05
to node      10 in steps of      4
* F,7,FY,-1.8859E6,9,1

Load case    1
Define force from node      7 with label FY    and value   -1.8859E+06
to node       9 in steps of      1
* F,1,FY,-1.0539E6,5,4

Load case    1
Define force from node      1 with label FY    and value   -1.0539E+06
to node       5 in steps of      4
* F,2,FY,-2.1078E6,4,1

Load case    1
Define force from node      2 with label FY    and value   -2.1078E+06
to node       4 in steps of      1
* F,16,FX,487500

Load case    1
Define force from node     16 with label FX    and value    4.8750E+05
to node      16 in steps of      1
* F,11,FX,1462500

Load case    1
Define force from node     11 with label FX    and value    1.4625E+06
to node      11 in steps of      1
* F,6,FX,2437500

Load case    1
Define force from node      6 with label FX    and value    2.4375E+06
to node       6 in steps of      1
* F,1,FX,3412500

Load case    1
Define force from node      1 with label FX    and value    3.4125E+06
to node       1 in steps of      1
* MPLIST

List Material properties
  MNUM     EXX           NUXY         DENS         ALPX         KXX
     1   5.760E+08    2.000E-01    8.640E-02    .000E+00     .000E+00
*
*
*
```

```
* NLIST,1,25

List nodes from    1 to    25 in steps of    1
 Node       X            Y            Z            SN
 ------  -----------  -----------  -----------  -----    - - - - - -
    1     .0000E+00    .0000E+00    .0000E+00       0    0 0 0 0 0 0
    2    1.2500E+02    .0000E+00    .0000E+00       0    0 0 0 0 0 0
    3    2.5000E+02    .0000E+00    .0000E+00       0    0 0 0 0 0 0
    4    3.7500E+02    .0000E+00    .0000E+00       0    0 0 0 0 0 0
    5    5.0000E+02    .0000E+00    .0000E+00       0    0 0 0 0 0 0
    6    2.5000E+01   1.2500E+02    .0000E+00       0    0 0 0 0 0 0
    7    1.3750E+02   1.2500E+02    .0000E+00       0    0 0 0 0 0 0
    8    2.5000E+02   1.2500E+02    .0000E+00       0    0 0 0 0 0 0
    9    3.6250E+02   1.2500E+02    .0000E+00       0    0 0 0 0 0 0
   10    4.7500E+02   1.2500E+02    .0000E+00       0    0 0 0 0 0 0
   11    5.0000E+01   2.5000E+02    .0000E+00       0    0 0 0 0 0 0
   12    1.5000E+02   2.5000E+02    .0000E+00       0    0 0 0 0 0 0
   13    2.5000E+02   2.5000E+02    .0000E+00       0    0 0 0 0 0 0
   14    3.5000E+02   2.5000E+02    .0000E+00       0    0 0 0 0 0 0
   15    4.5000E+02   2.5000E+02    .0000E+00       0    0 0 0 0 0 0
   16    7.5000E+01   3.7500E+02    .0000E+00       0    0 0 0 0 0 0
   17    1.6250E+02   3.7500E+02    .0000E+00       0    0 0 0 0 0 0
   18    2.5000E+02   3.7500E+02    .0000E+00       0    0 0 0 0 0 0
   19    3.3750E+02   3.7500E+02    .0000E+00       0    0 0 0 0 0 0
   20    4.2500E+02   3.7500E+02    .0000E+00       0    0 0 0 0 0 0
   21    1.0000E+02   5.0000E+02    .0000E+00       0    0 0 0 0 0 0
   22    1.7500E+02   5.0000E+02    .0000E+00       0    0 0 0 0 0 0
   23    2.5000E+02   5.0000E+02    .0000E+00       0    0 0 0 0 0 0
   24    3.2500E+02   5.0000E+02    .0000E+00       0    0 0 0 0 0 0
   25    4.0000E+02   5.0000E+02    .0000E+00       0    0 0 0 0 0 0
* ELIST,1,16

List element connectivities from    1 to    16 in steps of    1
ENUM   TP MN RN      CONNECTIVITY

    1    1  1  1    1    2    7    6    0    0    0    0
    2    1  1  1    2    3    8    7    0    0    0    0
    3    1  1  1    3    4    9    8    0    0    0    0
    4    1  1  1    4    5   10    9    0    0    0    0
    5    1  1  1    6    7   12   11    0    0    0    0
    6    1  1  1    7    8   13   12    0    0    0    0
    7    1  1  1    8    9   14   13    0    0    0    0
    8    1  1  1    9   10   15   14    0    0    0    0
    9    1  1  1   11   12   17   16    0    0    0    0
   10    1  1  1   12   13   18   17    0    0    0    0
   11    1  1  1   13   14   19   18    0    0    0    0
   12    1  1  1   14   15   20   19    0    0    0    0
   13    1  1  1   16   17   22   21    0    0    0    0
   14    1  1  1   17   18   23   22    0    0    0    0
   15    1  1  1   18   19   24   23    0    0    0    0
   16    1  1  1   19   20   25   24    0    0    0    0
* CHECK

Model checking routine . . .

     *** No errors detected ***
* EXIT


Statistics for job CONCDAM
```

CONCRETE DAM SOLUTION: PCTRAN PLUS

```
**********************************

Analysis type:  0 STATIC STRESS
Active coordinate system:    0    .0000E+00    .0000E+00    .0000E+00
Number of load cases:  1
Max node number:     25
Number of system nodes:    0
Max element number:    16
Number of applied displacements:    10
Half bandwidth:    0
Input file echo: ON
Reorder load cases in ascending order
Load Case, # of Loads
    1      24
Reorder displacements in ascending order
Copying "PCTX" files back to job CONCDAM
Exit preprocessor .....

C>STATICS

C>echo off

PCTRAN Plus 3.2 NODE RENUMBERING
(C) Copyright 1985, 1986  Brooks Scientific, Inc.

MAXN:          25
MAXE:          16
NSN:           25

NODE RENUMBERING [Y/N/(E to exit)]? Y
Enter number of starting nodes (30 max) NSTART =
Enter the  0 starting nodes [N1,N2, ... etc.]:

Nodes for element      1 renumbered
Nodes for element      2 renumbered
Nodes for element      5 renumbered
Nodes for element      6 renumbered
Nodes for element      3 renumbered
Nodes for element      7 renumbered
Nodes for element      9 renumbered
Nodes for element     10 renumbered
Nodes for element     11 renumbered
Nodes for element      4 renumbered
Nodes for element      8 renumbered
Nodes for element     12 renumbered
Nodes for element     13 renumbered
Nodes for element     14 renumbered
Nodes for element     15 renumbered
Nodes for element     16 renumbered

NUMBER OF NODES          =    25
NUMBER OF SYSTEM NODES   =    25
HALF BANDWIDTH           =    22
DEGREES OF FREEDOM       =    50

NODE RENUMBERING [Y/N/(E to exit)]? E

PCTRAN Plus 3.2 STIFFNESS ASSEMBLY
(C) Copyright 1985, 1986  Brooks Scientific, Inc.
```

```
Max node number:      25
Max element number:    16
Half bandwidth:      22
Total DOF:      50

Initializing stiffness file ...

Assembling element    1
Assembling element    2
Assembling element    3
Assembling element    4
Assembling element    5
Assembling element    6
Assembling element    7
Assembling element    8
Assembling element    9
Assembling element   10
Assembling element   11
Assembling element   12
Assembling element   13
Assembling element   14
Assembling element   15
Assembling element   16
EXIT STIFF . . .

PCTRAN Plus 3.2 BANDED EQUATION SOLVER
(C) Copyright 1985, 1986  Brooks Scientific, Inc.

NUMBER OF NODES:             25
NUMBER OF ELEMENTS:          16
NUMBER OF LOAD CASES:         1
# OF LOADS IN CASE  1:   24
DEGREES OF FREEDOM:          50
NUMBER OF DISPLACEMENTS:     10
ACTIVE DEGREES OF FREEDOM:   40
HALF BANDWIDTH:              22


LOADING GLOBAL STIFFNESS MATRIX ....


DECOMPOSITION ...
EQUATION SOLUTION: DOF =      5
EQUATION SOLUTION: DOF =     10
EQUATION SOLUTION: DOF =     15
EQUATION SOLUTION: DOF =     20
EQUATION SOLUTION: DOF =     25
EQUATION SOLUTION: DOF =     30
EQUATION SOLUTION: DOF =     35

BACKWARD SUBSTITUTION

REACTION FORCE CALCULATION


PCTRAN Plus 3.2 STRESS CALCULATION
(C) Copyright 1985, 1986  Brooks Scientific, Inc.

READ POST CODES FROM FILE "PCTRAN.CSD"
MAX NODE NUMBER:    25
```

CONCRETE DAM SOLUTION: PCTRAN PLUS

```
MAX ELEMENT NUMBER:     16
HALF BANDWIDTH:     22
TOTAL DOF:     50
NUMBER OF SYSTEM NODES:     25
STRESS CALCULATION: ELEMENT     1
STRESS CALCULATION: ELEMENT     6
STRESS CALCULATION: ELEMENT     11
STRESS CALCULATION: ELEMENT     16
EXIT STRESS ....

C>POST1

C>echo off

PCTRAN Plus 3.2 POSTPROCESSOR
(C) Copyright 1985, 1986  Brooks Scientific, Inc.

DEVICE DRIVER IS /PCTIBMG.DEV/     , MODE =  0
    ... Please wait ...

CONCDAM     IS AN OLD JOB

CALCULATE MIN,MAX POST DATA

    Statistics for CONCDAM
    ***************************

Analysis type:  0 STATIC STRESS
Active coordinate system:     0  .0000E+00    .0000E+00    .0000E+00
Current load case:  1
Max node number:     25
Max element number:     16
Number of displacemenets:     10
Half bandwidth:     22

Screen echo: OFF  , Input file echo: ON
Reading POST codes from file "PCTRAN.CSD"
* PRDISP,1,25
```

DISPLACEMENTS: LOAD CASE 1

NODE	UX	UY	UZ	ROTX	ROTY	ROTZ
1	.0000E+00	.0000E+00	.0000E+00	.0000E+00	.0000E+00	.0000E+00
2	.0000E+00	.0000E+00	.0000E+00	.0000E+00	.0000E+00	.0000E+00
3	.0000E+00	.0000E+00	.0000E+00	.0000E+00	.0000E+00	.0000E+00
4	.0000E+00	.0000E+00	.0000E+00	.0000E+00	.0000E+00	.0000E+00
5	.0000E+00	.0000E+00	.0000E+00	.0000E+00	.0000E+00	.0000E+00
6	6.5070E-04	-2.9739E-04	.0000E+00	.0000E+00	.0000E+00	.0000E+00
7	4.5336E-04	-6.8987E-04	.0000E+00	.0000E+00	.0000E+00	.0000E+00
8	4.1399E-04	-7.9360E-04	.0000E+00	.0000E+00	.0000E+00	.0000E+00
9	4.3847E-04	-8.5042E-04	.0000E+00	.0000E+00	.0000E+00	.0000E+00
10	5.3480E-04	-9.0505E-04	.0000E+00	.0000E+00	.0000E+00	.0000E+00
11	9.8362E-04	-8.0334E-04	.0000E+00	.0000E+00	.0000E+00	.0000E+00
12	8.8290E-04	-1.1147E-03	.0000E+00	.0000E+00	.0000E+00	.0000E+00
13	8.1418E-04	-1.3007E-03	.0000E+00	.0000E+00	.0000E+00	.0000E+00
14	8.0856E-04	-1.4214E-03	.0000E+00	.0000E+00	.0000E+00	.0000E+00
15	8.4776E-04	-1.5016E-03	.0000E+00	.0000E+00	.0000E+00	.0000E+00
16	1.1878E-03	-1.1665E-03	.0000E+00	.0000E+00	.0000E+00	.0000E+00
17	1.1544E-03	-1.3578E-03	.0000E+00	.0000E+00	.0000E+00	.0000E+00

```
18   1.1194E-03 =1.5340E-03  .0000E+00   .0000E+00   .0000E+00   .0000E+00
19   1.0995E-03 -1.6825E-03  .0000E+00   .0000E+00   .0000E+00   .0000E+00
20   1.1038E-03 -1.8071E-03  .0000E+00   .0000E+00   .0000E+00   .0000E+00
21   1.3690E-03 -1.2410E-03  .0000E+00   .0000E+00   .0000E+00   .0000E+00
22   1.3730E-03 -1.3675E-03  .0000E+00   .0000E+00   .0000E+00   .0000E+00
23   1.3614E-03 -1.5141E-03  .0000E+00   .0000E+00   .0000E+00   .0000E+00
24   1.3452E-03 -1.6536E-03  .0000E+00   .0000E+00   .0000E+00   .0000E+00
25   1.3350E-03 -1.7905E-03  .0000E+00   .0000E+00   .0000E+00   .0000E+00

                        ** MAX DEFLECTION **
NODE:      22         20          0           0           0           0
        1.3730E-03 -1.8071E-03  .0000E+00   .0000E+00   .0000E+00   .0000E+00
* PRRFOR
```

Reaction forces for load case 1

```
NODE  LABEL      RFORCE

  1    UX       1.4292E+05
  1    UY       5.5571E+05
  2    UX      -1.0485E+06
  2    UY       4.5267E+06
  3    UX      -1.1274E+06
  3    UY       5.6447E+06
  4    UX      -1.2856E+06
  4    UY       6.0200E+06
  5    UX      -1.0689E+06
  5    UY       3.2216E+06

TOTALS:

      UX          UY          UZ         ROTX        ROTY        ROTZ

  -4.3875E+06  1.9969E+07  .0000E+00   .0000E+00   .0000E+00   .0000E+00
* PRNSTR,1,25,1
```

Nodal Stresses: Load Case 1

```
NODE    SX         SY         SZ        SXY        SYZ        SXZ        SIGE

  1  -.2855E+03-.1427E+04  .0000E+00  .1249E+04  .0000E+00  .0000E+00  .0000E+00
  2  -.6623E+03-.3311E+04  .0000E+00  .8705E+03  .0000E+00  .0000E+00  .0000E+00
  3  -.7619E+03-.3809E+04  .0000E+00  .7949E+03  .0000E+00  .0000E+00  .0000E+00
  4  -.8164E+03-.4082E+04  .0000E+00  .8419E+03  .0000E+00  .0000E+00  .0000E+00
  5  -.8688E+03-.4344E+04  .0000E+00  .1027E+04  .0000E+00  .0000E+00  .0000E+00
  6  -.1354E+04-.1720E+04  .0000E+00  .1912E+03  .0000E+00  .0000E+00  .0000E+00
  7  -.1140E+04-.2669E+04  .0000E+00  .3436E+03  .0000E+00  .0000E+00  .0000E+00
  8  -.6641E+03-.3130E+04  .0000E+00  .6104E+03  .0000E+00  .0000E+00  .0000E+00
  9  -.3661E+03-.3377E+04  .0000E+00  .6702E+03  .0000E+00  .0000E+00  .0000E+00
 10  -.2186E+03-.3559E+04  .0000E+00  .7384E+03  .0000E+00  .0000E+00  .0000E+00
 11  -.9467E+03-.1833E+04  .0000E+00-.1832E+04  .0000E+00  .0000E+00  .0000E+00
 12  -.7991E+03-.1556E+04  .0000E+00  .9651E+02  .0000E+00  .0000E+00  .0000E+00
 13  -.5784E+03-.1822E+04  .0000E+00  .3092E+03  .0000E+00  .0000E+00  .0000E+00
 14  -.3107E+03-.2037E+04  .0000E+00  .3976E+03  .0000E+00  .0000E+00  .0000E+00
 15  -.2170E+03-.2214E+04  .0000E+00  .3727E+03  .0000E+00  .0000E+00  .0000E+00
 16  -.3866E+03-.8337E+03  .0000E+00-.1367E+04  .0000E+00  .0000E+00  .0000E+00
 17  -.3306E+03-.5276E+03  .0000E+00-.2421E+02  .0000E+00  .0000E+00  .0000E+00
 18  -.2906E+03-.5497E+03  .0000E+00  .8012E+02  .0000E+00  .0000E+00  .0000E+00
 19  -.1838E+03-.6615E+03  .0000E+00  .1386E+03  .0000E+00  .0000E+00  .0000E+00
```

CONCRETE DAM SOLUTION: PCTRAN PLUS

```
20  -.1436E+03-.8584E+03 .0000E+00 .1283E+03 .0000E+00 .0000E+00 .0000E+00
21   .1272E+01-.1488E+03 .0000E+00-.5953E+02 .0000E+00 .0000E+00 .0000E+00
22  -.1782E+02 .5700E+02 .0000E+00-.1602E+02 .0000E+00 .0000E+00 .0000E+00
23  -.9209E+02 .7365E+02 .0000E+00 .6759E+01 .0000E+00 .0000E+00 .0000E+00
24  -.9977E+02 .7079E+01 .0000E+00 .2519E+02 .0000E+00 .0000E+00 .0000E+00
25  -.1093E+03-.1557E+03 .0000E+00-.7702E+00 .0000E+00 .0000E+00 .0000E+00

                        ** MAX STRESS **
NODE:      6        5        0        1        0        0        0
        -.1354E+04-.4344E+04 .0000E+00 .1249E+04 .0000E+00 .0000E+00 .0000E+00
* DEFL,OVR

Plot label DEFL is: OVR
* PLOT

* MODE,1

Graphics mode set to   1
* PLOT
```

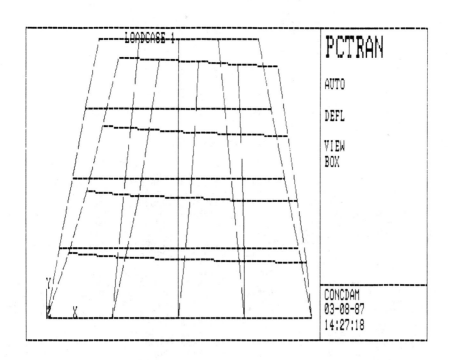

```
* PLOTSTAT


  Plot statistics for CONCDAM
**************************
Plot title:
Graphics mode is  1
Label color set:    2  0 10 12  7  2
Plot fill colors:   0  0  0  0  0  0  0  0  0  0  0
ELAB: 0 NLAB: 0 DEFL: 0 CONT: 1
CLAB: 1 MLAB: 0 TLAB: 0 RLAB: 0 HIDE: 0
View parameters: THETA,PHI =  -90.00    .00
Distance parameters: RHO,DIST =   1.0000E+05  1.0000E+05
Plot range:     .000E+00  5.000E+02   .000E+00  5.000E+02   .000E+00   .000E+00
Contour values = A: -1.354E+03 B: -1.204E+03 C: -1.053E+03
   D: -9.025E+02 E: -7.518E+02 F: -6.012E+02 G: -4.506E+02
   H: -3.000E+02 I: -1.494E+02 J:  1.273E+00
DBOX:   1.15
Deflection scale:    2.7669E+04
Plot mode:    AUTO
*
```

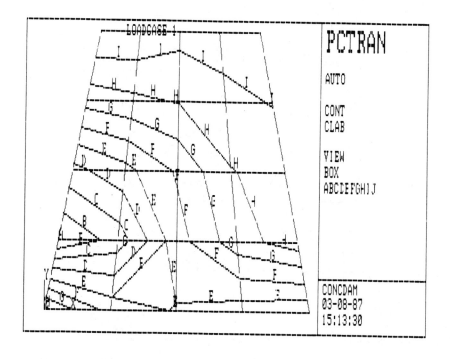

CONCRETE DAM SOLUTION: PCTRAN PLUS

```
* PLOTSTAT


   Plot statistics for CONCDAM
   ****************************

   Plot title:
   Graphics mode is  1
   Label color set:    2  0 10 12  7  2
   Plot fill colors.:  0  0  0  0  0  0  0  0  0  0
   ELAB: O NLAB: O DEFL: O CONT: 1
   CLAB: 1 MLAB: O TLAB: O RLAB: O HIDE: O
   View parameters: THETA,PHI =  -90.00     .00
   Distance parameters: RHO,DIST =    1.0000E+05  1.0000E+05
   Plot range:      .000E+00  5.000E+02    .000E+00  5.000E+02    .000E+00    .000E+00
   Contour values = A: -4.344E+03 B: -3.853E+03 C: -3.362E+03
      D: -2.872E+03 E: -2.381E+03 F: -1.890E+03 G: -1.399E+03
      H: -9.081E+02 I: -4.172E+02 J:  7.365E+01
   DBOX:   1.15
   Deflection scale:    2.7669E+04
   Plot mode:   AUTO
   *
```

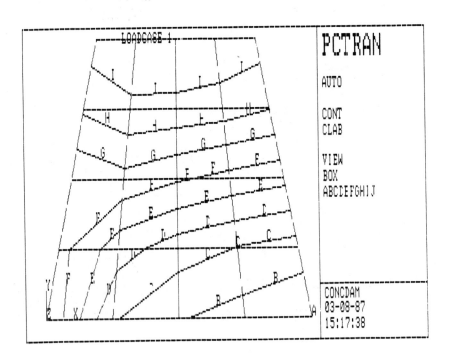

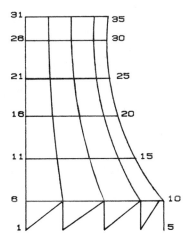

Figure 5.2 Hyperbolic cooling tower.

5.3 NATURAL FREQUENCY OF A HYPERBOLIC COOLING TOWER

The second problem addressed is that of determining the first natural frequency of a hyperbolic natural draft cooling tower. The natural draft cooling tower modeled is typical of that used at power plants to cool the plant circulating water system. A finite element model was developed that allowed several of the programs used in this book to be applied in the analysis. Problem size restraints among the various programs resulted in a "first cut" modeling effort.

The problem illustrated in Figure 5.2 illustrates the overall size of the cooling tower. Note that since the tower is symmetrical, only one-half is illustrated. This problem combines both the quadrilateral isoparametric solid and the three-dimensional beam element.

5.3.1 Input Data

In the analysis, only one-quarter of the tower is used to determine the first mode natural frequency. This is done to keep the problem size at a minimum and allow

Table 5.5   Key Node Data

Node	r	Θ	z
1	165.94	0	0
5	165.94	90	0
6	156.26	0	35
10	156.26	90	35
11	133.76	0	107.4
15	133.76	90	107.4
16	114.69	0	179.0
20	114.69	90	179.0
21	96.88	0	250.6
25	96.88	90	250.6
26	90.63	0	358.0
30	90.63	90	358.0
31	104.38	0	389.0
35	104.38	90	389.0

several programs to solve the problem. The data given for the input data for the node points is less detailed than for the concrete dam problem discussed earlier. Since the node and element generation capabilities of the finite element programs are used, only the minimum information is given at this level of presentation (see Table 5.5). Section 5.3.2 details all node and element data.

A natural frequency analysis requires fewer elements than a stress analysis in order to provide accurate results. Most finite element programs used a lumped mass approach in determining the natural frequency of a structure. As with the previous problems, element aspect ratios are within the limits described in Chapter 3.

There are two material requirements for this problem. One is for the concrete shell and the second is for the supporting columns. For this problem, the actual concrete and steel columns are replaced by equivalent steel columns. Table 5.6 lists the material properties.

Two element types are used. The first is the three-dimensional isoparametric quadrilateral element and the second is the three-dimensional beam element. The quad-plate thickness is assumed to be 12 inches.

Displacements or restraints are given in Table 5.7.

Table 5.6 Material Properties for Cooling Tower Problem

Concrete:	
Modulus of elasticity	4.0 E6 lb/in.^{2}
Poisson's ratio	0.2
Density	0.0868 lb/in.^{3}
Steel:	
Modulus of elasticity	30.0 E6 lb/in.^{2}
Poisson's ratio	0.3
Density	0.282 lb/in.^{3}

Table 5.7 Cooling Tower Model Restraints

Node	Ux	Uy	Uz	ROTX	ROTY	ROTZ
1	0	0	0	0	0	0
2	0	0	0	0	0	0
3	0	0	0	0	0	0
4	0	0	0	0	0	0
5	0	0	0	0	0	0
6		0				0
10		0	0			
11		0				0
15		0	0			
16		0				0
20		0	0			
21		0				0
25		0	0			
.26		0				0
31		0	0			
35		0				0

5.3.2 Solution

Two solutions are given to the cooling tower natural frequency problem. The first model is based on 35 nodes, 20 plate elements, and 9 beam elements. A natural frequency of 22.7 is calculated. In the second model, the number of nodes is increased to 120, plate elements to 90, and beam elements to 9. The calculated frequency is 23.5 Hz. For a first-cut effort, the additional amount of work to arrive at the second solution is unnecessary. Both solutions were run with the IMAGES3D program. The solutions illustrate the effect of increasing the number of elements and the effect on the answer. The first one is the 35-node model and the second is the increased-node model. The reader should compare the stress levels in the elements and the deflections for each case.

COOLING TOWER VIBRATION: IMAGES 3 D

```
?==============================================================================
=
================================= I M A G E S - 3 D ==========================
==============================================================================

02-13-1987                        Run ID=IQ92926                       17:55:39


----------------------------------------------------------------------------!
!                                                                            !
!  IIIIII    QQQQQQ    999999    222222    999999    222222    666666        !
!  IIIIII    QQQQQQ    999999    222222    999999    222222    666666        !
!    II     QQ    QQ 99     99 22      22 99     99 22      22 66             !
!    II     QQ    QQ 99     99 22      22 99     99 22      22 66             !
!    II     QQ    QQ 99     99         22 99     99         22 66             !
!    II     QQ    QQ 99     99         22 99     99         22 66             !
!    II     QQ    QQ 99999999       22    99999999       22 66666666         !
!    II     QQ    QQ 99999999       22    99999999       22 66666666         !
!    II     QQ QQ QQ        99 22           99 22      22 66       66 !
!    II     QQ QQ QQ        99 22           99 22      22 66       66 !
!    II     QQ    QQ        99 22           99 22      66       66 !
!    II     QQ    QQ        99 22           99 22      66       66 !
!  IIIIII    QQQQ QQ 999999 2222222222 999999 2222222222 666666             !
!  IIIIII    QQQQ QQ 999999 2222222222 999999 2222222222 666666             !
!                                                                            !
----------------------------------------------------------------------------!

                         IMAGES-3D   s/n:800189


        !---------------------------------------------------!
        !       J o b   I n f o r m a t i o n               !
        !---------------------------------------------------!
        !                                                   !
        !   Project    : _____             !
        !                                                   !
        !   Client     : _____             !
        !                                                   !
        !   Job Name   : _____             !
        !                                                   !
        !   Remarks    : _____             !
        !                                                   !
        !                                                   !
        !   Engineer   : _____/_____          !
        !                                                   !
        !   Chk'd by   : _____/_____          !
        !                                                   !
        !   Appr'd by  : _____/_____          !
        !                                                   !
        !   Comments   : _____                !
        !                                                   !
        !                _____                !
        !                                                   !
        !---------------------------------------------------!


==============================================================================
================================= I M A G E S - 3 D ==========================
==============================================================================
```

COOLING TOWER VIBRATION: IMAGES 3D

```
=============== I M A G E S   3 D ===============
= Copyright (c) 1984   Celestial Software Inc. =
================================================
```

Interactive Microcomputer Analysis & Graphics of Engineering Systems

IMAGES-3D   Version 1.4   12/01/86

RUN ID=IQ92926

```
================================================
=                   NOTICE                     =
=----------------------------------------------=
= Celestial Software Inc. assumes no responsi- =
= bility for the validity, accuracy, or        =
= applicability of the results obtained from   =
= IMAGES-3D.                                    =
================================================
```

```
================================================
= Any questions or comments concerning the use =
= of IMAGES-3D or the users manual should be   =
= addressed to:                                =
=                                              =
=           Celestial Software Inc.            =
=             125 University Ave.              =
=                 Berkeley,CA                  =
=                   94710                      =
=                                              =
=               415-420-0300                   =
================================================
```

STRUCTURAL MODELS

```
================ I M A G E S   3 D ================
= Copyright (c) 1984   Celestial Software Inc. =
==================================================
```

CHECK GEOMETRY   Version 1.4   12/01/86

COOLING TOWER VIBRATION

MATERIAL PROPERTIES

Material No	Modulus of Elasticity	Weight Density	Coeff of Thermal Exp.	Poisson's Ratio	Shear Web Modulus
1	4.00000E+06	8.68000E-02	0.00000E+00	2.00E-01	0.00000E+00
2	3.00000E+07	2.82000E-01	0.00000E+00	3.00E-01	0.00000E+00

NODE COORDINATES

Node	X-Coord.	Y-Coord.	Z-Coord.
1	1.65940E+02	0.00000E+00	0.00000E+00
2	1.53309E+02	0.00000E+00	-6.35025E+01
3	1.17337E+02	0.00000E+00	-1.17337E+02
4	6.35025E+01	0.00000E+00	-1.53309E+02
5	0.00000E+00	0.00000E+00	-1.65940E+02
6	1.56260E+02	3.50000E+01	0.00000E+00
7	1.44365E+02	3.50000E+01	-5.97981E+01
8	1.10493E+02	3.50000E+01	-1.10493E+02
9	5.97981E+01	3.50000E+01	-1.44365E+02
10	0.00000E+00	3.50000E+01	-1.56260E+02
11	1.33760E+02	1.07400E+02	0.00000E+00
12	1.23578E+02	1.07400E+02	-5.11877E+01
13	9.45826E+01	1.07400E+02	-9.45826E+01
14	5.11877E+01	1.07400E+02	-1.23578E+02
15	0.00000E+00	1.07400E+02	-1.33760E+02
16	1.14690E+02	1.79000E+02	0.00000E+00
17	1.05960E+02	1.79000E+02	-4.38900E+01
18	8.10981E+01	1.79000E+02	-8.10981E+01
19	4.38900E+01	1.79000E+02	-1.05960E+02
20	0.00000E+00	1.79000E+02	-1.14690E+02
21	9.68800E+01	2.50600E+02	0.00000E+00
22	8.95054E+01	2.50600E+02	-3.70744E+01
23	6.85045E+01	2.50600E+02	-6.85045E+01
24	3.70744E+01	2.50600E+02	-8.95054E+01
25	0.00000E+00	2.50600E+02	-9.68800E+01
26	9.06300E+01	3.58000E+02	0.00000E+00
27	8.37312E+01	3.58000E+02	-3.46826E+01

COOLING TOWER VIBRATION: IMAGES 3D

```
IMAGES-3D  s/n:800189                                      02-13-1987
                                                           PAGE 2
        ================ I M A G E S  3 D ===============
        = Copyright (c) 1984   Celestial Software Inc. =
        ================================================

        CHECK GEOMETRY                 Version 1.4  12/01/86

        COOLING TOWER VIBRATION

            Node    X-Coord.       Y-Coord.       Z-Coord.
            ----   ------------   ------------   ------------
             28    6.40851E+01    3.58000E+02    -6.40851E+01
             29    3.46826E+01    3.58000E+02    -8.37312E+01
             30    0.00000E+00    3.58000E+02    -9.06300E+01
             31    1.04380E+02    3.89000E+02     0.00000E+00
             32    9.64345E+01    3.89000E+02    -3.99445E+01
             33    7.38078E+01    3.89000E+02    -7.38078E+01
             34    3.99445E+01    3.89000E+02    -9.64345E+01
             35    0.00000E+00    3.89000E+02    -1.04380E+02


                            BEAM PROPERTIES

        Multiplier = 1  (For AISC database properties only)

            Prop   X-Section    Moment of Inertia     Torsional
            No       Area        Iy   /   Iz         Const.- J
            ----   ----------   ---------- ----------  ----------
             1     5.857E+02    2.433E+04  2.433E+04   2.433E+04

            Prop   Max. Fiber Dist     Shear Shape Fact
            No       Cy   /   Cz        SSFy  /  SSFz        Ctors
            ----   -------- --------   -------- --------   ----------
             1     1.00E+00 1.00E+00   0.00E+00 0.00E+00   1.00E+00


                            BEAM CONNECTIVITY

        Beam    Nodes     Prop Mat   Pincodes                Y Dir Cosines    Beam
        No   From/ To/Ref  No   No    I  /  J    Length      X     Y     Z    Type
        ----  ------------  ---- ---  ------------- ---------- ----- ----- ----- -----
          1    1   6   0    1    2               3.631E+01   0.96  0.27  0.00 Beam
          2    1   7   0    1    2               7.257E+01   0.16  0.88  0.45 Beam
          3    2   7   0    1    2               3.631E+01   0.89  0.27 -0.37 Beam
          4    2   8   0    1    2               7.257E+01   0.32  0.88  0.36 Beam
          5    3   8   0    1    2               3.631E+01   0.68  0.27 -0.68 Beam
          6    3   9   0    1    2               7.257E+01   0.44  0.88  0.21 Beam
          7    4   9   0    1    2               3.631E+01   0.37  0.27 -0.89 Beam
          8    4  10   0    1    2               7.257E+01   0.48  0.88  0.02 Beam
          9    5  10   0    1    2               3.631E+01   0.00  0.27 -0.96 Beam
```

IMAGES-3D s/n:800189   02=13=1987
                                                         PAGE 3
         ================ I M A G E S   3 D ================
         = Copyright (c) 1984   Celestial Software Inc. =
         ==

           CHECK GEOMETRY   Version 1.4 12/01/86

      COOLING TOWER VIBRATION

                     PLATE ELEMENT CONNECTIVITY

     Plate   N o d e s   Mat   Shear Web Aspect   Plate
     No.   I   J   K   L   No. Thickness   Area   Thickness Ratio   Type
     ----- --------------- --- --------- --------- ---------- --------- --------
QUAD   1   6   7 12 11   1 1.200E+01 4.282E+03   7.475E-01 Mem+Bend
QUAD   2   7   8 13 12   1 1.200E+01 4.282E+03   7.475E-01 Mem+Bend
QUAD   3   8   9 14 13   1 1.200E+01 4.282E+03   7.475E-01 Mem+Bend
QUAD   4   9 10 15 14   1 1.200E+01 4.282E+03   7.475E-01 Mem+Bend
QUAD   5   11 12 17 16   1 1.200E+01 3.587E+03   6.550E-01 Mem+Bend
QUAD   6   12 13 18 17   1 1.200E+01 3.587E+03   6.550E-01 Mem+Bend
QUAD   7   13 14 19 18   1 1.200E+01 3.587E+03   6.550E-01 Mem+Bend
QUAD   8   14 15 20 19   1 1.200E+01 3.587E+03   6.550E-01 Mem+Bend
QUAD   9   16 17 22 21   1 1.200E+01 3.042E+03   5.600E-01 Mem+Bend
QUAD 10   17 18 23 22   1 1.200E+01 3.042E+03   5.600E-01 Mem+Bend
QUAD 11   18 19 24 23   1 1.200E+01 3.042E+03   5.600E-01 Mem+Bend
QUAD 12   19 20 25 24   1 1.200E+01 3.042E+03   5.600E-01 Mem+Bend
QUAD 13   21 22 27 26   1 1.200E+01 3.935E+03   3.401E-01 Mem+Bend
QUAD 14   22 23 28 27   1 1.200E+01 3.935E+03   3.401E-01 Mem+Bend
QUAD 15   23 24 29 28   1 1.200E+01 3.935E+03   3.401E-01 Mem+Bend
QUAD 16   24 25 30 29   1 1.200E+01 3.935E+03   3.401E-01 Mem+Bend
QUAD 17   26 27 32 31   1 1.200E+01 1.286E+03   1.125E+00 Mem+Bend
QUAD 18   27 28 33 32   1 1.200E+01 1.286E+03   1.125E+00 Mem+Bend
QUAD 19   28 29 34 33   1 1.200E+01 1.286E+03   1.125E+00 Mem+Bend
QUAD 20   29 30 35 34   1 1.200E+01 1.286E+03   1.125E+00 Mem+Bend

                         RESTRAINTS

                Node   Restraint
                No   Directions
                ----   ----------------

                 1   X Y Z RX RY RZ
                 2   X Y Z RX RY RZ
                 3   X Y Z RX RY RZ
                 4   X Y Z RX RY RZ
                 5   X Y Z RX RY RZ
                 6   - - Z - - RZ
                10   - - Z RX - -
                11   - - Z - - RZ
                15   - - Z RX - -
                16   - - Z - - RZ
                20   - - Z RX - -

COOLING TOWER VIBRATION: IMAGES 3D

```
IMAGES-3D   s/n:800189
                              02-13-1987
                                                    PAGE 4
=============== I M A G E S   3 D ===============
= Copyright (c) 1984   Celestial Software Inc. =
================================================

    CHECK GEOMETRY                Version 1.4  12/01/86

   COOLING TOWER VIBRATION


                    Node       Restraint
                    No         Directions
                    ----       -----------------

                    21      - - Z -  -  RZ
                    25      - - Z RX -  -
                    26      - - Z -  -  RZ
                    30      - - Z RX -  -
                    31      - - Z -  -  RZ
                    35      - - Z RX -  -



IMAGES-3D   s/n:800189
                                                    02=13=1987
                                                    PAGE 1
=============== I M A G E S   3 D ===============
= Copyright (c) 1984   Celestial Software Inc. =
================================================

    RENUMBER NODES                Version 1.4  12/01/86

   COOLING TOWER VIBRATION

            Node Renumbering Cross Reference List

             Was    Is     Was    Is     Was    Is
             ----   ----    ----   ----    ----   ----
               1     1        2     2        3     3
               4     4        5     5        6     6
               7     7        8     8        9     9
              10    10       11    11       12    12
              13    13       14    14       15    15
              16    16       17    17       18    18
              19    19       20    20       21    21
              22    22       23    23       24    24
              25    25       26    26       27    27
              28    28       29    29       30    30
              31    31       32    32       33    33
              34    34       35    35

                 Original Nodal Band    7
                 Final Nodal band       7
```

IMAGES-3D   s/n:800189   02-13-87
                                                          PAGE   1
```
================ I M A G E S   3 D ================
= Copyright (c) 1984   Celestial Software Inc. =
================================================
```

ASSEMBLE STIFFNESS MATRIX   Version 1.4   12/01/86

COOLING TOWER VIBRATION

                STIFFNESS ASSEMBLY SUMMARY

        Number of Node Points.................   35
        Number of Truss and Beam Elements.....   9
        Number of Plate Elements..............   20
        Number of Spring Elements.............   0
        Number of Solid Elements..............   0
        Number of Axisymmetric Elements.......   0
        Number of Nodes with Restraints.......   17
        Number of Equations to Be Solvesd..... 156
        Number of Blocks in the Matrix........   1

            B L O C K   I N F O R M A T I O N

BLCK   SIZE   BLCK   SIZE   BLCK   SIZE   BLCK   SIZE
 NO   (Byte)   NO   (Byte)   NO   (Byte)   NO   (Byte)
---- ------   ---- ------   ---- ------   ---- ------
  1   36784

COOLING TOWER VIBRATION: IMAGES 3D

```
IMAGES-3D   s/n:800189                                          02-13-87
                                                               PAGE    2
================= I M A G E S   3 D ================
= Copyright (c) 1984   Celestial Software Inc. =
===================================================

ASSEMBLE STIFFNESS MATRIX  Version 1.4  12/01/86

COOLING TOWER VIBRATION


        E Q U A T I O N   N U M B E R   L I S T


        NODE         TRANSLATION         ROTATION
        WAS  IS    X     Y     Z      X     Y     Z
        ---  ---  ---   ---   ---    ---   ---   ---
         1    1    0     0     0      0     0     0
         2    2    0     0     0      0     0     0
         3    3    0     0     0      0     0     0
         4    4    0     0     0      0     0     0
         5    5    0     0     0      0     0     0
         6    6    1     2     0      3     4     0
         7    7    5     6     7      8     9    10
         8    8   11    12    13     14    15    16
         9    9   17    18    19     20    21    22
        10   10   23    24     0      0    25    26
        11   11   27    28     0     29    30     0
        12   12   31    32    33     34    35    36
        13   13   37    38    39     40    41    42
        14   14   43    44    45     46    47    48
        15   15   49    50     0      0    51    52
        16   16   53    54     0     55    56     0
        17   17   57    58    59     60    61    62
        18   18   63    64    65     66    67    68
        19   19   69    70    71     72    73    74
        20   20   75    76     0      0    77    78
        21   21   79    80     0     81    82     0
        22   22   83    84    85     86    87    88
        23   23   89    90    91     92    93    94
        24   24   95    96    97     98    99   100
        25   25  101   102     0      0   103   104
        26   26  105   106     0    107   108     0
        27   27  109   110   111    112   113   114
        28   28  115   116   117    118   119   120
        29   29  121   122   123    124   125   126
        30   30  127   128     0      0   129   130
        31   31  131   132     0    133   134     0
        32   32  135   136   137    138   139   140
        33   33  141   142   143    144   145   146
        34   34  147   148   149    150   151   152
        35   35  153   154     0      0   155   156
```

STRUCTURAL MODELS

```
IMAGES-3D   s/n:800189                                    02-13-87
                                                         PAGE    3
============== I M A G E S   3 D ===============
= Copyright (c) 1984   Celestial Software Inc. =
================================================

ASSEMBLE STIFFNESS MATRIX  Version 1.4  12/01/86

COOLING TOWER VIBRATION


            STIFFNESS SUMMARY IN  1 BLOCKS

            Minimum Diagonal Stiffness.....  .9980D+07
            Eq No of Minimum Diagonal......   27
            Maximum Diagonal Stiffness.....  .1155D+12
            Eq No of Maximum Diagonal......   26

IMAGES-3D   s/n:800189                                    02-13-87
                                                         PAGE    1
============== I M A G E S   3 D ===============
= Copyright (c) 1984   Celestial Software Inc. =
================================================

WEIGHTS                          Version 1.4  12/01/86

COOLING TOWER VIBRATION

                         Weight Matrix

        W e i g h t s       /   R o t a r y   I n e r t i a s
Node     X          Y          Z    /   X          Y          Z
----  ---------  ---------  --------- / ---------  ---------  ---------
  1   .8992E+04  .8992E+04  .8992E+04 / .0000E+00  .0000E+00  .0000E+00
  2   .8992E+04  .8992E+04  .8992E+04 / .0000E+00  .0000E+00  .0000E+00
  3   .8992E+04  .8992E+04  .8992E+04 / .0000E+00  .0000E+00  .0000E+00
  4   .8992E+04  .8992E+04  .8992E+04 / .0000E+00  .0000E+00  .0000E+00
  5   .2999E+04  .2999E+04  .2999E+04 / .0000E+00  .0000E+00  .0000E+00
  6   .4143E+04  .4143E+04  .4143E+04 / .0000E+00  .0000E+00  .0000E+00
  7   .1128E+05  .1128E+05  .1128E+05 / .0000E+00  .0000E+00  .0000E+00
  8   .1128E+05  .1128E+05  .1128E+05 / .0000E+00  .0000E+00  .0000E+00
  9   .1128E+05  .1128E+05  .1128E+05 / .0000E+00  .0000E+00  .0000E+00
 10   .1014E+05  .1014E+05  .1014E+05 / .0000E+00  .0000E+00  .0000E+00
 11   .2044E+04  .2044E+04  .2044E+04 / .0000E+00  .0000E+00  .0000E+00
 12   .4088E+04  .4088E+04  .4088E+04 / .0000E+00  .0000E+00  .0000E+00
 13   .4088E+04  .4088E+04  .4088E+04 / .0000E+00  .0000E+00  .0000E+00
 14   .4088E+04  .4088E+04  .4088E+04 / .0000E+00  .0000E+00  .0000E+00
 15   .2044E+04  .2044E+04  .2044E+04 / .0000E+00  .0000E+00  .0000E+00
 16   .1724E+04  .1724E+04  .1724E+04 / .0000E+00  .0000E+00  .0000E+00
 17   .3449E+04  .3449E+04  .3449E+04 / .0000E+00  .0000E+00  .0000E+00
 18   .3449E+04  .3449E+04  .3449E+04 / .0000E+00  .0000E+00  .0000E+00
 19   .3449E+04  .3449E+04  .3449E+04 / .0000E+00  .0000E+00  .0000E+00
 20   .1724E+04  .1724E+04  .1724E+04 / .0000E+00  .0000E+00  .0000E+00
 21   .1806E+04  .1806E+04  .1806E+04 / .0000E+00  .0000E+00  .0000E+00
 22   .3612E+04  .3612E+04  .3612E+04 / .0000E+00  .0000E+00  .0000E+00
 23   .3612E+04  .3612E+04  .3612E+04 / .0000E+00  .0000E+00  .0000E+00
 24   .3612E+04  .3612E+04  .3612E+04 / .0000E+00  .0000E+00  .0000E+00
 25   .1806E+04  .1806E+04  .1806E+04 / .0000E+00  .0000E+00  .0000E+00
 26   .1340E+04  .1340E+04  .1340E+04 / .0000E+00  .0000E+00  .0000E+00
 27   .2681E+04  .2681E+04  .2681E+04 / .0000E+00  .0000E+00  .0000E+00
 28   .2681E+04  .2681E+04  .2681E+04 / .0000E+00  .0000E+00  .0000E+00
 29   .2681E+04  .2681E+04  .2681E+04 / .0000E+00  .0000E+00  .0000E+00
 30   .1340E+04  .1340E+04  .1340E+04 / .0000E+00  .0000E+00  .0000E+00
 31   .3428E+03  .3428E+03  .3428E+03 / .0000E+00  .0000E+00  .0000E+00
 32   .6856E+03  .6856E+03  .6856E+03 / .0000E+00  .0000E+00  .0000E+00
 33   .6856E+03  .6856E+03  .6856E+03 / .0000E+00  .0000E+00  .0000E+00
 34   .6856E+03  .6856E+03  .6856E+03 / .0000E+00  .0000E+00  .0000E+00
 35   .3428E+03  .3428E+03  .3428E+03 / .0000E+00  .0000E+00  .0000E+00
```

COOLING TOWER VIBRATION: IMAGES 3D

```
IMAGES-3D   s/n:800189                                    02-13-87
                                                         PAGE    2
            ================ I M A G E S  3 D ================
            = Copyright (c) 1984  Celestial Software Inc. =
            ================================================

            WEIGHTS                        Version 1.4  12/01/86

         COOLING TOWER VIBRATION

                  W e i g h t s      /  R o t a r y   I n e r t i a s
         Node      X         Y        Z  /    X         Y         Z
         ----   --------- --------- --------- / --------- --------- ---------

         Total:  .1451E+06 .1451E+06 .1451E+06 / .0000E+00 .0000E+00 .0000E+00

         Total:  .1062E+06 .1062E+06 .7739E+05 / .0000E+00 .0000E+00 .0000E+00
         (used)

                        Center of Gravity Based on X-Weights

            X = .877213E+02   Y = .994587E+02   Z = -.873216E+02

                        Center of Gravity Based on Y-Weights

            X = .877213E+02   Y = .994587E+02   Z = -.873216E+02

                        Center of Gravity Based on Z-Weights

            X = .877213E+02   Y = .994587E+02   Z = -.873216E+02


   IMAGES-3D   s/n:800189                                 02-13-87
                                                         PAGE    1
            ================ I M A G E S  3 D ================
            = Copyright (c) 1984  Celestial Software Inc. =
            ================================================

            SOLVE FREQUENCIES              Version 1.4  12/01/86

         COOLING TOWER VIBRATION

                    Number of frequencies requested    1
                    Number of frequencies printed      1
                    Acceleration of gravity        386.40

            Mode   Eigenvalue     Frequency      Period
            ----   -----------   -----------    -----------
              1   .203639E+05   .227117E+02    .440301E-01
```

```
IMAGES-3D   s/n:800189                            02-13-87
                                                 PAGE   1
          =============== I M A G E S   3 D ===============
          = Copyright (c) 1984   Celestial Software Inc. =
          ================================================

             SOLVE MODE SHAPES           Version 1.4  12/01/86

         COOLING TOWER VIBRATION


                     Number of modes        1

               ***Mode  1***     Eigenvalue= .203639E+05
```

Node	T r a n s l a t i o n s			/	R o t a t i o n s		
	X	Y	Z	/	X	Y	Z
1	.000000	.000000	.000000	/	.000000	.000000	.000000
2	.000000	.000000	.000000	/	.000000	.000000	.000000
3	.000000	.000000	.000000	/	.000000	.000000	.000000
4	.000000	.000000	.000000	/	.000000	.000000	.000000
5	.000000	.000000	.000000	/	.000000	.000000	.000000
6	.000674	-.001584	.000000	/	.000001	-.000003	.000000
7	.002370	-.002025	-.001544	/	-.000060	-.000010	-.000076
8	.001649	-.000422	-.001131	/	-.000052	.000003	-.000065
9	.001418	.002523	.000990	/	.000002	.000021	-.000070
10	.002584	.003536	.000000	/	.000000	.000010	-.000102
11	.043088	-.035347	.000000	/	-.004109	-.001666	.000000
12	.063266	-.023570	-.008422	/	-.000237	-.000326	-.001264
13	.071605	.002737	-.018087	/	-.000525	-.000120	-.001456
14	.071978	.045595	-.018717	/	-.000614	.000221	-.001565
15	.079035	.114515	.000000	/	.000000	.001148	-.004118
16	.153963	-.051242	.000000	/	-.008636	-.003470	.000000
17	.202752	-.026657	-.015956	/	.000163	-.000957	-.002452
18	.232222	.024805	-.040617	/	-.000453	-.000586	-.002485
19	.234904	.102810	-.043508	/	-.000360	.000630	-.002846
20	.232562	.201689	.000000	/	.000000	.001657	-.004040
21	.332529	-.042969	.000000	/	-.011709	-.004073	.000000
22	.407634	-.008396	-.016100	/	.000355	-.002130	-.003853
23	.462152	.058818	-.054335	/	-.000127	-.001261	-.003461
24	.468159	.155121	-.062228	/	-.000055	.000906	-.003436
25	.459664	.270361	.000000	/	.000000	.002293	-.003922
26	.788833	-.024509	.000000	/	-.007334	-.000470	.000000
27	.844204	-.008317	-.019687	/	-.000219	-.001211	-.003387
28	.876212	.065252	-.044641	/	-.000020	-.000644	-.003713
29	.878382	.178461	-.045067	/	.000020	.000720	-.003863
30	.871969	.319334	.000000	/	.000000	.001527	-.003906
31	.886271	-.063565	.000000	/	-.020770	.006568	.000000
32	.964777	-.057953	-.007545	/	-.000329	-.001476	-.003751
33	1.000000	.027713	-.035068	/	.000205	-.000830	-.003966
34	.995728	.159318	-.041357	/	.000229	.000558	-.004062
35	.984152	.322834	.000000	/	.000000	.001530	-.003922

COOLING TOWER VIBRATION: IMAGES 3D

```
IMAGES-3D  s/n:800189                                    02-13-87
                                                        PAGE    1
          =============== I M A G E S  3 D ===============
          = Copyright (c) 1984   Celestial Software Inc. =
          ================================================

          SOLVE PARTICIPATION        Version 1.4  12/01/86

       COOLING TOWER VIBRATION

                                     PARTICIPATION  FACTORS
     Mode    Generalized Weight        X         Y          Z
     ----    ------------------     ---------  ---------  ----------
       1          .1471E+05         .1516E+01  .2215E+00  -.9378E-01


                 EFFECTIVE MODAL WEIGHTS        % TOTAL SYSTEM WEIGHTS
     Mode      X         Y          Z          X        Y        Z
     ----    --------  --------  ---------   --------  ------  --------
       1     .3379E+05 .7216E+03 .1293E+03    31.82     .68      .17

                                Summation:    31.82     .68      .17
```

```
IMAGES-3D   s/n:800189                              03-05-87
                                                    PAGE    1
          =============== I M A G E S  3 D ===============
          = Copyright (c) 1984   Celestial Software Inc. =
          ================================================

          SOLVE SEISMIC RESPONSE       Version 1.4  12/01/86

      COOLING TOWER VIBRATION


                         Spectrum Multipliers

                    X-Direction =  .1000E+01
                    Y-Direction =  .1000E+01
                    Z-Direction =  .1000E+01

                      INPUT RESPONSE SPECTRA

              X-Direction    /     Y-Direction    /    Z-Direction
      Point Frequency Acceleration Frequency Acceleration Frequency Acceleration
      ----- --------- ------------ --------- ------------ --------- ------------
        1  .227E+02  .100000E+01

                     Interpolated Accelerations

              Mode  Frequency  X-Direction  Y-Direction  Z-Direction
              ----  ---------  -----------  -----------  -----------
                1  .2271E+02  .10000E+01   .00000E+00   .00000E+00

                     Generalized Displacements

                                                            A B S
      Mode  Frequency  X-Direction  Y-Direction  Z-Direction  Combination
      ----  ---------  -----------  -----------  -----------  -----------
        1  .2271E+02  .28760E-01   .00000E+00   .00000E+00   .28760E-01

                     Generalized Accelerations

                                                            A B S
      Mode  Frequency  X-Direction  Y-Direction  Z-Direction  Combination
      ----  ---------  -----------  -----------  -----------  -----------
        1  .2271E+02  .58568E+03   .00000E+00   .00000E+00   .58568E+03

                    DISPLACEMENTS for MODE  1

               T r a n s l a t i o n s   /    R o t a t i o n s
      Node     X          Y          Z     /    X          Y          Z
      ----  ---------- ---------- ---------- / ---------- ---------- ----------
        1  .0000E+00  .0000E+00  .0000E+00 / .0000E+00  .0000E+00  .0000E+00
        2  .0000E+00  .0000E+00  .0000E+00 / .0000E+00  .0000E+00  .0000E+00
        3  .0000E+00  .0000E+00  .0000E+00 / .0000E+00  .0000E+00  .0000E+00
```

COOLING TOWER VIBRATION: IMAGES 3D

```
IMAGES-3D   s/n:800189                                      03-05-87
                                                           PAGE    2
================ I M A G E S   3 D ================
= Copyright (c) 1984   Celestial Software Inc. =
==================================================

         SOLVE SEISMIC RESPONSE      Version 1.4  12/01/86

   COOLING TOWER VIBRATION


          T r a n s l a t i o n s   /      R o t a t i o n s
   Node   X          Y          Z      /   X          Y          Z
   ----   ----------  ----------  ----------  /  ----------  ----------  ----------
    4   .0000E+00   .0000E+00   .0000E+00  /  .0000E+00   .0000E+00   .0000E+00
    5   .0000E+00   .0000E+00   .0000E+00  /  .0000E+00   .0000E+00   .0000E+00
    6   .1937E-04  -.4555E-04   .0000E+00  /  .1580E-07  -.8863E-07   .0000E+00
    7   .6818E-04  -.5825E-04  -.4441E-04  / -.1721E-05  -.2936E-06  -.2176E-05
    8   .4744E-04  -.1215E-04  -.3252E-04  / -.1488E-05   .9464E-07  -.1859E-05
    9   .4077E-04   .7257E-04   .2848E-04  /  .5766E-07   .6094E-06  -.2024E-05
   10   .7431E-04   .1017E-03   .0000E+00  /  .0000E+00   .2737E-06  -.2941E-05
   11   .1239E-02  -.1017E-02   .0000E+00  / -.1182E-03  -.4792E-04   .0000E+00
   12   .1820E-02  -.6779E-03  -.2422E-03  / -.6815E-05  -.9364E-05  -.3635E-04
   13   .2059E-02   .7871E-04  -.5202E-03  / -.1509E-04  -.3449E-05  -.4188E-04
   14   .2070E-02   .1311E-02  -.5383E-03  / -.1765E-04   .6355E-05  -.4502E-04
   15   .2273E-02   .3293E-02   .0000E+00  /  .0000E+00   .3301E-04  -.1184E-03
   16   .4428E-02  -.1474E-02   .0000E+00  / -.2484E-03  -.9978E-04   .0000E+00
   17   .5831E-02  -.7667E-03  -.4589E-03  /  .4683E-05  -.2751E-04  -.7053E-04
   18   .6679E-02   .7134E-03  -.1168E-02  / -.1303E-04  -.1685E-04  -.7146E-04
   19   .6756E-02   .2957E-02  -.1251E-02  / -.1036E-04   .1811E-04  -.8185E-04
   20   .6689E-02   .5801E-02   .0000E+00  /  .0000E+00   .4766E-04  -.1162E-03
   21   .9564E-02  -.1236E-02   .0000E+00  / -.3368E-03  -.1172E-03   .0000E+00
   22   .1172E-01  -.2415E-03  -.4630E-03  /  .1021E-04  -.6126E-04  -.1108E-03
   23   .1329E-01   .1692E-02  -.1563E-02  / -.3662E-05  -.3626E-04  -.9954E-04
   24   .1346E-01   .4461E-02  -.1790E-02  / -.1585E-05   .2606E-04  -.9882E-04
   25   .1322E-01   .7776E-02   .0000E+00  /  .0000E+00   .6596E-04  -.1128E-03
   26   .2269E-01  -.7049E-03   .0000E+00  / -.2109E-03  -.1351E-04   .0000E+00
   27   .2428E-01  -.2392E-03  -.5662E-03  / -.6311E-05  -.3483E-04  -.9742E-04
   28   .2520E-01   .1877E-02  -.1284E-02  / -.5752E-06  -.1854E-04  -.1068E-03
   29   .2526E-01   .5133E-02  -.1296E-02  /  .5609E-06   .2071E-04  -.1111E-03
   30   .2508E-01   .9184E-02   .0000E+00  /  .0000E+00   .4392E-04  -.1123E-03
   31   .2549E-01  -.1828E-02   .0000E+00  / -.5973E-03   .1889E-03   .0000E+00
   32   .2775E-01  -.1667E-02  -.2170E-03  / -.9469E-05  -.4245E-04  -.1079E-03
   33   .2876E-01   .7970E-03  -.1009E-02  /  .5883E-05  -.2387E-04  -.1141E-03
   34   .2864E-01   .4582E-02  -.1189E-02  /  .6594E-05   .1605E-04  -.1168E-03
   35   .2830E-01   .9285E-02   .0000E+00  /  .0000E+00   .4399E-04  -.1128E-03
```

```
                                                      03-05-87
                                                      PAGE   3
       =============== I M A G E S   3 D ===============
       = Copyright (c) 1984   Celestial Software Inc. =
       ================================================

          SOLVE SEISMIC RESPONSE      Version 1.4  12/01/86

       COOLING TOWER VIBRATION
```

ACCELERATIONS for MODE 1

Node	Translational X	Y	Z	/	Rotational X	Y	Z
1	.0000E+00	.0000E+00	.0000E+00	/	.0000E+00	.0000E+00	.0000E+00
2	.0000E+00	.0000E+00	.0000E+00	/	.0000E+00	.0000E+00	.0000E+00
3	.0000E+00	.0000E+00	.0000E+00	/	.0000E+00	.0000E+00	.0000E+00
4	.0000E+00	.0000E+00	.0000E+00	/	.0000E+00	.0000E+00	.0000E+00
5	.0000E+00	.0000E+00	.0000E+00	/	.0000E+00	.0000E+00	.0000E+00
6	.1021E-02	-.2401E-02	.0000E+00	/	.8326E-06	-.4671E-05	.0000E+00
7	.3593E-02	-.3070E-02	-.2341E-02	/	-.9067E-04	-.1548E-04	-.1147E-03
8	.2500E-02	-.6402E-03	-.1714E-02	/	-.7844E-04	.4988E-05	-.9798E-04
9	.2149E-02	.3824E-02	.1501E-02	/	.3039E-05	.3212E-04	-.1066E-03
10	.3916E-02	.5359E-02	.0000E+00	/	.0000E+00	.1442E-04	-.1550E-03
11	.6531E-01	-.5358E-01	.0000E+00	/	-.6228E-02	-.2526E-02	.0000E+00
12	.9589E-01	-.3573E-01	-.1277E-01	/	-.3591E-03	-.4935E-03	-.1916E-02
13	.1085E+00	.4148E-02	-.2741E-01	/	-.7953E-03	-.1818E-03	-.2207E-02
14	.1091E+00	.6911E-01	-.2837E-01	/	-.9304E-03	.3349E-03	-.2373E-02
15	.1198E+00	.1736E+00	.0000E+00	/	.0000E+00	.1740E-02	-.6241E-02
16	.2334E+00	-.7767E-01	.0000E+00	/	-.1309E-01	-.5259E-02	.0000E+00
17	.3073E+00	-.4040E-01	-.2418E-01	/	.2468E-03	-.1450E-02	-.3717E-02
18	.3520E+00	.3760E-01	-.6156E-01	/	-.6868E-03	-.8881E-03	-.3766E-02
19	.3560E+00	.1558E+00	-.6595E-01	/	-.5460E-03	.9542E-03	-.4314E-02
20	.3525E+00	.3057E+00	.0000E+00	/	.0000E+00	.2512E-02	-.6124E-02
21	.5040E+00	-.6513E-01	.0000E+00	/	-.1775E-01	-.6174E-02	.0000E+00
22	.6179E+00	-.1273E-01	-.2440E-01	/	.5381E-03	-.3229E-02	-.5840E-02
23	.7005E+00	.8915E-01	-.8236E-01	/	-.1930E-03	-.1911E-02	-.5246E-02
24	.7096E+00	.2351E+00	-.9432E-01	/	-.8351E-04	.1373E-02	-.5208E-02
25	.6967E+00	.4098E+00	.0000E+00	/	.0000E+00	.3476E-02	-.5945E-02
26	.1196E+01	-.3715E-01	.0000E+00	/	-.1112E-01	-.7122E-03	.0000E+00
27	.1280E+01	-.1261E-01	-.2984E-01	/	-.3326E-03	-.1836E-02	-.5134E-02
28	.1328E+01	.9890E-01	-.6766E-01	/	-.3032E-04	-.9769E-03	-.5628E-02
29	.1331E+01	.2705E+00	-.6831E-01	/	.2956E-04	.1092E-02	-.5855E-02
30	.1322E+01	.4840E+00	.0000E+00	/	.0000E+00	.2315E-02	-.5920E-02
31	.1343E+01	-.9635E-01	.0000E+00	/	-.3148E-01	.9955E-02	.0000E+00
32	.1462E+01	-.8784E-01	-.1144E-01	/	-.4990E-03	-.2237E-02	-.5685E-02
33	.1516E+01	.4201E-01	-.5315E-01	/	.3100E-03	-.1258E-02	-.6011E-02
34	.1509E+01	.2415E+00	-.6269E-01	/	.3475E-03	.8458E-03	-.6157E-02
35	.1492E+01	.4893E+00	.0000E+00	/	.0000E+00	.2318E-02	-.5944E-02

ABS DISPLACEMENTS

Node	Translations X	Y	Z	/	Rotations X	Y	Z
1	.0000E+00	.0000E+00	.0000E+00	/	.0000E+00	.0000E+00	.0000E+00

COOLING TOWER VIBRATION: IMAGES 3D

```
IMAGES-3D   s/n:800189                              03-05-87
                                                    PAGE    4
            ================ I M A G E S  3 D ================
            = Copyright (c) 1984   Celestial Software Inc. =
            ===================================================

            SOLVE SEISMIC RESPONSE        Version 1.4  12/01/86

        COOLING TOWER VIBRATION
```

Node	Translations X	Y	Z	/	Rotations X	Y	Z
2	.0000E+00	.0000E+00	.0000E+00	/	.0000E+00	.0000E+00	.0000E+00
3	.0000E+00	.0000E+00	.0000E+00	/	.0000E+00	.0000E+00	.0000E+00
4	.0000E+00	.0000E+00	.0000E+00	/	.0000E+00	.0000E+00	.0000E+00
5	.0000E+00	.0000E+00	.0000E+00	/	.0000E+00	.0000E+00	.0000E+00
6	.1937E-04	.4555E-04	.0000E+00	/	.1580E-07	.8863E-07	.0000E+00
7	.6818E-04	.5825E-04	.4441E-04	/	.1721E-05	.2936E-06	.2176E-05
8	.4744E-04	.1215E-04	.3252E-04	/	.1488E-05	.9464E-07	.1859E-05
9	.4077E-04	.7257E-04	.2848E-04	/	.5766E-07	.6094E-06	.2024E-05
10	.7431E-04	.1017E-03	.0000E+00	/	.0000E+00	.2737E-06	.2941E-05
11	.1239E-02	.1017E-02	.0000E+00	/	.1182E-03	.4792E-04	.0000E+00
12	.1820E-02	.6779E-03	.2422E-03	/	.6815E-05	.9364E-05	.3635E-04
13	.2059E-02	.7871E-04	.5202E-03	/	.1509E-04	.3449E-05	.4188E-04
14	.2070E-02	.1311E-02	.5383E-03	/	.1765E-04	.6355E-05	.4502E-04
15	.2273E-02	.3293E-02	.0000E+00	/	.0000E+00	.3301E-04	.1184E-03
16	.4428E-02	.1474E-02	.0000E+00	/	.2484E-03	.9978E-04	.0000E+00
17	.5831E-02	.7667E-03	.4589E-03	/	.4683E-05	.2751E-04	.7053E-04
18	.6679E-02	.7134E-03	.1168E-02	/	.1303E-04	.1685E-04	.7146E-04
19	.6756E-02	.2957E-02	.1251E-02	/	.1036E-04	.1811E-04	.8185E-04
20	.6689E-02	.5801E-02	.0000E+00	/	.0000E+00	.4766E-04	.1162E-03
21	.9564E-02	.1236E-02	.0000E+00	/	.3368E-03	.1172E-03	.0000E+00
22	.1172E-01	.2415E-01	.4630E-03	/	.1021E-04	.6126E-04	.1108E-03
23	.1329E-01	.1692E-02	.1563E-02	/	.3662E-05	.3626E-04	.9954E-04
24	.1346E-01	.4461E-02	.1790E-02	/	.1585E-04	.2606E-04	.9882E-04
25	.1322E-01	.7776E-02	.0000E+00	/	.0000E+00	.6596E-04	.1128E-03
26	.2269E-01	.7049E-03	.0000E+00	/	.2109E-03	.1351E-04	.0000E+00
27	.2428E-01	.2392E-03	.5662E-03	/	.6311E-05	.3483E-04	.9742E-04
28	.2520E-01	.1877E-02	.1284E-02	/	.5752E-06	.1854E-04	.1068E-03
29	.2526E-01	.5133E-02	.1296E-02	/	.5609E-06	.2071E-04	.1111E-03
30	.2508E-01	.9184E-02	.0000E+00	/	.0000E+00	.4392E-04	.1123E-03
31	.2549E-01	.1828E-02	.0000E+00	/	.5973E-03	.1889E-03	.0000E+00
32	.2775E-01	.1667E-02	.2170E-02	/	.9469E-05	.4245E-04	.1079E-03
33	.2876E-01	.7970E-03	.1009E-02	/	.5883E-05	.2387E-04	.1141E-03
34	.2864E-01	.4582E-02	.1189E-02	/	.6594E-05	.1605E-04	.1168E-03
35	.2830E-01	.9285E-02	.0000E+00	/	.0000E+00	.4399E-04	.1128E-03

```
=============== I M A G E S   3 D ===============
= Copyright (c) 1984   Celestial Software Inc. =
================================================
```

SOLVE SEISMIC RESPONSE   Version 1.4   12/01/86

COOLING TOWER VIBRATION

ABS ACCELERATIONS

Node	T r a n s l a t i o n a l			/	R o t a t i o n a l		
	X	Y	Z	/	X	Y	Z
1	.0000E+00	.0000E+00	.0000E+00	/	.0000E+00	.0000E+00	.0000E+00
2	.0000E+00	.0000E+00	.0000E+00	/	.0000E+00	.0000E+00	.0000E+00
3	.0000E+00	.0000E+00	.0000E+00	/	.0000E+00	.0000E+00	.0000E+00
4	.0000E+00	.0000E+00	.0000E+00	/	.0000E+00	.0000E+00	.0000E+00
5	.0000E+00	.0000E+00	.0000E+00	/	.0000E+00	.0000E+00	.0000E+00
6	.1021E-02	.2401E-02	.0000E+00	/	.8326E-06	.4671E-05	.0000E+00
7	.3593E-02	.3070E-02	.2341E-02	/	.9067E-04	.1548E-04	.1147E-03
8	.2500E-02	.6402E-03	.1714E-02	/	.7844E-04	.4988E-05	.9798E-04
9	.2149E-02	.3824E-02	.1501E-02	/	.3039E-05	.3212E-04	.1066E-03
10	.3916E-02	.5359E-02	.0000E+00	/	.0000E+00	.1442E-04	.1550E-03
11	.6531E-01	.5358E-01	.0000E+00	/	.6228E-02	.2526E-02	.0000E+00
12	.9589E-01	.3573E-01	.1277E-01	/	.3591E-02	.4935E-03	.1916E-02
13	.1085E+00	.4148E-02	.2741E-01	/	.7953E-03	.1818E-03	.2207E-02
14	.1091E+00	.6911E-01	.2837E-01	/	.9304E-03	.3349E-03	.2373E-02
15	.1198E+00	.1736E+00	.0000E+00	/	.0000E+00	.1740E-02	.6241E-02
16	.2334E+00	.7767E-01	.0000E+00	/	.1309E-01	.5259E-02	.0000E+00
17	.3073E+00	.4040E-01	.2418E-01	/	.2468E-03	.1450E-02	.3717E-02
18	.3520E+00	.3760E-01	.6156E-01	/	.6868E-03	.8881E-03	.3766E-02
19	.3560E+00	.1558E+00	.6595E-01	/	.5460E-03	.9542E-03	.4314E-02
20	.3525E+00	.3057E+00	.0000E+00	/	.0000E+00	.2512E-02	.6124E-02
21	.5040E+00	.6513E-01	.0000E+00	/	.1775E-01	.6174E-02	.0000E+00
22	.6179E+00	.1273E-01	.2440E-01	/	.5381E-02	.3229E-02	.5840E-02
23	.7005E+00	.8915E-01	.8236E-01	/	.1930E-03	.1911E-02	.5246E-02
24	.7096E+00	.2351E+00	.9432E-01	/	.8351E-04	.1373E-02	.5208E-02
25	.6967E+00	.4098E+00	.0000E+00	/	.0000E+00	.3476E-02	.5945E-02
26	.1196E+01	.3715E-01	.0000E+00	/	.1112E-01	.7122E-03	.0000E+00
27	.1280E+01	.1261E-01	.2984E-01	/	.3326E-03	.1836E-02	.5134E-02
28	.1328E+01	.9890E-01	.6766E-01	/	.3032E-04	.9769E-03	.5628E-02
29	.1331E+01	.2705E+00	.6831E-01	/	.2956E-04	.1092E-02	.5855E-02
30	.1322E+01	.4840E+00	.0000E+00	/	.0000E+00	.2315E-02	.5920E-02
31	.1343E+01	.9635E-01	.0000E+00	/	.3148E-01	.9955E-02	.0000E+00
32	.1462E+01	.8784E-01	.1144E-01	/	.4990E-02	.2237E-02	.5685E-02
33	.1516E+01	.4201E-01	.5315E-01	/	.3100E-03	.1258E-02	.6011E-02
34	.1509E+01	.2415E+00	.6269E-01	/	.3475E-03	.8458E-03	.6157E-02
35	.1492E+01	.4893E+00	.0000E+00	/	.0000E+00	.2318E-02	.5944E-02

COOLING TOWER VIBRATION: IMAGES 3D

```
IMAGES-3D   s/n:800189                                    03-05-87
                                                         PAGE   1
       =============== I M A G E S  3 D ===============
       = Copyright (c) 1984   Celestial Software Inc. =
       ================================================

          SOLVE BEAM LOADS/STRESSES    Version 1.4  12/01/86

       COOLING TOWER VIBRATION

Mode  1 -


                        BEAM LOADS AND/OR STRESSES

GLoads Node    Fx          Fy          Fz          Mx          My          Mz
LLoads Node    Axial       Y-Shear     Z-Shear     Torsion     Y-Bending   Z-Bending
/Stress
------ ----  ----------  ----------  ----------  ----------  ----------  ----------
                        ***BEAM   NO.    1***
GLoads    1 -.7479E+04  .2256E+05 -.2789E+02 -.5101E+03  .5778E+03  .2167E+05
GLoads    6  .7479E+04 -.2256E+05  .2789E+02 -.4661E+03 -.8478E+03  .2167E+05
LLoads    1  .2374E+05 -.1194E+04  .2789E+02  .6929E+03 -.3376E+03 -.2167E+05
LLoads    6 -.2374E+05  .1194E+04 -.2789E+02 -.6929E+03 -.6752E+03 -.2167E+05
Stress    1 -.4053E+02 -.2038E+01  .4762E-01 -.2848E-01  .8908E+00  .1388E-01
Stress    6 -.4053E+02 -.2038E+01  .4762E-01 -.2848E-01 -.8908E+00 -.2775E-01
                        ***BEAM   NO.    2***
GLoads    1 -.1255E+04  .1938E+04 -.1870E+04  .3854E+05  .2677E+05  .1192E+05
GLoads    7  .1255E+04 -.1938E+04  .1870E+04  .1189E+05  .7949E+04 -.9793E+04
LLoads    1  .2849E+04  .6438E+03 -.5461E+03 -.8368E+04  .3517E+05  .3221E+05
LLoads    7 -.2849E+04 -.6438E+03  .5461E+03  .8368E+04  .4467E+04  .1451E+05
Stress    1 -.4864E+01  .1099E+01 -.9325E+00  .3439E+00 -.1324E+01 -.1445E+01
Stress    7 -.4864E+01  .1099E+01 -.9325E+00  .3439E+00  .5964E+00  .1836E+00
                        ***BEAM   NO.    3***
GLoads    2 -.1135E+05  .3549E+05  .4754E+04  .5230E+05  .5179E+04  .8350E+05
GLoads    7  .1135E+05 -.3549E+05 -.4754E+04 -.1736E+05 -.4688E+04 -.3774E+04
LLoads    2  .3748E+05 -.2397E+04 -.5073E+02  .6284E+03  .1716E+05 -.9716E+05
LLoads    7 -.3748E+05  .2397E+04  .5073E+02 -.6284E+03 -.1531E+05  .1013E+05
Stress    2 -.6399E+02 -.4092E+01 -.8661E-01 -.2583E-01  .3993E+01 -.7051E+00
Stress    7 -.6399E+02 -.4092E+01 -.8661E-01 -.2583E-01  .4163E+00 -.6294E+00
                        ***BEAM   NO.    4***
GLoads    2 -.2046E+04  .1741E+04 -.1620E+04  .1975E+05  .1879E+05  .8710E+04
GLoads    8  .2046E+04 -.1741E+04  .1620E+04  .5350E+04  .7992E+04 -.1163E+05
LLoads    2  .3096E+04  .2828E+03 -.4213E+03 -.8231E+04  .2598E+05  .8733E+04
LLoads    8 -.3096E+04 -.2828E+03  .4213E+03  .8231E+04  .4591E+04  .1179E+05
Stress    2 -.5287E+01  .4829E+00 -.7193E+00  .3383E+00 -.3590E+00 -.1068E+01
Stress    8 -.5287E+01  .4829E+00 -.7193E+00  .3383E+00  .4846E+00  .1887E+00
                        ***BEAM   NO.    5***
GLoads    3 -.4185E+04  .1209E+05  .2762E+04  .3685E+05 -.6518E+04  .6930E+05
GLoads    8  .4185E+04 -.1209E+05 -.2762E+04 -.2289E+05 -.3222E+04 -.5534E+04
LLoads    3  .1296E+05 -.1513E+04  .1006E+04 -.1650E+03 -.2385E+05 -.7506E+05
LLoads    8 -.1296E+05  .1513E+04 -.1006E+04  .1650E+03 -.1268E+05  .2010E+05
Stress    3 -.2212E+02 -.2584E+01  .1718E+01  .6783E-02  .3085E+01  .9805E+00
Stress    8 -.2212E+02 -.2584E+01  .1718E+01  .6783E-02  .8260E+00 -.5214E+00
                        ***BEAM   NO.    6***
GLoads    3 -.1691E+04  .5281E+03 -.8757E+03 -.1368E+05 -.5479E+04  .3244E+05
```

IMAGES-3D s/n:800189

```
=============== I M A G E S  3 D ===============
= Copyright (c) 1984   Celestial Software Inc. =
================================================
```

SOLVE BEAM LOADS/STRESSES   Version 1.4 12/01/86

COOLING TOWER VIBRATION

Mode 1 -

GLoads	Node	Fx	Fy	Fz	Mx	My	Mz
LLoads	Node	Axial	Y-Shear	Z-Shear	Torsion	Y-Bending	Z-Bending
/Stress							
GLoads	9	.1691E+04	-.5281E+03	.8757E+03	-.2691E+04	.7983E+03	-.3643E+04
LLoads	3	.1922E+04	-.4552E+03	.7363E+02	-.3876E+04	-.4121E+04	-.3518E+05
LLoads	9	-.1922E+04	.4552E+03	-.7363E+02	.3876E+04	-.1222E+04	.2153E+04
Stress	3	-.3281E+01	-.7771E+00	.1257E+00	.1593E+00	.1446E+01	.1694E+00
Stress	9	-.3281E+01	-.7771E+00	.1257E+00	.1593E+00	.8848E-01	-.5025E-01
				BEAM NO. 7			
GLoads	4	.1753E+04	-.3375E+05	-.1021E+05	-.2908E+05	-.2234E+05	.7276E+05
GLoads	9	-.1753E+04	.3375E+05	.1021E+05	-.2655E+05	.1792E+03	-.9089E+04
LLoads	4	-.3522E+05	.7443E+03	.2289E+04	-.6427E+03	-.8147E+05	-.9792E+03
LLoads	9	.3522E+05	-.7443E+03	-.2289E+04	.6427E+03	-.1652E+04	.2801E+05
Stress	4	.6014E+02	.1271E+01	.3908E+01	.2642E-01	.4025E-01	.3349E+01
Stress	9	.6014E+02	.1271E+01	.3908E+01	.2642E-01	.1151E+01	-.6788E-01
				BEAM NO. 8			
GLoads	4	-.3599E+04	.1499E+04	.5668E+02	.1841E+04	.5109E+04	.4489E+05
GLoads	10	.3599E+04	-.1499E+04	-.5668E+02	.4567E+04	.9112E+04	-.1413E+05
LLoads	4	.3870E+04	-.4193E+03	-.2237E+03	-.9733E+03	.6367E+04	-.4476E+05
LLoads	10	-.3870E+04	.4193E+03	.2237E+03	.9733E+03	.9866E+04	.1433E+05
Stress	4	-.6607E+01	-.7159E+00	-.3819E+00	.4000E-01	.1840E+01	-.2617E+00
Stress	10	-.6607E+01	-.7159E+00	-.3819E+00	.4000E-01	.5890E+00	.4055E+00
				BEAM NO. 9			
GLoads	5	-.3937E+04	-.4703E+05	-.7863E+04	.9002E+05	-.3076E+05	.1263E+06
GLoads	10	.3937E+04	.4703E+05	.7863E+04	.9002E+05	-.7353E+04	.1150E+05
LLoads	5	-.4742E+05	-.4958E+04	.3937E+04	.4021E+04	-.1299E+06	-.9002E+05
LLoads	10	.4742E+05	.4958E+04	-.3937E+04	-.4021E+04	-.1305E+05	.9002E+05
Stress	5	.8097E+02	-.8465E+01	.6722E+01	-.1653E+00	.3700E+01	.5340E+01
Stress	10	.8097E+02	-.8465E+01	.6722E+01	-.1653E+00	-.3700E+01	-.5363E+00

IMAGES-3D s/n:800189

```
=============== I M A G E S  3 D ===============
= Copyright (c) 1984   Celestial Software Inc. =
================================================
```

SOLVE BEAM LOADS/STRESSES   Version 1.4 12/01/86

COOLING TOWER VIBRATION

Mode 1 -

MAXIMUM STRESS SUMMARY FOR BEAMS/TRUSSES
WITHIN SPECIFIED RANGE   1-   9

Maximum (absolute) Stress = .8097E+02 at BEAM   9

Beam	Axial	Y-Shear	Z-Shear	Torsion	Y-Bending	Z-Bending
9	.8097E+02	-.8465E+01	.6722E+01	-.1653E+00	.3700E+01	.5340E+01

COOLING TOWER VIBRATION: IMAGES 3D

```
IMAGES-3D   s/n:800189                          03-05-87
                                                PAGE    1
         =============== I M A G E S  3 D ===============
         = Copyright (c) 1984   Celestial Software Inc. =
         ================================================

         SOLVE PLATE LOADS/STRESSES    Version 1.4  12/01/86

     COOLING TOWER VIBRATION

 Mode  1 -

                    PLATE LOADS AND/OR STRESSES

GLoads Node    Fx          Fy          Fz          Mx          My          Mz
Stress Surf  Sigma  X    Sigma  Y    Tau XY     Sigma  1    Sigma  2     Angle
Stress       Shear XZ    Shear YZ
------ ----  ----------  ----------  ----------  ----------  ----------  ----------
                         ***PLATE    1***
Loads    6 -.7475E+04  .2255E+05 -.3283E+04  .4661E+03  .8478E+03  .3196E+04
Loads    7 -.5392E+04  .1927E+05  .4082E+04  .2861E+04 -.2463E+04  .1008E+05
Loads   12  .7440E+04 -.2136E+05  .2168E+04 -.2584E+04  .3344E+04 -.8202E+04
Loads   11  .5426E+04 -.2047E+05 -.2967E+04 -.2098E+04  .4050E+04 -.2859E+04
Stress  TOP -.8489E+01 -.7378E+02 -.5387E+01 -.8048E+01 -.7422E+02    -4.7
Stress  MID -.7283E+01 -.6441E+02 -.2543E+01 -.7170E+01 -.6452E+02    -2.5
Stress  BOT -.6077E+01 -.5504E+02  .3011E+00 -.6075E+01 -.5505E+02     .4
                         ***PLATE    2***
Loads    7 -.7168E+04  .1812E+05 -.1224E+04  .2603E+04 -.7977E+03  .3485E+04
Loads    8 -.2303E+04  .8714E+04  .1957E+04  .9768E+04 -.5304E+04  .1174E+05
Loads   13  .6020E+04 -.1442E+05  .6490E+03 -.4525E+04  .5061E+03 -.6517E+04
Loads   12  .3452E+04 -.1241E+05 -.1382E+04 -.1901E+04 -.1346E+04 -.3615E+04
Stress  TOP -.5559E+01 -.5295E+02 -.8922E+01 -.3935E+01 -.5457E+02   -10.3
Stress  MID -.4755E+01 -.4136E+02 -.6853E+01 -.3514E+01 -.4260E+02   -10.3
Stress  BOT -.3950E+01 -.2977E+02 -.4783E+01 -.3093E+01 -.3063E+02   -10.2
                         ***PLATE    3***
Loads    8 -.3901E+04  .5106E+04 -.8344E+03  .7768E+04  .5343E+03  .5432E+04
Loads    9 -.1874E+04 -.1153E+05 -.4997E+04  .1569E+05 -.5351E+04  .8561E+04
Loads   14  .2745E+04  .1753E+04  .2035E+04 -.1966E+04  .2927E+04 -.2909E+03
Loads   13  .3030E+04  .4671E+04  .3796E+04 -.1746E+04 -.1650E+04 -.1808E+04
Stress  TOP  .6458E+00 -.1756E+01 -.1294E+02  .1244E+02 -.1355E+02   -42.3
Stress  MID  .9894E+00  .9754E+01 -.1184E+02  .1800E+02 -.7256E+01   -55.2
Stress  BOT  .1333E+01  .2126E+02 -.1075E+02  .2596E+02 -.3361E+01   -66.4
                         ***PLATE    4***
Loads    9  .1960E+04 -.2165E+05 -.6075E+04  .1356E+05  .4374E+04  .4171E+04
Loads   10 -.7496E+04 -.4548E+05 -.1536E+05  .1540E+05 -.1759E+04  .2629E+04
Loads   15  .4580E+03  .3411E+05  .1054E+05  .5805E+04 -.3036E+04 -.1990E+02
Loads   14  .5078E+04  .3301E+05  .1089E+05  .2469E+04 -.5148E+04 -.1106E+04
```

              SOLVE PLATE LOADS/STRESSES   Version 1.4 12/01/86

          COOLING TOWER VIBRATION

 Mode   1 -

GLoads	Node	Fx	Fy	Fz	Mx	My	Mz
Stress	Surf	Sigma X	Sigma Y	Tau XY	Sigma 1	Sigma 2	Angle
Stress		Shear XZ	Shear YZ				
Stress	TOP	.7962E+01	.9529E+02	-.1441E+02	.9760E+02	.5646E+01	-80.9
Stress	MID	.8911E+01	.1031E+03	-.1415E+02	.1052E+03	.6830E+01	-81.6
Stress	BOT	.9859E+01	.1110E+03	-.1390E+02	.1129E+03	.7983E+01	-82.3
			PLATE 5				
Loads	11	-.5293E+04	.2036E+05	-.9580E+03	.2098E+04	-.4050E+04	.5492E+04
Loads	12	-.4458E+04	.1497E+05	-.1135E+03	.2334E+04	-.1279E+04	.8902E+04
Loads	17	.5698E+04	-.1923E+05	-.6027E+02	-.2109E+04	.4595E+04	-.5727E+04
Loads	16	.4053E+04	-.1610E+05	.1132E+04	-.1371E+04	.3847E+04	.2886E+03
Stress	TOP	.9231E+00	-.7181E+02	-.1126E+02	.2626E+01	-.7351E+02	-8.6
Stress	MID	-.1598E+00	-.6283E+02	.5077E+01	.2488E+00	-.6324E+02	-4.6
Stress	BOT	-.1243E+01	-.5386E+02	.1104E+01	-.1220E+01	-.5388E+02	1.2
			PLATE 6				
Loads	12	-.6042E+04	.1865E+05	-.7247E+03	.2150E+04	-.7194E+03	.2915E+04
Loads	13	-.3510E+04	.2561E+04	-.3392E+04	.4071E+04	.1996E+04	.7075E+04
Loads	18	.5782E+04	-.1513E+05	.3432E+03	-.1644E+04	.9863E+04	.2161E+04
Loads	17	.3769E+04	-.6085E+04	.3774E+04	-.2372E+04	-.7024E+03	-.3944E+04
Stress	TOP	.5167E+01	-.4437E+02	-.1991E+02	.1218E+02	-.5138E+02	-19.4
Stress	MID	.1393E+01	-.3775E+02	-.1501E+02	.6486E+01	-.4284E+02	-18.7
Stress	BOT	-.2381E+01	-.3113E+02	-.1011E+02	.8160E+00	-.3432E+02	-17.6
			PLATE 7				
Loads	13	-.5096E+04	.7202E+04	-.1165E+04	.2201E+04	-.8525E+03	.1249E+04
Loads	14	-.3859E+04	-.1685E+05	-.7801E+04	.1024E+04	.5595E+04	.2478E+04
Loads	19	.4621E+04	-.9542E+03	.2325E+04	-.1093E+04	.1801E+05	.4860E+04
Loads	18	.4334E+04	.1060E+05	.6640E+04	-.1606E+04	-.9427E+04	-.4050E+04
Stress	TOP	.1218E+02	.1430E+02	-.2327E+02	.3654E+02	-.1006E+02	-46.3
Stress	MID	.2734E+01	.1713E+02	-.2137E+02	.3248E+02	-.1261E+02	-54.3
Stress	BOT	-.6714E+01	.1996E+02	-.1946E+02	.3022E+02	-.1697E+02	-62.2
			PLATE 8				
Loads	14	-.3518E+04	-.1763E+05	-.5237E+04	-.1527E+04	-.3374E+04	-.1081E+04
Loads	15	-.2131E+03	-.3376E+05	-.9292E+04	-.3604E+04	.3036E+04	.1989E+02
Loads	20	.1905E+04	.2458E+05	.7749E+04	-.5707E+04	-.1717E+04	-.1668E+04
Loads	19	.1825E+04	.2680E+05	.6779E+04	.5067E+04	-.1974E+05	-.4224E+04

COOLING TOWER VIBRATION: IMAGES 3D

```
IMAGES-3D  s/n:800189                                    03-05-87
                                                         PAGE    3
         ================ I M A G E S   3 D ================
         = Copyright (c) 1984   Celestial Software Inc. =
         ===================================================

         SOLVE PLATE LOADS/STRESSES   Version 1.4  12/01/86

      COOLING TOWER VIBRATION

   Mode  1 -
```

GLoads Stress Stress	Node Surf	Fx Sigma X Shear XZ	Fy Sigma Y Shear YZ	Fz Tau XY	Mx Sigma 1	My Sigma 2	Mz Angle
------	----	----------	----------	----------	----------	----------	----------
Stress	TOP	.5170E+01	.9278E+02	-.9376E+01	.9377E+02	.4178E+01	-84.0
Stress	MID	-.1625E+01	.9136E+02	-.1116E+02	.9268E+02	-.2946E+01	-83.2
Stress	BOT	-.8419E+01	.8993E+02	-.1295E+02	.9161E+02	-.1010E+02	-82.6
				PLATE 9			
Loads	16	-.3651E+04	.1597E+05	-.4622E+03	.1371E+04	-.3847E+04	.3262E+04
Loads	17	-.2860E+04	.7808E+04	-.2531E+04	.2403E+04	-.2731E+04	.7058E+04
Loads	22	.5714E+04	-.1476E+05	.1894E+04	-.4044E+04	.6581E+04	-.1376E+05
Loads	21	.7966E+03	-.9013E+04	.1099E+04	-.9541E+04	.4113E+05	.5559E+04
Stress	TOP	-.8713E+01	-.6026E+02	-.1805E+02	-.3023E+01	-.6595E+02	-17.5
Stress	MID	.2487E+00	-.4940E+02	-.8492E+01	.1661E+01	-.5082E+02	-9.4
Stress	BOT	.9210E+01	-.3855E+02	.1061E+01	.9233E+01	-.3857E+02	1.3
				PLATE 10			
Loads	17	-.5546E+04	.1737E+05	-.1266E+04	.2078E+04	-.1162E+04	.2613E+04
Loads	18	-.3292E+04	-.3774E+04	-.6102E+04	-.2229E+04	.1092E+05	.1510E+04
Loads	23	.6776E+04	-.1500E+05	.1514E+04	.1956E+04	.2332E+05	.1316E+05
Loads	22	.2062E+04	.1405E+04	.5854E+04	-.1013E+05	-.4129E+04	-.1703E+05
Stress	TOP	.1335E+02	-.3390E+02	-.2768E+02	.2612E+02	-.4667E+02	-24.8
Stress	MID	.2210E+01	-.2822E+02	-.2228E+02	.1397E+02	-.3993E+02	-27.8
Stress	BOT	-.8932E+01	-.2255E+02	-.1688E+02	.2466E+01	-.3394E+02	-34.0
				PLATE 11			
Loads	18	-.5611E+04	.8429E+04	-.1094E+04	.5480E+04	-.1135E+05	.3780E+03
Loads	19	-.3117E+04	-.1476E+05	-.6304E+04	-.2709E+04	.2099E+05	.4353E+04
Loads	24	.5451E+04	-.4480E+04	.1399E+04	.7176E+04	.3966E+05	.1639E+05
Loads	23	.3276E+04	.1081E+05	.5998E+04	.1184E+04	-.2480E+05	-.6494E+04
Stress	TOP	.2841E+02	.1693E+02	-.2464E+02	.4797E+02	-.2635E+01	-38.4
Stress	MID	.8898E+00	.1304E+02	-.2295E+02	.3070E+02	-.1677E+02	-52.4
Stress	BOT	-.2663E+02	.9156E+01	-.2126E+02	.1905E+02	-.3652E+02	-65.0
				PLATE 12			
Loads	19	-.2101E+04	-.1055E+05	-.3029E+04	-.1265E+04	-.1926E+05	-.4989E+04
Loads	20	-.1298E+04	-.2406E+05	-.6011E+04	.6399E+04	.1717E+04	.1668E+04
Loads	25	.9147E+03	.1519E+05	.5735E+04	-.1079E+05	-.7280E+04	-.3991E+04
Loads	24	.2484E+04	.1942E+05	.3305E+04	.2593E+05	-.4610E+05	-.6296E+04

```
IMAGES-3D   s/n:800189                          03-05-87
                                                PAGE    4
        ================ I M A G E S  3 D ================
        = Copyright (c) 1984   Celestial Software Inc. =
        =================================================

        SOLVE PLATE LOADS/STRESSES      Version 1.4  12/01/86

        COOLING TOWER VIBRATION

Mode  1 -

GLoads Node     Fx         Fy         Fz         Mx         My         Mz
Stress Surf  Sigma  X   Sigma  Y    Tau XY    Sigma  1   Sigma  2    Angle
Stress       Shear XZ   Shear YZ
------ ----  ---------- ---------- ---------- ---------- ---------- ----------
Stress TOP   .1749E+02  .7619E+02 -.8526E+01  .7741E+02  .1628E+02    -81.9
Stress MID   .4249E+00  .7181E+02 -.1029E+02  .7326E+02 -.1029E+01    -82.0
Stress BOT  -.1664E+02  .6743E+02 -.1206E+02  .6912E+02 -.1834E+02    -82.0
                       ***PLATE    13***
Loads   21   .1136E+03  .8896E+04  .3298E+04  .9541E+04 -.4113E+05  .3865E+05
Loads   22  -.2370E+04 -.2669E+04 -.7261E+04  .3391E+04 -.5130E+03  .1490E+05
Loads   27   .5641E+03 -.1004E+05 -.8740E+04 -.2683E+04  .1356E+05 -.1129E+05
Loads   26   .1692E+04  .3814E+04  .1270E+05  .1130E+05  .4887E+05  .7216E+05
Stress TOP   .1357E+02 -.1025E+02 -.1730E+02  .2266E+02 -.1935E+02    -27.7
Stress MID   .1230E+02 -.1435E+02 -.9857E+01  .1555E+02 -.1760E+02    -18.2
Stress BOT   .1103E+02 -.1844E+02 -.2409E+01  .1122E+02 -.1864E+02     -4.6
                       ***PLATE    14***
Loads   22  -.3175E+04  .1598E+05 -.5753E+03  .1079E+05 -.1939E+04  .1589E+05
Loads   23  -.1833E+04 -.8428E+04 -.5756E+04 -.7311E+04  .3948E+05 -.6862E+04
Loads   28   .1853E+04 -.1506E+05  .1467E+03  .1259E+04  .3018E+05  .4974E+04
Loads   27   .3156E+04  .7512E+04  .6184E+04  .5025E+04 -.1423E+04  .7413E+04
Stress TOP   .1776E+02 -.1468E+02 -.1664E+02  .2478E+02 -.2170E+02    -22.9
Stress MID   .3576E+01 -.1725E+02 -.1833E+02  .1424E+02 -.2792E+02    -30.2
Stress BOT  -.1061E+02 -.1983E+02 -.2002E+02  .5321E+01 -.3576E+02    -38.5
                       ***PLATE    15***
Loads   23  -.5689E+04  .1294E+05 -.2054E+04  .4171E+04 -.3800E+05  .1997E+03
Loads   24  -.1362E+04 -.1382E+05 -.2726E+04 -.9710E+04  .6114E+05 -.2283E+04
Loads   29   .5968E+04 -.1107E+05  .2892E+04  .9545E+03  .4567E+05  .3748E+04
Loads   28   .1083E+04  .1195E+05  .1888E+04  .2831E+04 -.3180E+05 -.2964E+03
Stress TOP   .3139E+02  .8174E+01 -.1879E+02  .4186E+02 -.2303E+01    -29.1
Stress MID  -.3020E+01  .2017E+01 -.1940E+02  .1907E+02 -.2007E+02    -48.7
Stress BOT  -.3743E+02 -.4139E+01 -.2002E+02  .5253E+01 -.4682E+02    -64.9
                       ***PLATE    16***
Loads   24  -.4010E+04 -.2705E+03 -.2319E+04 -.2340E+05 -.5470E+05 -.7809E+04
Loads   25   .3435E+03 -.1445E+05  .6890E+03  .1800E+05  .7280E+04  .3991E+04
Loads   30   .3368E+04  .2909E+04  .1746E+04 -.1190E+05  .4461E+04 -.2118E+04
Loads   29   .2993E+03  .1181E+05 -.1161E+03  .1373E+05 -.4243E+05  .2576E+03
```

COOLING TOWER VIBRATION: IMAGES 3D

```
IMAGES-3D   s/n:800189                                    03-05-87
                                                         PAGE    5
         =============== I M A G E S   3 D ===============
         = Copyright (c) 1984   Celestial Software Inc. =
         ================================================

         SOLVE PLATE LOADS/STRESSES    Version 1.4  12/01/86

      COOLING TOWER VIBRATION

Mode  1 -
```

GLoads	Node	Fx	Fy	Fz	Mx	My	Mz
Stress	Surf	Sigma X	Sigma Y	Tau XY	Sigma 1	Sigma 2	Angle
Stress		Shear XZ	Shear YZ				
------	----	---------	---------	---------	---------	---------	---------
Stress	TOP	.1802E+02	.3781E+02	-.1049E+02	.4233E+02	.1350E+02	-66.7
Stress	MID	-.3207E+01	.3358E+02	-.8917E+01	.3563E+02	-.5254E+01	-77.1
Stress	BOT	-.2444E+02	.2936E+02	-.7346E+01	.3035E+02	-.2542E+02	-82.4
			PLATE 17				
Loads	26	-.8919E+02	-.3864E+04	.4213E+04	-.1130E+05	-.4887E+05	.5216E+05
Loads	27	-.1473E+04	.2126E+04	-.2409E+02	-.5840E+04	-.1345E+05	-.3375E+03
Loads	32	.1101E+04	.1770E+04	-.9725E+03	.2546E+04	.5532E+04	-.5215E+03
Loads	31	.4605E+03	-.3302E+02	-.3217E+04	-.1229E+02	-.2626E-01	.1963E+04
Stress	TOP	.2662E+02	-.2015E+02	.6185E+01	.2743E+02	-.2095E+02	7.4
Stress	MID	.2001E+01	.5542E+01	.8333E+01	.1229E+02	-.4747E+01	51.0
Stress	BOT	-.2262E+02	.3123E+02	.1048E+02	.3320E+02	-.2459E+02	79.4
			PLATE 18				
Loads	27	.1183E+04	.3680E+03	.2500E+04	.3498E+04	.1313E+04	.4217E+04
Loads	28	.5859E+03	.1559E+04	-.1313E+04	.1749E+04	.9816E+04	-.5064E+04
Loads	33	-.1670E+04	-.9597E+02	-.2151E+04	.1653E+04	.5171E+04	-.1589E+04
Loads	32	-.9883E+02	-.1831E+04	.9646E+03	-.2546E+04	-.5532E+04	.5214E+03
Stress	TOP	.2157E+02	-.6559E+01	.3140E+01	.2191E+02	-.6905E+01	6.3
Stress	MID	.8582E+01	-.4578E+01	.4314E+01	.9870E+01	-.5867E+01	16.6
Stress	BOT	-.4404E+01	-.2598E+01	.5488E+01	.2061E+01	-.9063E+01	49.7
			PLATE 19				
Loads	28	.3890E+02	.1820E+04	-.9033E+03	-.5840E+04	-.8197E+04	.3860E+03
Loads	29	-.1318E+04	-.5206E+03	-.1160E+04	-.2988E+04	.1418E+05	-.9401E+04
Loads	34	-.1430E+04	-.1424E+04	-.5189E+02	.1595E+04	.9301E+04	-.3799E+04
Loads	33	.2709E+04	.1248E+03	.2115E+04	-.1653E+04	-.5171E+04	.1589E+04
Stress	TOP	.3258E+02	.2633E+01	-.5464E+01	.3354E+02	.1667E+01	-10.0
Stress	MID	.7585E+01	-.3487E+01	-.4841E+01	.9403E+01	-.5305E+01	-20.6
Stress	BOT	-.1741E+02	-.9606E+01	-.4218E+01	-.7762E+01	-.1925E+02	-66.4
			PLATE 20				
Loads	29	-.1380E+04	.5032E+03	-.1799E+04	-.1169E+05	-.1741E+05	.5396E+04
Loads	30	-.1596E+04	-.2260E+04	.1651E+04	.6713E+03	-.4461E+04	.2118E+04
Loads	35	.5113E+03	.1677E+03	.1390E+03	-.2696E+02	.2402E-02	.1754E-02
Loads	34	.2464E+04	.1589E+04	.8922E+01	-.1595E+04	-.9301E+04	.3799E+04

```
IMAGES-3D   s/n:800189                                    03-05-87
                                                         PAGE    6
            =============== I M A G E S  3 D ===============
            = Copyright (c) 1984   Celestial Software Inc. =
            ================================================

            SOLVE PLATE LOADS/STRESSES    Version 1.4  12/01/86

       COOLING TOWER VIBRATION

Mode  1 -

GLoads Node    Fx          Fy          Fz          Mx          My          Mz
Stress Surf  Sigma  X    Sigma  Y    Tau XY      Sigma  1    Sigma  2     Angle
Stress       Shear XZ    Shear YZ
------ ----  ----------  ----------  ----------  ----------  ----------  ----------
Stress TOP   .1718E+02   .8623E+01  -.7501E+01   .2154E+02   .4266E+01   -30.2
Stress MID   .1966E+01   .3910E+01  -.6456E+01   .9466E+01  -.3591E+01   -49.3
Stress BOT  -.1325E+02  -.8042E+00  -.5411E+01   .1219E+01  -.1527E+02   -69.5


IMAGES-3D   s/n:800189                                    03-05-87
                                                         PAGE    7
            =============== I M A G E S  3 D ===============
            = Copyright (c) 1984   Celestial Software Inc. =
            ================================================

            SOLVE PLATE LOADS/STRESSES    Version 1.4  12/01/86

       COOLING TOWER VIBRATION

  Mode  1 -


              MAXIMUM STRESS SUMMARY FOR PLATES
              WITHIN SPECIFIED RANGE     1-   20

      Maximum (absolute) Stress =  .1110E+03 at Plate    4

          Plate   Sigma  X   Sigma  Y    Tau XY
          -----   ---------- ---------- ----------
             4    .9859E+01   .1110E+03  -.1390E+02
```

COOLING TOWER VIBRATION: IMAGES 3D

```
IMAGES-3D  s/n:800189                              03-05-87
                                                   PAGE    1
      =============== I M A G E S   3 D ===============
      = Copyright (c) 1984   Celestial Software Inc. =
      ================================================

          SOLVE REACTIONS              Version 1.4  12/01/86

       COOLING TOWER VIBRATION

  Mode  1 -


                               REACTIONS

     Node     Fx          Fy          Fz          Mx          My          Mz
     ----  ----------  ----------  ----------  ----------  ----------  ----------
        1  -.8734E+04  .2450E+05  -.1898E+04   .3803E+05   .2735E+05   .3359E+05
        2  -.1339E+05  .3723E+05   .3134E+04   .7205E+05   .2397E+05   .9221E+05
        3  -.5876E+04  .1261E+05   .1887E+04   .2316E+05  -.1200E+05   .1017E+06
        4  -.1846E+04 -.3225E+05  -.1016E+05  -.2724E+05  -.1723E+05   .1177E+06
        5  -.3937E+04 -.4703E+05  -.7863E+04   .9002E+05  -.3076E+05   .1263E+06
        6   .0000E+00  .0000E+00  -.3255E+04   .0000E+00   .0000E+00   .2487E+05
       10   .0000E+00  .0000E+00  -.7551E+04   .1100E+06   .0000E+00   .0000E+00
       11   .0000E+00  .0000E+00  -.3925E+04   .0000E+00   .0000E+00   .2633E+04
       15   .0000E+00  .0000E+00   .1253E+04   .2200E+04   .0000E+00   .0000E+00
       16   .0000E+00  .0000E+00   .6696E+03   .0000E+00   .0000E+00   .3550E+04
       20   .0000E+00  .0000E+00   .1738E+04   .6926E+03   .0000E+00   .0000E+00
       21   .0000E+00  .0000E+00   .4397E+04   .0000E+00   .0000E+00   .4421E+05
       25   .0000E+00  .0000E+00   .6424E+04   .7217E+04   .0000E+00   .0000E+00
       26   .0000E+00  .0000E+00   .1692E+05   .0000E+00   .0000E+00   .1243E+06
       30   .0000E+00  .0000E+00   .3396E+04  -.1123E+05   .0000E+00   .0000E+00
       31   .0000E+00  .0000E+00  -.3217E+04   .0000E+00   .0000E+00   .1963E+04
       35   .0000E+00  .0000E+00   .1390E+03  -.2696E+02   .0000E+00   .0000E+00
```

IMAGES-3D   s/n:800189   03-05-87
                                                    PAGE   1
        =============== I M A G E S   3 D ===============
        = Copyright (c) 1984   Celestial Software Inc. =
        ===

        SOLVE BEAM LOADS/STRESSES   Version 1.4 12/01/86

     COOLING TOWER VIBRATION

ABS

                  BEAM LOADS AND/OR STRESSES

GLoads Node	Fx	Fy	Fz	Mx	My	Mz
LLoads Node /Stress	Axial	Y-Shear	Z-Shear	Torsion	Y-Bending	Z-Bending
		BEAM NO. 1				
GLoads 1	.7479E+04	.2256E+05	.2789E+02	.5101E+03	.5778E+03	.2167E+05
GLoads 6	.7479E+04	.2256E+05	.2789E+02	.4661E+03	.8478E+03	.2167E+05
LLoads 1	.2374E+05	.1194E+04	.2789E+02	.6929E+03	.3376E+03	.2167E+05
LLoads 6	.2374E+05	.1194E+04	.2789E+02	.6929E+03	.6752E+03	.2167E+05
Stress 1	.4053E+02	.2038E+01	.4762E-01	.2848E-01	.8908E+00	.1388E-01
Stress 6	.4053E+02	.2038E+01	.4762E-01	.2848E-01	.8908E+00	.2775E-01
		BEAM NO. 2				
GLoads 1	.1255E+04	.1938E+04	.1870E+04	.3854E+05	.2677E+05	.1192E+05
GLoads 7	.1255E+04	.1938E+04	.1870E+04	.1189E+05	.7949E+04	.9793E+04
LLoads 1	.2849E+04	.6438E+03	.5461E+03	.8368E+04	.3517E+05	.3221E+05
LLoads 7	.2849E+04	.6438E+03	.5461E+03	.8368E+04	.4467E+04	.1451E+05
Stress 1	.4864E+01	.1099E+01	.9325E+00	.3439E+00	.1324E+01	.1445E+01
Stress 7	.4864E+01	.1099E+01	.9325E+00	.3439E+00	.5964E+00	.1836E+00
		BEAM NO. 3				
GLoads 2	.1135E+05	.3549E+05	.4754E+04	.5230E+05	.5179E+04	.8350E+05
GLoads 7	.1135E+05	.3549E+05	.4754E+04	.1736E+05	.4688E+04	.3774E+04
LLoads 2	.3748E+05	.2397E+04	.5073E+02	.6284E+03	.1716E+05	.9716E+05
LLoads 7	.3748E+05	.2397E+04	.5073E+02	.6284E+03	.1531E+05	.1013E+05
Stress 2	.6399E+02	.4092E+01	.8661E-01	.2583E-01	.3993E+01	.7051E+00
Stress 7	.6399E+02	.4092E+01	.8661E-01	.2583E-01	.4163E+00	.6294E+00
		BEAM NO. 4				
GLoads 2	.2046E+04	.1741E+04	.1620E+04	.1975E+05	.1879E+05	.8710E+04
GLoads 8	.2046E+04	.1741E+04	.1620E+04	.5350E+04	.7992E+04	.1163E+05
LLoads 2	.3096E+04	.2828E+03	.4213E+03	.8231E+04	.2598E+05	.8733E+04
LLoads 8	.3096E+04	.2828E+03	.4213E+03	.8231E+04	.4591E+04	.1179E+05
Stress 2	.5287E+01	.4829E+00	.7193E+00	.3383E+00	.3590E+00	.1068E+01
Stress 8	.5287E+01	.4829E+00	.7193E+00	.3383E+00	.4846E+00	.1887E+00
		BEAM NO. 5				
GLoads 3	.4185E+04	.1209E+05	.2762E+04	.3685E+05	.6518E+04	.6930E+05
GLoads 8	.4185E+04	.1209E+05	.2762E+04	.2289E+05	.3222E+04	.5534E+04
LLoads 3	.1296E+05	.1513E+04	.1006E+04	.1650E+03	.2385E+05	.7506E+05
LLoads 8	.1296E+05	.1513E+04	.1006E+04	.1650E+03	.1268E+05	.2010E+05
Stress 3	.2212E+02	.2584E+01	.1718E+01	.6783E-02	.3085E+01	.9805E+00
Stress 8	.2212E+02	.2584E+01	.1718E+01	.6783E-02	.8260E+00	.5214E+00
		BEAM NO. 6				
GLoads 3	.1691E+04	.5281E+03	.8757E+03	.1368E+05	.5479E+04	.3244E+05

COOLING TOWER VIBRATION: IMAGES 3D

```
IMAGES-3D  s/n:800189                                      03-05-87
                                                          PAGE    2
          =============== I M A G E S  3 D ===============
          = Copyright (c) 1984  Celestial Software Inc. =
          ===============================================

          SOLVE BEAM LOADS/STRESSES     Version 1.4  12/01/86

      COOLING TOWER VIBRATION

ABS

GLoads Node    Fx          Fy          Fz          Mx          My          Mz
LLoads Node    Axial       Y-Shear     Z-Shear     Torsion     Y-Bending   Z-Bending
/Stress
------ ----  ----------  ----------  ----------  ----------  ----------  ----------
GLoads   9  .1691E+04   .5281E+03   .8757E+03   .2691E+04   .7983E+03   .3643E+04
LLoads   3  .1922E+04   .4552E+03   .7363E+02   .3876E+04   .4121E+04   .3518E+05
LLoads   9  .1922E+04   .4552E+03   .7363E+02   .3876E+04   .1222E+04   .2153E+04
Stress   3  .3281E+01   .7771E+00   .1257E+00   .1593E+00   .1446E+01   .1694E+00
Stress   9  .3281E+01   .7771E+00   .1257E+00   .1593E+00   .8848E-01   .5025E-01
                        ***BEAM   NO.   7***
GLoads   4  .1753E+04   .3375E+05   .1021E+05   .2908E+05   .2234E+05   .7276E+05
GLoads   9  .1753E+04   .3375E+05   .1021E+05   .2655E+05   .1792E+03   .9089E+04
LLoads   4  .3522E+05   .7443E+03   .2289E+04   .6427E+03   .8147E+05   .9792E+03
LLoads   9  .3522E+05   .7443E+03   .2289E+04   .6427E+03   .1652E+04   .2801E+05
Stress   4  .6014E+02   .1271E+01   .3908E+01   .2642E-01   .4025E-01   .3349E+01
Stress   9  .6014E+02   .1271E+01   .3908E+01   .2642E-01   .1151E+01   .6788E-01
                        ***BEAM   NO.   8***
GLoads   4  .3599E+04   .1499E+04   .5668E+02   .1841E+04   .5109E+04   .4489E+05
GLoads  10  .3599E+04   .1499E+04   .5668E+02   .4567E+04   .9112E+04   .1413E+05
LLoads   4  .3870E+04   .4193E+03   .2237E+03   .9733E+03   .6367E+04   .4476E+05
LLoads  10  .3870E+04   .4193E+03   .2237E+03   .9733E+03   .9866E+04   .1433E+05
Stress   4  .6607E+01   .7159E+00   .3819E+00   .4000E-01   .1840E+01   .2617E+00
Stress  10  .6607E+01   .7159E+00   .3819E+00   .4000E-01   .5890E+00   .4055E+00
                        ***BEAM   NO.   9***
GLoads   5  .3937E+04   .4703E+05   .7863E+04   .9002E+05   .3076E+05   .1263E+06
GLoads  10  .3937E+04   .4703E+05   .7863E+04   .9002E+05   .7353E+04   .1150E+05
LLoads   5  .4742E+05   .4958E+04   .3937E+04   .4021E+04   .1299E+06   .9002E+05
LLoads  10  .4742E+05   .4958E+04   .3937E+04   .4021E+04   .1305E+04   .9002E+05
Stress   5  .8097E+02   .8465E+01   .6722E+01   .1653E+00   .3700E+01   .5340E+01
Stress  10  .8097E+02   .8465E+01   .6722E+01   .1653E+00   .3700E+01   .5363E+00


    IMAGES-3D  s/n:800189                                  03-05-87
                                                          PAGE    3
          =============== I M A G E S  3 D ===============
          = Copyright (c) 1984  Celestial Software Inc. =
          ===============================================

          SOLVE BEAM LOADS/STRESSES     Version 1.4  12/01/86

      COOLING TOWER VIBRATION

ABS


              MAXIMUM STRESS SUMMARY FOR BEAMS/TRUSSES
                 WITHIN SPECIFIED RANGE    1-     9


        Maximum (absolute) Stress =  .8097E+02 at BEAM     9

   Beam     Axial       Y-Shear     Z-Shear     Torsion     Y-Bending   Z-Bending
   ----   ----------  ----------  ----------  ----------  ----------  ----------
     9    .8097E+02   .8465E+01   .6722E+01   .1653E+00   .3700E+01   .5340E+01
```

IMAGES-3D   s/n:800189   03-05-87
                                                         PAGE   1
          =============== I M A G E S   3 D ===============
          = Copyright (c) 1984   Celestial Software Inc. =
          ==

          SOLVE PLATE LOADS/STRESSES   Version 1.4   12/01/86

          COOLING TOWER VIBRATION

ABS

                    PLATE LOADS AND/OR STRESSES

GLoads	Node	Fx	Fy	Fz	Mx	My	Mz
Stress	Surf	Sigma X	Sigma Y	Tau XY	Sigma 1	Sigma 2	Angle
Stress		Shear XZ	Shear YZ				
			PLATE 1				
Loads	6	.7475E+04	.2255E+05	.3283E+04	.4661E+03	.8478E+03	.3196E+04
Loads	7	.5392E+04	.1927E+05	.4082E+04	.2861E+04	.2463E+04	.1008E+05
Loads	12	.7440E+04	.2136E+05	.2168E+04	.2584E+04	.3344E+04	.8202E+04
Loads	11	.5426E+04	.2047E+05	.2967E+04	.2098E+04	.4050E+04	.2859E+04
Stress	TOP	.8489E+01	.7378E+02	.5387E+01	.7422E+02	.8048E+01	85.3
Stress	MID	.7283E+01	.6441E+02	.2543E+01	.6452E+02	.7170E+01	87.5
Stress	BOT	.6077E+01	.5504E+02	-.3011E+00	.5505E+02	.6075E+01	-89.6
			PLATE 2				
Loads	7	.7168E+04	.1812E+05	.1224E+04	.2603E+04	.7977E+03	.3485E+04
Loads	8	.2303E+04	.8714E+04	.1957E+04	.9768E+04	.5304E+04	.1174E+05
Loads	13	.6020E+04	.1442E+05	.6490E+03	.4525E+04	.5061E+03	.6517E+04
Loads	12	.3452E+04	.1241E+05	.1382E+04	.1901E+04	.1346E+04	.3615E+04
Stress	TOP	.5559E+01	.5295E+02	.8922E+01	.5457E+02	.3935E+01	79.7
Stress	MID	.4755E+01	.4136E+02	.6853E+01	.4260E+02	.3514E+01	79.7
Stress	BOT	.3950E+01	.2977E+02	.4783E+01	.3063E+02	.3093E+01	79.8
			PLATE 3				
Loads	8	.3901E+04	.5106E+04	.8344E+03	.7768E+04	.5343E+03	.5432E+04
Loads	9	.1874E+04	.1153E+05	.4997E+04	.1569E+05	.5351E+04	.8561E+04
Loads	14	.2745E+04	.1753E+04	.2035E+04	.1966E+04	.2927E+04	.2909E+03
Loads	13	.3030E+04	.4671E+04	.3796E+04	.1746E+04	.1650E+04	.1808E+04
Stress	TOP	.1333E+01	.2126E+02	.1294E+02	.2763E+02	.3935E+01	63.8
Stress	MID	.9894E+00	.9754E+01	.1184E+02	.1800E+02	-.7256E+01	55.2
Stress	BOT	.6458E+00	-.1756E+01	.1075E+02	.1026E+02	-.1137E+02	41.8
			PLATE 4				
Loads	9	.1960E+04	.2165E+05	.6075E+04	.1356E+05	.4374E+04	.4171E+04
Loads	10	.7496E+04	.4548E+05	.1536E+05	.1540E+05	.1759E+04	.2629E+04
Loads	15	.4580E+03	.3411E+05	.1054E+05	.5805E+04	.3036E+04	.1990E+02
Loads	14	.5078E+04	.3301E+05	.1089E+05	.2469E+04	.5148E+04	.1106E+0⁴

COOLING TOWER VIBRATION: IMAGES 3D

```
IMAGES-3D   s/n:800189                                    03-05-87
                                                         PAGE    2
        =============== I M A G E S   3 D ===============
        = Copyright (c) 1984   Celestial Software Inc. =
        ================================================

            SOLVE PLATE LOADS/STRESSES    Version 1.4  12/01/86

        COOLING TOWER VIBRATION

ABS
```

GLoads Stress Stress	Node Surf	Fx Sigma X Shear XZ	Fy Sigma Y Shear YZ	Fz Tau XY	Mx Sigma 1	My Sigma 2	Mz Angle
Stress	TOP	.9859E+01	.1110E+03	.1441E+02	.1130E+03	.7847E+01	82.0
Stress	MID	.8911E+01	.1031E+03	.1415E+02	.1052E+03	.6830E+01	81.6
Stress	BOT	.7962E+01	.9529E+02	.1390E+02	.9745E+02	.5802E+01	81.2
			***PLATE	5***			
Loads	11	.5293E+04	.2036E+05	.9580E+03	.2098E+04	.4050E+04	.5492E+04
Loads	12	.4458E+04	.1497E+05	.1135E+03	.2334E+04	.1279E+04	.8902E+04
Loads	17	.5698E+04	.1923E+05	.6027E+02	.2109E+04	.4595E+04	.5727E+04
Loads	16	.4053E+04	.1610E+05	.1132E+04	.1371E+04	.3847E+04	.2886E+03
Stress	TOP	.1243E+01	.7181E+02	.1126E+02	.7356E+02	-.5101E+00	81.2
Stress	MID	.1598E+00	.6283E+02	.5077E+01	.6324E+02	-.2488E+00	85.4
Stress	BOT	-.9231E+00	.5386E+02	-.1104E+01	.5388E+02	-.9454E+00	-88.8
			***PLATE	6***			
Loads	12	.6042E+04	.1865E+05	.7247E+03	.2150E+04	.7194E+03	.2915E+04
Loads	13	.3510E+04	.2561E+04	.3392E+04	.4071E+04	.1996E+04	.7075E+04
Loads	18	.5782E+04	.1513E+05	.3432E+03	.1644E+04	.9863E+04	.2161E+04
Loads	17	.3769E+04	.6085E+04	.3774E+04	.2372E+04	.7024E+03	.3944E+04
Stress	TOP	.5167E+01	.4437E+02	.1991E+02	.5271E+02	-.3173E+01	67.3
Stress	MID	.1393E+01	.3775E+02	.1501E+02	.4314E+02	-.4002E+01	70.2
Stress	BOT	-.2381E+01	.3113E+02	.1011E+02	.3394E+02	-.5192E+01	74.5
			***PLATE	7***			
Loads	13	.5096E+04	.7202E+04	.1165E+04	.2201E+04	.8525E+03	.1249E+04
Loads	14	.3859E+04	.1685E+05	.7801E+04	.1024E+04	.5595E+04	.2478E+04
Loads	19	.4621E+04	.9542E+03	.2325E+04	.1093E+04	.1801E+05	.4860E+04
Loads	18	.4334E+04	.1060E+05	.6640E+04	.1606E+04	.9427E+04	.4050E+04
Stress	TOP	.1218E+02	.1996E+02	.2327E+02	.3967E+02	-.7524E+01	49.7
Stress	MID	.2734E+01	.1713E+02	.2137E+02	.3248E+02	-.1261E+02	54.3
Stress	BOT	-.6714E+01	.1430E+02	.1946E+02	.2591E+02	-.1832E+02	59.2
			***PLATE	8***			
Loads	14	.3518E+04	.1763E+05	.5237E+06	.1527E+04	.3374E+04	.1081E+04
Loads	15	.2131E+03	.3376E+05	.9292E+04	.3604E+04	.3036E+04	.1989E+02
Loads	20	.1905E+04	.2458E+05	.7749E+04	.5707E+04	.1717E+04	.1668E+04
Loads	19	.1825E+04	.2680E+05	.6779E+04	.5067E+04	.1974E+05	.4224E+04

IMAGES-3D   s/n:800189   03-05-87
                                                            PAGE   3
```
=============== I M A G E S   3 D ===============
= Copyright (c) 1984   Celestial Software Inc. =
================================================
```

          SOLVE PLATE LOADS/STRESSES   Version 1.4 12/01/86

       COOLING TOWER VIBRATION

ABS

GLoads Stress Stress	Node Surf	Fx Sigma X Shear XZ	Fy Sigma Y Shear YZ	Fz Tau XY	Mx Sigma 1	My Sigma 2	Mz Angle
Stress	TOP	.8419E+01	.9278E+02	.1295E+02	.9473E+02	.6475E+01	81.5
Stress	MID	.1625E+01	.9136E+02	.1116E+02	.9273E+02	.2566E+00	83.0
Stress	BOT	-.5170E+01	.8993E+02	.9376E+01	.9085E+02	-.6085E+01	84.4
			***PLATE	9***			
Loads	16	.3651E+04	.1597E+05	.4622E+03	.1371E+04	.3847E+04	.3262E+04
Loads	17	.2860E+04	.7808E+04	.2531E+04	.2403E+04	.2731E+04	.7058E+04
Loads	22	.5714E+04	.1476E+05	.1894E+04	.4044E+04	.6581E+04	.1376E+05
Loads	21	.7966E+03	.9013E+04	.1099E+04	.9541E+04	.4113E+05	.5559E+04
Stress	TOP	.9210E+01	.6026E+02	.1805E+02	.6599E+02	.3476E+01	72.4
Stress	MID	.2487E+00	.4940E+02	.5083E+02	.5083E+02	-.1177E+01	80.5
Stress	BOT	-.8713E+01	.3855E+02	-.1061E+01	.3857E+02	-.8736E+01	-88.7
			***PLATE	10***			
Loads	17	.5546E+04	.1737E+05	.1266E+04	.2078E+04	.1162E+04	.2613E+04
Loads	18	.3292E+04	.3774E+04	.6102E+04	.2229E+04	.1092E+05	.1510E+04
Loads	23	.6776E+04	.1500E+05	.1514E+04	.1956E+04	.2332E+05	.1316E+05
Loads	22	.2062E+04	.1405E+04	.5854E+04	.1013E+05	.4129E+04	.1703E+05
Stress	TOP	.1335E+02	.3390E+02	.2768E+02	.5315E+02	-.5899E+01	55.2
Stress	MID	.2210E+01	.2822E+02	.2228E+02	.4102E+02	-.1058E+02	60.1
Stress	BOT	-.8932E+01	.2255E+02	.1688E+02	.2989E+02	-.1627E+02	66.5
			***PLATE	11***			
Loads	18	.5611E+04	.8429E+04	.1094E+04	.5480E+04	.1135E+05	.3780E+03
Loads	19	.3117E+04	.1476E+05	.6304E+04	.2709E+04	.2099E+05	.4353E+04
Loads	24	.5451E+04	.4480E+04	.1399E+04	.7176E+04	.3966E+04	.1639E+05
Loads	23	.3276E+04	.1081E+05	.5998E+04	.1184E+04	.2480E+05	.6494E+04
Stress	TOP	.2841E+02	.1693E+02	.2464E+02	.4797E+02	-.2635E+01	38.4
Stress	MID	.8898E+00	.1304E+02	.2295E+02	.3070E+02	-.1677E+02	52.4
Stress	BOT	-.2663E+02	.9156E+01	.2126E+02	.1905E+02	-.3652E+02	65.0
			***PLATE	12***			
Loads	19	.2101E+04	.1055E+05	.3029E+04	.1265E+04	.1926E+05	.4989E+04
Loads	20	.1298E+04	.2406E+05	.6011E+04	.6399E+04	.1717E+04	.1668E+04
Loads	25	.9147E+03	.1519E+05	.5735E+04	.1079E+05	.7280E+04	.3991E+04
Loads	24	.2484E+04	.1942E+05	.3305E+04	.2593E+05	.4610E+05	.6296E+04

COOLING TOWER VIBRATION: IMAGES 3D

```
IMAGES-3D  s/n:800189                                    03-05-87
                                                         PAGE   4
          =============== I M A G E S  3 D ===============
          = Copyright (c) 1984  Celestial Software Inc. =
          ================================================

              SOLVE PLATE LOADS/STRESSES   Version 1.4  12/01/86

       COOLING TOWER VIBRATION

  ABS
```

GLoads Stress Stress	Node Surf	Fx Sigma X Shear XZ	Fy Sigma Y Shear YZ	Fz Tau XY	Mx Sigma 1	My Sigma 2	Mz Angle
Stress	TOP	.1749E+02	.7619E+02	.1206E+02	.7857E+02	.1511E+02	78.8
Stress	MID	.4249E+00	.7181E+02	.1029E+02	.7326E+02	-.1029E+01	82.0
Stress	BOT	-.1664E+02	.6743E+02	.8526E+02	.6828E+02	-.1750E+02	84.3
				PLATE 13			
Loads	21	.1136E+03	.8896E+04	.3298E+04	.9541E+04	.4113E+05	.3865E+05
Loads	22	.2370E+04	.2669E+04	.7261E+04	.3391E+04	.5130E+03	.1490E+05
Loads	27	.5641E+03	.1004E+05	.8740E+04	.2683E+04	.1356E+05	.1129E+05
Loads	26	.1692E+04	.3814E+04	.1270E+05	.1130E+05	.4887E+05	.7216E+05
Stress	TOP	.1357E+02	.1844E+02	.1730E+02	.3348E+02	-.1471E+01	49.0
Stress	MID	.1230E+02	.1435E+02	.9857E+01	.2323E+02	.3411E+02	48.0
Stress	BOT	.1103E+02	.1025E+02	.2409E+01	.1308E+02	.8198E+01	40.4
				PLATE 14			
Loads	22	.3175E+04	.1598E+05	.5753E+03	.1079E+05	.1939E+04	.1589E+05
Loads	23	.1833E+04	.8428E+04	.5756E+04	.7311E+04	.3948E+05	.6862E+04
Loads	28	.1853E+04	.1506E+05	.1467E+03	.1259E+04	.3018E+05	.4974E+04
Loads	27	.3156E+04	.7512E+04	.6184E+04	.5025E+04	.1423E+04	.7413E+04
Stress	TOP	.1776E+02	.1983E+02	.2002E+02	.3884E+02	-.1248E+01	46.5
Stress	MID	.3576E+01	.1725E+02	.1833E+02	.2998E+02	-.9149E+01	55.2
Stress	BOT	-.1061E+02	.1468E+02	.1664E+02	.2294E+02	-.1887E+02	63.6
				PLATE 15			
Loads	23	.5689E+04	.1294E+05	.2054E+04	.4171E+04	.3800E+05	.1997E+03
Loads	24	.1362E+04	.1382E+05	.2726E+04	.9710E+04	.6114E+05	.2283E+04
Loads	29	.5968E+04	.1107E+05	.2892E+04	.9545E+03	.4567E+05	.3748E+04
Loads	28	.1083E+04	.1195E+05	.1888E+04	.2831E+04	.3180E+05	.2964E+03
Stress	TOP	.3743E+02	.8174E+01	.2002E+02	.4760E+02	-.1995E+02	26.9
Stress	MID	.3020E+02	.2017E+01	.1940E+02	.2193E+02	-.1689E+02	44.3
Stress	BOT	-.3139E+02	-.4139E+01	.1879E+02	.5444E+01	-.4097E+02	63.0
				PLATE 16			
Loads	24	.4010E+04	.2705E+03	.2319E+04	.2340E+05	.5470E+05	.7809E+04
Loads	25	.3435E+03	.1445E+05	.6890E+03	.1800E+05	.7280E+04	.3991E+04
Loads	30	.3368E+04	.2909E+04	.1746E+04	.1190E+05	.4461E+04	.2118E+04
Loads	29	.2993E+03	.1181E+05	.1161E+03	.1373E+05	.4243E+05	.2576E+03

```
IMAGES-3D   s/n:800189                              03-05-87
                                                   PAGE    5
        =============== I M A G E S  3 D ===============
        = Copyright (c) 1984  Celestial Software Inc. =
        ===============================================

        SOLVE PLATE LOADS/STRESSES    Version 1.4  12/01/86

      COOLING TOWER VIBRATION

ABS
```

GLoads	Node	Fx	Fy	Fz	Mx	My	Mz
Stress	Surf	Sigma X	Sigma Y	Tau XY	Sigma 1	Sigma 2	Angle
Stress		Shear XZ	Shear YZ				
Stress	TOP	.2444E+02	.3781E+02	.1049E+02	.4356E+02	.1869E+02	61.3
Stress	MID	.3207E+01	.3358E+02	.8917E+01	.3601E+02	.7828E+00	74.8
Stress	BOT	-.1802E+02	.2936E+02	.7346E+01	.3047E+02	-.1914E+02	81.4
			***PLATE	17***			
Loads	26	.8919E+02	.3864E+04	.4213E+04	.1130E+05	.4887E+05	.5216E+05
Loads	27	.1473E+04	.2126E+04	.2409E+02	.5840E+04	.1345E+05	.3375E+03
Loads	32	.1101E+04	.1770E+04	.9725E+03	.2546E+04	.5532E+04	.5215E+03
Loads	31	.4605E+03	.3302E+02	.3217E+04	.1229E-02	.2626E-01	.1963E+04
Stress	TOP	.2662E+02	.3123E+02	.1048E+02	.3966E+02	.1820E+02	51.2
Stress	MID	.2001E+01	.5542E+01	.8333E+01	.1229E+02	-.4747E+01	51.0
Stress	BOT	-.2262E+02	-.2015E+02	.6185E+01	-.1508E+02	-.2769E+02	50.6
			***PLATE	18***			
Loads	27	.1183E+04	.3680E+03	.2500E+04	.3498E+04	.1313E+04	.4217E+04
Loads	28	.5859E+03	.1559E+04	.1313E+04	.1749E+04	.9816E+04	.5064E+04
Loads	33	.1670E+04	.9597E+02	.2151E+04	.1653E+04	.5171E+04	.1589E+04
Loads	32	.9883E+02	.1831E+04	.9646E+03	.2546E+04	.5532E+04	.5214E+03
Stress	TOP	.2157E+02	.6559E+01	.5488E+01	.2336E+02	.4766E+01	18.1
Stress	MID	.8582E+01	.4578E+01	.4314E+01	.1134E+02	.1824E+01	32.6
Stress	BOT	-.4404E+01	.2598E+01	.3140E+01	.3800E+01	-.5606E+01	69.1
			***PLATE	19***			
Loads	28	.3890E+02	.1820E+04	.9033E+03	.5840E+04	.8197E+04	.3860E+03
Loads	29	.1318E+04	.5206E+03	.1160E+04	.2988E+04	.1418E+05	.9401E+04
Loads	34	.1430E+04	.1424E+04	.5189E+02	.1595E+04	.9301E+04	.3799E+04
Loads	33	.2709E+04	.1248E+03	.2115E+04	.1653E+04	.5171E+04	.1589E+04
Stress	TOP	.3258E+02	.9606E+01	.5464E+01	.3381E+02	.8373E+01	12.7
Stress	MID	.7585E+01	.3487E+01	.4841E+01	.1079E+02	.2793E+00	33.5
Stress	BOT	-.1741E+02	-.2633E+01	.4218E+01	-.1513E+01	-.1853E+02	75.1
			***PLATE	20***			
Loads	29	.1380E+04	.5032E+03	.1799E+04	.1169E+05	.1741E+05	.5396E+04
Loads	30	.1596E+04	.2260E+04	.1651E+04	.6713E+03	.4461E+04	.2118E+04
Loads	35	.5113E+03	.1677E+03	.1390E+03	.2696E+02	.2402E-02	.1754E-02
Loads	34	.2464E+04	.1589E+04	.8922E+01	.1595E+04	.9301E+04	.3799E+04

COOLING TOWER VIBRATION: IMAGES 3D

```
IMAGES-3D   s/n:800189                              03-05-87
                                                   PAGE    6
            =============== I M A G E S  3 D ===============
            = Copyright (c) 1984   Celestial Software Inc. =
            ================================================

            SOLVE PLATE LOADS/STRESSES    Version 1.4   12/01/86

            COOLING TOWER VIBRATION

ABS

GLoads Node    Fx          Fy          Fz          Mx          My          Mz
Stress Surf  Sigma  X    Sigma  Y    Tau XY      Sigma  1    Sigma  2    Angle
Stress       Shear XZ    Shear YZ
------ ----  ---------- ---------- ---------- ---------- ---------- ----------
Stress TOP   .1718E+02  .8623E+01  .7501E+01  .2154E+02  .4266E+01     30.2
Stress MID   .1966E+01  .3910E+01  .6456E+01  .9466E+01 -.3591E+01     49.3
Stress BOT  -.1325E+02 -.8042E+00  .5411E+01  .1219E+01 -.1527E+02     69.5


IMAGES-3D   s/n:800189                              03-05-87
                                                   PAGE    7
            =============== I M A G E S  3 D ===============
            = Copyright (c) 1984   Celestial Software Inc. =
            ================================================

            SOLVE PLATE LOADS/STRESSES    Version 1.4   12/01/86

            COOLING TOWER VIBRATION

ABS


                MAXIMUM STRESS SUMMARY FOR PLATES
                WITHIN SPECIFIED RANGE    1-   20

        Maximum (absolute) Stress =  .1110E+03 at Plate    4

            Plate    Sigma  X    Sigma  Y    Tau XY
            -----   ---------- ---------- ----------
              4     .9859E+01  .1110E+03  .1441E+02
```

IMAGES-3D   s/n:800189   03-05-87
                                                          PAGE   1
       =============== I M A G E S   3 D ===============
       = Copyright (c) 1984   Celestial Software Inc. =
       ==

          SOLVE REACTIONS   Version 1.4   12/01/86

       COOLING TOWER VIBRATION

ABS

                           REACTIONS

Node	Fx	Fy	Fz	Mx	My	Mz
1	.8734E+04	.2450E+05	.1898E+04	.3803E+05	.2735E+05	.3359E+05
2	.1339E+05	.3723E+05	.3134E+04	.7205E+05	.2397E+05	.9221E+05
3	.5876E+04	.1261E+05	.1887E+04	.2316E+05	.1200E+05	.1017E+06
4	.1846E+04	.3225E+05	.1016E+05	.2724E+05	.1723E+05	.1177E+06
5	.3937E+04	.4703E+05	.7863E+04	.9002E+05	.3076E+05	.1263E+06
6	.0000E+00	.0000E+00	.3255E+04	.0000E+00	.0000E+00	.2487E+05
10	.0000E+00	.0000E+00	.7551E+04	.1100E+06	.0000E+00	.0000E+00
11	.0000E+00	.0000E+00	.3925E+04	.0000E+00	.0000E+00	.2633E+04
15	.0000E+00	.0000E+00	.1253E+04	.2200E+04	.0000E+00	.0000E+00
16	.0000E+00	.0000E+00	.6696E+03	.0000E+00	.0000E+00	.3550E+04
20	.0000E+00	.0000E+00	.1738E+04	.6926E+03	.0000E+00	.0000E+00
21	.0000E+00	.0000E+00	.4397E+04	.0000E+00	.0000E+00	.4421E+05
25	.0000E+00	.0000E+00	.6424E+04	.7217E+04	.0000E+00	.0000E+00
26	.0000E+00	.0000E+00	.1692E+05	.0000E+00	.0000E+00	.1243E+06
30	.0000E+00	.0000E+00	.3396E+04	.1123E+05	.0000E+00	.0000E+00
31	.0000E+00	.0000E+00	.3217E+04	.0000E+00	.0000E+00	.1963E+04
35	.0000E+00	.0000E+00	.1390E+03	.2696E+02	.0000E+00	.0000E+00

COOLING TOWER VIBRATION: INCREASED NODE MODEL

```
?===============================================================================
=
============================== I M A G E S - 3 D ==============================
===============================================================================

02-14-1987                    Run ID=JA43945                         12:34:45

-------------------------------------------------------------------------------
!                                                                             !
!        JJ   AAAAAA  44   44      333333     999999   44     44   5555555555  !
!        JJ   AAAAAA  44   44      333333     999999   44     44   5555555555  !
!        JJ AA    AA  44   44   33      33 99      99 44     44   55           !
!        JJ AA    AA  44   44   33      33 99      99 44     44   55           !
!        JJ AA    AA  44   44            33 99      99 44     44   555555       !
!        JJ AA    AA  44   44            33 99      99 44     44   555555       !
!        JJ AA    AA 4444444444       33     99999999 4444444444         55    !
!        JJ AA    AA 4444444444       33     99999999 4444444444         55    !
! JJ     JJ AAAAAAAAAA      44        33           99       44           55    !
! JJ     JJ AAAAAAAAAA      44        33           99       44           55    !
! JJ     JJ AA    AA       44   33    33           99       44   55      55    !
! JJ     JJ AA    AA       44   33    33           99       44   55      55    !
!  JJJJJJ   AA    AA       44      333333     999999        44      555555     !
!  JJJJJJ   AA    AA       44      333333     999999        44      555555     !
!                                                                             !
-------------------------------------------------------------------------------
```

IMAGES-3D   s/n:800189

```
!-------------------------------------------------!
!          J o b   I n f o r m a t i o n          !
!-------------------------------------------------!
!                                                 !
!     Project    : FEA BOOK                       !
!                                                 !
!     Client     : MARCEL DEKKER, INC.            !
!                                                 !
!     Job Name   : COOLING TOWER 2                !
!                                                 !
!     Remarks    : _____           !
!                                                 !
!                                                 !
!                                                 !
!     Engineer   : _____/_____      !
!                   CHAMPION                       !
!                                                 !
!     Chk'd by   : _____/_____      !
!                                                 !
!                                                 !
!     Appr'd by  : _____/_____      !
!                                                 !
!                                                 !
!     Comments   : _____           !
!                                                 !
!                  _____            !
!-------------------------------------------------!
```

```
===============================================================================
============================== I M A G E S - 3 D ==============================
===============================================================================
```

```
=============== I M A G E S  3 D ===============
= Copyright (c) 1984   Celestial Software Inc. =
================================================
```

Interactive Microcomputer Analysis & Graphics of Engineering Systems

IMAGES-3D Version 1.4 12/01/86

RUN ID=JA43945

```
================================================
=                    NOTICE                    =
=----------------------------------------------=
= Celestial Software Inc. assumes no responsi- =
= bility for the validity, accuracy, or        =
= applicability of the results obtained from   =
= IMAGES-3D.                                    =
================================================


================================================
= Any questions or comments concerning the use =
= of IMAGES-3D or the users manual should be    =
= addressed to:                                 =
=                                               =
=              Celestial Software Inc.          =
=                 125 University Ave.           =
=                   Berkeley,CA                 =
=                     94710                     =
=                                               =
=                 415-420-0300                  =
================================================
```

COOLING TOWER VIBRATION: INCREASED NODE MODEL

IMAGES-3D   s/n:800189
                                                            02-14-1987
                                                            PAGE 1
          =============== I M A G E S   3 D ===============
          = Copyright (c) 1984   Celestial Software Inc. =
          ==

              CHECK GEOMETRY   Version 1.4 12/01/86

        COOLING TOWER VIBRATION - NODE INCREASE

                           MATERIAL PROPERTIES

Material No	Modulus of Elasticity	Weight Density	Coeff of Thermal Exp.	Poisson's Ratio	Shear Web Modulus
1	4.00000E+06	8.68000E-02	0.00000E+00	2.00E-01	0.00000E+00
2	3.00000E+07	2.82000E-01	0.00000E+00	3.00E-01	0.00000E+00

                           NODE COORDINATES

Node	X-Coord.	Y-Coord.	Z-Coord.
1	1.65940E+02	0.00000E+00	0.00000E+00
2	1.63419E+02	0.00000E+00	-2.88152E+01
3	1.55933E+02	0.00000E+00	-5.67548E+01
4	1.43708E+02	0.00000E+00	-8.29700E+01
5	1.27117E+02	0.00000E+00	-1.06664E+02
6	1.06664E+02	0.00000E+00	-1.27117E+02
7	8.29700E+01	0.00000E+00	-1.43708E+02
8	5.67548E+01	0.00000E+00	-1.55933E+02
9	2.88152E+01	0.00000E+00	-1.63419E+02
10	0.00000E+00	0.00000E+00	-1.65940E+02
11	1.56260E+02	3.58000E+01	0.00000E+00
12	1.53886E+02	3.58000E+01	-2.71343E+01
13	1.46836E+02	3.58000E+01	-5.34441E+01
14	1.35325E+02	3.58000E+01	-7.81300E+01
15	1.19702E+02	3.58000E+01	-1.00442E+02
16	1.00442E+02	3.58000E+01	-1.19702E+02
17	7.81300E+01	3.58000E+01	-1.35325E+02
18	5.34441E+01	3.58000E+01	-1.46836E+02
19	2.71343E+01	3.58000E+01	-1.53886E+02
20	0.00000E+00	3.58000E+01	-1.56260E+02
21	1.43760E+02	7.16000E+01	0.00000E+00
22	1.41576E+02	7.16000E+01	-2.49637E+01
23	1.35090E+02	7.16000E+01	-4.91688E+01
24	1.24500E+02	7.16000E+01	-7.18800E+01
25	1.10127E+02	7.16000E+01	-9.24072E+01
26	9.24072E+01	7.16000E+01	-1.10127E+02
27	7.18800E+01	7.16000E+01	-1.24500E+02

IMAGES-3D   s/n:800189

02-14-1987
PAGE 2

=============== I M A G E S   3 D ===============
= Copyright (c) 1984   Celestial Software Inc. =
==

CHECK GEOMETRY   Version 1.4 12/01/86

COOLING TOWER VIBRATION - NODE INCREASE

Node	X-Coord.	Y-Coord.	Z-Coord.
28	4.91688E+01	7.16000E+01	-1.35090E+02
29	2.49637E+01	7.16000E+01	-1.41576E+02
30	0.00000E+00	7.16000E+01	-1.43760E+02
31	1.33760E+02	1.07400E+02	0.00000E+00
32	1.31728E+02	1.07400E+02	-2.32272E+01
33	1.25693E+02	1.07400E+02	-4.57486E+01
34	1.15840E+02	1.07400E+02	-6.68800E+01
35	1.02466E+02	1.07400E+02	-8.59793E+01
36	8.59793E+01	1.07400E+02	-1.02466E+02
37	6.68800E+01	1.07400E+02	-1.15840E+02
38	4.57486E+01	1.07400E+02	-1.25693E+02
39	2.32272E+01	1.07400E+02	-1.31728E+02
40	0.00000E+00	1.07400E+02	-1.33760E+02
41	1.24070E+02	1.43200E+02	0.00000E+00
42	1.22185E+02	1.43200E+02	-2.15445E+01
43	1.16588E+02	1.43200E+02	-4.24344E+01
44	1.07448E+02	1.43200E+02	-6.20350E+01
45	9.50432E+01	1.43200E+02	-7.97507E+01
46	7.97507E+01	1.43200E+02	-9.50432E+01
47	6.20350E+01	1.43200E+02	-1.07448E+02
48	4.24345E+01	1.43200E+02	-1.16588E+02
49	2.15445E+01	1.43200E+02	-1.22185E+02
50	0.00000E+00	1.43200E+02	-1.24070E+02
51	1.14690E+02	1.79000E+02	0.00000E+00
52	1.12948E+02	1.79000E+02	-1.99157E+01
53	1.07773E+02	1.79000E+02	-3.92263E+01
54	9.93245E+01	1.79000E+02	-5.73450E+01
55	8.78577E+01	1.79000E+02	-7.37213E+01
56	7.37213E+01	1.79000E+02	-8.78577E+01
57	5.73450E+01	1.79000E+02	-9.93245E+01
58	3.92263E+01	1.79000E+02	-1.07773E+02
59	1.99157E+01	1.79000E+02	-1.12948E+02
60	0.00000E+00	1.79000E+02	-1.14690E+02
61	1.05320E+02	2.14800E+02	0.00000E+00
62	1.03720E+02	2.14800E+02	-1.82886E+01
63	9.89684E+01	2.14800E+02	-3.60216E+01
64	9.12098E+01	2.14800E+02	-5.26600E+01
65	8.06798E+01	2.14800E+02	-6.76984E+01
66	6.76984E+01	2.14800E+02	-8.06798E+01
67	5.26600E+01	2.14800E+02	-9.12098E+01
68	3.60216E+01	2.14800E+02	-9.89685E+01

COOLING TOWER VIBRATION: INCREASED NODE MODEL

```
IMAGES-3D  s/n:800189                                        02-14-1987
                                                            PAGE 3
       =============== I M A G E S  3 D ===============
       = Copyright (c) 1984   Celestial Software Inc. =
       ================================================

       CHECK GEOMETRY              Version 1.4  12/01/86

   COOLING TOWER VIBRATION - NODE INCREASE

        Node    X-Coord.       Y-Coord.       Z-Coord.
        ----    ------------   ------------   ------------
          69    1.82886E+01    2.14800E+02   -1.03720E+02
          70    0.00000E+00    2.14800E+02   -1.05320E+02
          71    9.68750E+01    2.50600E+02    0.00000E+00
          72    9.54033E+01    2.50600E+02   -1.68222E+01
          73    9.10327E+01    2.50600E+02   -3.31332E+01
          74    8.38962E+01    2.50600E+02   -4.84375E+01
          75    7.42106E+01    2.50600E+02   -6.22701E+01
          76    6.22701E+01    2.50600E+02   -7.42106E+01
          77    4.84375E+01    2.50600E+02   -8.38962E+01
          78    3.31332E+01    2.50600E+02   -9.10328E+01
          79    1.68222E+01    2.50600E+02   -9.54033E+01
          80    0.00000E+00    2.50600E+02   -9.68750E+01
          81    9.21900E+01    2.86400E+02    0.00000E+00
          82    9.07894E+01    2.86400E+02   -1.60086E+01
          83    8.66303E+01    2.86400E+02   -3.15308E+01
          84    7.98389E+01    2.86400E+02   -4.60950E+01
          85    7.06216E+01    2.86400E+02   -5.92586E+01
          86    5.92586E+01    2.86400E+02   -7.06216E+01
          87    4.60950E+01    2.86400E+02   -7.98389E+01
          88    3.15308E+01    2.86400E+02   -8.66303E+01
          89    1.60086E+01    2.86400E+02   -9.07894E+01
          90    0.00000E+00    2.86400E+02   -9.21900E+01
          91    9.06300E+01    3.22200E+02    0.00000E+00
          92    8.92531E+01    3.22200E+02   -1.57377E+01
          93    8.51644E+01    3.22200E+02   -3.09973E+01
          94    7.84879E+01    3.22200E+02   -4.53150E+01
          95    6.94266E+01    3.22200E+02   -5.82558E+01
          96    5.82559E+01    3.22200E+02   -6.94266E+01
          97    4.53150E+01    3.22200E+02   -7.84879E+01
          98    3.09973E+01    3.22200E+02   -8.51644E+01
          99    1.57378E+01    3.22200E+02   -8.92532E+01
         100    0.00000E+00    3.22200E+02   -9.06300E+01
         101    9.50050E+01    3.58000E+02    0.00000E+00
         102    9.35617E+01    3.58000E+02   -1.64974E+01
         103    8.92755E+01    3.58000E+02   -3.24936E+01
         104    8.22767E+01    3.58000E+02   -4.75025E+01
         105    7.27781E+01    3.58000E+02   -6.10680E+01
         106    6.10680E+01    3.58000E+02   -7.27781E+01
         107    4.75025E+01    3.58000E+02   -8.22768E+01
         108    3.24936E+01    3.58000E+02   -8.92755E+01
         109    1.64974E+01    3.58000E+02   -9.35617E+01
```

```
IMAGES-3D  s/n:800189                                02-14-1987
                                                     PAGE 4
       =============== I M A G E S  3 D ===============
       = Copyright (c) 1984   Celestial Software Inc. =
       ================================================

       CHECK GEOMETRY               Version 1.4  12/01/86

  COOLING TOWER VIBRATION - NODE INCREASE
```

Node	X-Coord.	Y-Coord.	Z-Coord.
110	0.00000E+00	3.58000E+02	-9.50050E+01
111	1.04380E+02	3.89000E+02	0.00000E+00
112	1.02794E+02	3.89000E+02	-1.81254E+01
113	9.80851E+01	3.89000E+02	-3.57001E+01
114	9.03958E+01	3.89000E+02	-5.21900E+01
115	7.99597E+01	3.89000E+02	-6.70942E+01
116	6.70942E+01	3.89000E+02	-7.99597E+01
117	5.21900E+01	3.89000E+02	-9.03958E+01
118	3.57001E+01	3.89000E+02	-9.80851E+01
119	1.81254E+01	3.89000E+02	-1.02794E+02
120	0.00000E+00	3.89000E+02	-1.04380E+02

```
                          BEAM PROPERTIES

       Multiplier = 1  (For AISC database properties only)
```

Prop No	X-Section Area	Moment of Inertia Iy / Iz		Torsional Const.- J
1	5.857E+02	2.433E+04	2.433E+04	2.433E+04

Prop No	Max. Fiber Dist Cy / Cz		Shear Shape Fact SSFy / SSFz		Ctors
1	1.00E+00	1.00E+00	0.00E+00	0.00E+00	1.00E+00

```
                        BEAM CONNECTIVITY
```

Beam No	Nodes From/ To/Ref		Prop No	Mat No	Pincodes I / J	Length	Y Dir Cosines X	Y	Z	Beam Type
1	1	11	0	1	2		3.709E+01	0.97	0.26	0.00	Beam
2	1	12	0	1	2		4.651E+01	0.31	0.64	0.70	Beam
3	2	12	0	1	2		3.709E+01	0.95	0.26	-0.17	Beam
4	2	13	0	1	2		4.651E+01	0.43	0.64	0.64	Beam
5	3	13	0	1	2		3.709E+01	0.91	0.26	-0.33	Beam
6	3	14	0	1	2		4.651E+01	0.53	0.64	0.55	Beam
7	4	14	0	1	2		3.709E+01	0.84	0.26	-0.48	Beam

COOLING TOWER VIBRATION: INCREASED NODE MODEL

```
IMAGES-3D  s/n:800189                                    02-14-1987
                                                         PAGE 5
            ================ I M A G E S   3 D ================
            = Copyright (c) 1984   Celestial Software Inc. =
            ====================================================

            CHECK GEOMETRY            Version 1.4  12/01/86

     COOLING TOWER VIBRATION - NODE INCREASE
```

Beam No	Nodes From/ To/Ref	Prop No	Mat No	Pincodes I / J	Length	X	Y Dir Cosines Y	Z	Beam Type
8	4 15	0	1	2	4.651E+01	0.62	0.64	0.45	Beam
9	5 15	0	1	2	3.709E+01	0.74	0.26	-0.62	Beam
10	5 16	0	1	2	4.651E+01	0.69	0.64	0.34	Beam
11	6 16	0	1	2	3.709E+01	0.62	0.26	-0.74	Beam
12	6 17	0	1	2	4.651E+01	0.74	0.64	0.21	Beam
13	7 17	0	1	2	3.709E+01	0.48	0.26	-0.84	Beam
14	7 18	0	1	2	4.651E+01	0.77	0.64	0.08	Beam
15	8 18	0	1	2	3.709E+01	0.33	0.26	-0.91	Beam
16	8 19	0	1	2	4.651E+01	0.77	0.64	-0.05	Beam
17	9 19	0	1	2	3.709E+01	0.17	0.26	-0.95	Beam
18	9 20	0	1	2	4.651E+01	0.75	0.64	-0.19	Beam
19	10 20	0	1	2	3.709E+01	0.00	0.26	-0.97	Beam

```
                        PLATE ELEMENT CONNECTIVITY
```

Plate No.	N o d e s I	J	K	L	Mat No.	Thickness	Area	Shear Web Thickness	Aspect Ratio	Plate Type
QUAD 1	11	12	22	21	1	1.200E+01	9.911E+02		6.899E-01	Mem+Bend
QUAD 2	12	13	23	22	1	1.200E+01	9.911E+02		6.899E-01	Mem+Bend
QUAD 3	13	14	24	23	1	1.200E+01	9.911E+02		6.899E-01	Mem+Bend
QUAD 4	14	15	25	24	1	1.200E+01	9.911E+02		6.899E-01	Mem+Bend
QUAD 5	15	16	26	25	1	1.200E+01	9.911E+02		6.899E-01	Mem+Bend
QUAD 6	16	17	27	26	1	1.200E+01	9.911E+02		6.899E-01	Mem+Bend
QUAD 7	17	18	28	27	1	1.200E+01	9.911E+02		6.899E-01	Mem+Bend
QUAD 8	18	19	29	28	1	1.200E+01	9.911E+02		6.899E-01	Mem+Bend
QUAD 9	19	20	30	29	1	1.200E+01	9.911E+02		6.899E-01	Mem+Bend
QUAD 10	21	22	32	31	1	1.200E+01	8.988E+02		6.509E-01	Mem+Bend
QUAD 11	22	23	33	32	1	1.200E+01	8.988E+02		6.509E-01	Mem+Bend
QUAD 12	23	24	34	33	1	1.200E+01	8.988E+02		6.509E-01	Mem+Bend
QUAD 13	24	25	35	34	1	1.200E+01	8.988E+02		6.509E-01	Mem+Bend
QUAD 14	25	26	36	35	1	1.200E+01	8.988E+02		6.509E-01	Mem+Bend
QUAD 15	26	27	37	36	1	1.200E+01	8.988E+02		6.509E-01	Mem+Bend
QUAD 16	27	28	38	37	1	1.200E+01	8.988E+02		6.509E-01	Mem+Bend
QUAD 17	28	29	39	38	1	1.200E+01	8.988E+02		6.509E-01	Mem+Bend
QUAD 18	29	30	40	39	1	1.200E+01	8.988E+02		6.509E-01	Mem+Bend
QUAD 19	31	32	42	41	1	1.200E+01	8.332E+02		6.060E-01	Mem+Bend
QUAD 20	32	33	43	42	1	1.200E+01	8.332E+02		6.060E-01	Mem+Bend
QUAD 21	33	34	44	43	1	1.200E+01	8.332E+02		6.060E-01	Mem+Bend

IMAGES-3D s/n:800189   02-14-1987
                                                         PAGE 6
         =============== I M A G E S   3 D ===============
         = Copyright (c) 1984   Celestial Software Inc. =
         ===

                 CHECK GEOMETRY   Version 1.4 12/01/86

         COOLING TOWER VIBRATION - NODE INCREASE

Plate No.	N	o	d	e	s	Mat No.	Thickness	Area	Shear Web Thickness	Aspect Ratio	Plate Type
		I	J	K	L						
QUAD 22	34	35	45	44		1	1.200E+01	8.332E+02		6.060E-01	Mem+Bend
QUAD 23	35	36	46	45		1	1.200E+01	8.332E+02		6.060E-01	Mem+Bend
QUAD 24	36	37	47	46		1	1.200E+01	8.332E+02		6.060E-01	Mem+Bend
QUAD 25	37	38	48	47		1	1.200E+01	8.332E+02		6.060E-01	Mem+Bend
QUAD 26	38	39	49	48		1	1.200E+01	8.332E+02		6.060E-01	Mem+Bend
QUAD 27	39	40	50	49		1	1.200E+01	8.332E+02		6.060E-01	Mem+Bend
QUAD 28	41	42	52	51		1	1.200E+01	7.699E+02		5.624E-01	Mem+Bend
QUAD 29	42	43	53	52		1	1.200E+01	7.699E+02		5.624E-01	Mem+Bend
QUAD 30	43	44	54	53		1	1.200E+01	7.699E+02		5.624E-01	Mem+Bend
QUAD 31	44	45	55	54		1	1.200E+01	7.699E+02		5.624E-01	Mem+Bend
QUAD 32	45	46	56	55		1	1.200E+01	7.699E+02		5.624E-01	Mem+Bend
QUAD 33	46	47	57	56		1	1.200E+01	7.699E+02		5.624E-01	Mem+Bend
QUAD 34	47	48	58	57		1	1.200E+01	7.699E+02		5.624E-01	Mem+Bend
QUAD 35	48	49	59	58		1	1.200E+01	7.699E+02		5.624E-01	Mem+Bend
QUAD 36	49	50	60	59		1	1.200E+01	7.699E+02		5.624E-01	Mem+Bend
QUAD 37	51	52	62	61		1	1.200E+01	7.094E+02		5.183E-01	Mem+Bend
QUAD 38	52	53	63	62		1	1.200E+01	7.094E+02		5.183E-01	Mem+Bend
QUAD 39	53	54	64	63		1	1.200E+01	7.094E+02		5.183E-01	Mem+Bend
QUAD 40	54	55	65	64		1	1.200E+01	7.094E+02		5.183E-01	Mem+Bend
QUAD 41	55	56	66	65		1	1.200E+01	7.094E+02		5.183E-01	Mem+Bend
QUAD 42	56	57	67	66		1	1.200E+01	7.094E+02		5.183E-01	Mem+Bend
QUAD 43	57	58	68	67		1	1.200E+01	7.094E+02		5.183E-01	Mem+Bend
QUAD 44	58	59	69	68		1	1.200E+01	7.094E+02		5.183E-01	Mem+Bend
QUAD 45	59	60	70	69		1	1.200E+01	7.094E+02		5.183E-01	Mem+Bend
QUAD 46	61	62	72	71		1	1.200E+01	6.481E+02		4.792E-01	Mem+Bend
QUAD 47	62	63	73	72		1	1.200E+01	6.481E+02		4.792E-01	Mem+Bend
QUAD 48	63	64	74	73		1	1.200E+01	6.481E+02		4.792E-01	Mem+Bend
QUAD 49	64	65	75	74		1	1.200E+01	6.481E+02		4.792E-01	Mem+Bend
QUAD 50	65	66	76	75		1	1.200E+01	6.481E+02		4.792E-01	Mem+Bend
QUAD 51	66	67	77	76		1	1.200E+01	6.481E+02		4.792E-01	Mem+Bend
QUAD 52	67	68	78	77		1	1.200E+01	6.481E+02		4.792E-01	Mem+Bend
QUAD 53	68	69	79	78		1	1.200E+01	6.481E+02		4.792E-01	Mem+Bend
QUAD 54	69	70	80	79		1	1.200E+01	6.481E+02		4.792E-01	Mem+Bend
QUAD 55	71	72	82	81		1	1.200E+01	5.949E+02		4.564E-01	Mem+Bend
QUAD 56	72	73	83	82		1	1.200E+01	5.949E+02		4.564E-01	Mem+Bend
QUAD 57	73	74	84	83		1	1.200E+01	5.949E+02		4.564E-01	Mem+Bend
QUAD 58	74	75	85	84		1	1.200E+01	5.949E+02		4.564E-01	Mem+Bend
QUAD 59	75	76	86	85		1	1.200E+01	5.949E+02		4.564E-01	Mem+Bend
QUAD 60	76	77	87	86		1	1.200E+01	5.949E+02		4.564E-01	Mem+Bend
QUAD 61	77	78	88	87		1	1.200E+01	5.949E+02		4.564E-01	Mem+Bend
QUAD 62	78	79	89	88		1	1.200E+01	5.949E+02		4.564E-01	Mem+Bend

COOLING TOWER VIBRATION: INCREASED NODE MODEL

```
IMAGES-3D  s/n:800189                                    02-14-1987
                                                         PAGE 7
           =============== I M A G E S  3 D ===============
           = Copyright (c) 1984   Celestial Software Inc. =
           ================================================

              CHECK GEOMETRY              Version 1.4  12/01/86

           COOLING TOWER VIBRATION - NODE INCREASE
```

Plate No.	Nodes I	J	K	L	Mat No.	Thickness	Area	Shear Web Thickness	Aspect Ratio	Plate Type
QUAD 63	79	80	90	89	1	1.200E+01	5.949E+02		4.564E-01	Mem+Bend
QUAD 64	81	82	92	91	1	1.200E+01	5.710E+02		4.447E-01	Mem+Bend
QUAD 65	82	83	93	92	1	1.200E+01	5.710E+02		4.447E-01	Mem+Bend
QUAD 66	83	84	94	93	1	1.200E+01	5.710E+02		4.447E-01	Mem+Bend
QUAD 67	84	85	95	94	1	1.200E+01	5.710E+02		4.447E-01	Mem+Bend
QUAD 68	85	86	96	95	1	1.200E+01	5.710E+02		4.447E-01	Mem+Bend
QUAD 69	86	87	97	96	1	1.200E+01	5.710E+02		4.447E-01	Mem+Bend
QUAD 70	87	88	98	97	1	1.200E+01	5.710E+02		4.447E-01	Mem+Bend
QUAD 71	88	89	99	98	1	1.200E+01	5.710E+02		4.447E-01	Mem+Bend
QUAD 72	89	90	100	99	1	1.200E+01	5.710E+02		4.447E-01	Mem+Bend
QUAD 73	91	92	102	101	1	1.200E+01	5.835E+02		4.486E-01	Mem+Bend
QUAD 74	92	93	103	102	1	1.200E+01	5.835E+02		4.486E-01	Mem+Bend
QUAD 75	93	94	104	103	1	1.200E+01	5.835E+02		4.486E-01	Mem+Bend
QUAD 76	94	95	105	104	1	1.200E+01	5.835E+02		4.486E-01	Mem+Bend
QUAD 77	95	96	106	105	1	1.200E+01	5.835E+02		4.486E-01	Mem+Bend
QUAD 78	96	97	107	106	1	1.200E+01	5.835E+02		4.486E-01	Mem+Bend
QUAD 79	97	98	108	107	1	1.200E+01	5.835E+02		4.486E-01	Mem+Bend
QUAD 80	98	99	109	108	1	1.200E+01	5.835E+02		4.486E-01	Mem+Bend
QUAD 81	99	100	110	109	1	1.200E+01	5.835E+02		4.486E-01	Mem+Bend
QUAD 82	101	102	112	111	1	1.200E+01	5.626E+02		5.367E-01	Mem+Bend
QUAD 83	102	103	113	112	1	1.200E+01	5.626E+02		5.367E-01	Mem+Bend
QUAD 84	103	104	114	113	1	1.200E+01	5.626E+02		5.367E-01	Mem+Bend
QUAD 85	104	105	115	114	1	1.200E+01	5.626E+02		5.367E-01	Mem+Bend
QUAD 86	105	106	116	115	1	1.200E+01	5.626E+02		5.367E-01	Mem+Bend
QUAD 87	106	107	117	116	1	1.200E+01	5.626E+02		5.367E-01	Mem+Bend
QUAD 88	107	108	118	117	1	1.200E+01	5.626E+02		5.367E-01	Mem+Bend
QUAD 89	108	109	119	118	1	1.200E+01	5.626E+02		5.367E-01	Mem+Bend
QUAD 90	109	110	120	119	1	1.200E+01	5.626E+02		5.367E-01	Mem+Bend

```
                          RESTRAINTS

                 Node        Restraint
                 No          Directions
                 ----        ----------------
                  1      X Y Z RX RY RZ
                  2      X Y Z RX RY RZ
                  3      X Y Z RX RY RZ
                  4      X Y Z RX RY RZ
```

```
IMAGES-3D   s/n:800189
                          02-14-1987
                                                    PAGE 8
    =============== I M A G E S  3 D ===============
    = Copyright (c) 1984   Celestial Software Inc. =
    ================================================

        CHECK GEOMETRY            Version 1.4  12/01/86

    COOLING TOWER VIBRATION - NODE INCREASE

                    Node      Restraint
                    No        Directions
                    ----      -----------------

                      5     X Y Z RX RY RZ
                      6     X Y Z RX RY RZ
                      7     X Y Z RX RY RZ
                      8     X Y Z RX RY RZ
                      9     X Y Z RX RY RZ
                     10     X Y Z RX RY RZ
                     11     - - Z -  -  RZ
                     20     - - Z RX -  -
                     21     - - Z -  -  RZ
                     30     - - Z RX -  -
                     31     - - Z -  -  RZ
                     40     - - Z RX -  -
                     41     - - Z -  -  RZ
                     50     - - Z RX -  -
                     51     - - Z -  -  RZ
                     60     - - Z RX -  -
                     61     - - Z -  -  RZ
                     70     - - Z RX -  -
                     71     - - Z -  -  RZ
                     80     - - Z RX -  -
                     81     - - Z -  -  RZ
                     90     - - Z RX -  -
                     91     - - Z -  -  RZ
                    100     - - Z RX -  -
                    101     - - Z -  -  RZ
                    110     - - Z RX -  -
                    111     - - Z -  -  RZ
                    120     - - Z RX -  -
```

COOLING TOWER VIBRATION: INCREASED NODE MODEL

```
IMAGES-3D  s/n:800189                                    02=14=1987
                                                         PAGE 1
        =============== I M A G E S  3 D ===============
        = Copyright (c) 1984   Celestial Software Inc. =
        ================================================

        RENUMBER NODES                 Version 1.4  12/01/86

     COOLING TOWER VIBRATION - NODE INCREASE

              Node Renumbering Cross Reference List

              Was    Is      Was    Is      Was    Is
              ----   ----    ----   ----    ----   ----
               1      1       2      2       3      3
               4      4       5      5       6      6
               7      7       8      8       9      9
              10     10      11     11      12     12
              13     13      14     14      15     15
              16     16      17     17      18     18
              19     19      20     20      21     21
              22     22      23     23      24     24
              25     25      26     26      27     27
              28     28      29     29      30     30
              31     31      32     32      33     33
              34     34      35     35      36     36
              37     37      38     38      39     39
              40     40      41     41      42     42
              43     43      44     44      45     45
              46     46      47     47      48     48
              49     49      50     50      51     51
              52     52      53     53      54     54
              55     55      56     56      57     57
              58     58      59     59      60     60
              61     61      62     62      63     63
              64     64      65     65      66     66
              67     67      68     68      69     69
              70     70      71     71      72     72
              73     73      74     74      75     75
              76     76      77     77      78     78
              79     79      80     80      81     81
              82     82      83     83      84     84
              85     85      86     86      87     87
              88     88      89     89      90     90
              91     91      92     92      93     93
              94     94      95     95      96     96
              97     97      98     98      99     99
             100    100     101    101     102    102
             103    103     104    104     105    105
             106    106     107    107     108    108
```

```
=============== I M A G E S  3 D ===============
= Copyright (c) 1984   Celestial Software Inc. =
================================================
```

RENUMBER NODES   Version 1.4 12/01/86

COOLING TOWER VIBRATION - NODE INCREASE

Was	Is	Was	Is	Was	Is
109	109	110	110	111	111
112	112	113	113	114	114
115	115	116	116	117	117
118	118	119	119	120	120

Original Nodal Band   12
Final Nodal band   12

IMAGES-3D s/n:800189   02-14-87
                                                           PAGE   1
```
=============== I M A G E S  3 D ===============
= Copyright (c) 1984   Celestial Software Inc. =
================================================
```

ASSEMBLE STIFFNESS MATRIX Version 1.4 12/01/86

COOLING TOWER VIBRATION - NODE INCREASE

STIFFNESS ASSEMBLY SUMMARY

```
Number of Node Points..................  120
Number of Truss and Beam Elements.....   19
Number of Plate Elements..............   90
Number of Spring Elements.............    0
Number of Solid Elements..............    0
Number of Axisymmetric Elements.......    0
Number of Nodes with Restraints.......   32
Number of Equations to Be Solvesd.....  616
Number of Blocks in the Matrix........    3
```

B L O C K   I N F O R M A T I O N

BLCK NO	SIZE (Byte)	BLCK NO	SIZE (Byte)	BLCK NO	SIZE (Byte)	BLCK NO	SIZE (Byte)
1	127936	2	128000	3	37888		

COOLING TOWER VIBRATION: INCREASED NODE MODEL

IMAGES-3D   s/n:800189

```
=============== I M A G E S   3 D ===============
= Copyright (c) 1984   Celestial Software Inc. =
================================================
```

ASSEMBLE STIFFNESS MATRIX   Version 1.4   12/01/86

COOLING TOWER VIBRATION - NODE INCREASE

E Q U A T I O N   N U M B E R   L I S T

NODE		TRANSLATION			ROTATION		
WAS	IS	X	Y	Z	X	Y	Z
1	1	0	0	0	0	0	0
2	2	0	0	0	0	0	0
3	3	0	0	0	0	0	0
4	4	0	0	0	0	0	0
5	5	0	0	0	0	0	0
6	6	0	0	0	0	0	0
7	7	0	0	0	0	0	0
8	8	0	0	0	0	0	0
9	9	0	0	0	0	0	0
10	10	0	0	0	0	0	0
11	11	1	2	0	3	4	0
12	12	5	6	7	8	9	10
13	13	11	12	13	14	15	16
14	14	17	18	19	20	21	22
15	15	23	24	25	26	27	28
16	16	29	30	31	32	33	34
17	17	35	36	37	38	39	40
18	18	41	42	43	44	45	46
19	19	47	48	49	50	51	52
20	20	53	54	0	0	55	56
21	21	57	58	0	59	60	0
22	22	61	62	63	64	65	66
23	23	67	68	69	70	71	72
24	24	73	74	75	76	77	78
25	25	79	80	81	82	83	84
26	26	85	86	87	88	89	90
27	27	91	92	93	94	95	96
28	28	97	98	99	100	101	102
29	29	103	104	105	106	107	108
30	30	109	110	0	0	111	112
31	31	113	114	0	115	116	0
32	32	117	118	119	120	121	122
33	33	123	124	125	126	127	128
34	34	129	130	131	132	133	134
35	35	135	136	137	138	139	140
36	36	141	142	143	144	145	146
37	37	147	148	149	150	151	152
38	38	153	154	155	156	157	158
39	39	159	160	161	162	163	164
40	40	165	166	0	0	167	168

IMAGES-3D s/n:800189
02-14-87
PAGE 3
============= I M A G E S 3 D =============
= Copyright (c) 1984 Celestial Software Inc. =
===

ASSEMBLE STIFFNESS MATRIX Version 1.4 12/01/86

COOLING TOWER VIBRATION - NODE INCREASE

NODE		TRANSLATION			ROTATION		
WAS	IS	X	Y	Z	X	Y	Z
41	41	169	170	0	171	172	0
42	42	173	174	175	176	177	178
43	43	179	180	181	182	183	184
44	44	185	186	187	188	189	190
45	45	191	192	193	194	195	196
46	46	197	198	199	200	201	202
47	47	203	204	205	206	207	208
48	48	209	210	211	212	213	214
49	49	215	216	217	218	219	220
50	50	221	222	0	0	223	224
51	51	225	226	0	227	228	0
52	52	229	230	231	232	233	234
53	53	235	236	237	238	239	240
54	54	241	242	243	244	245	246
55	55	247	248	249	250	251	252
56	56	253	254	255	256	257	258
57	57	259	260	261	262	263	264
58	58	265	266	267	268	269	270
59	59	271	272	273	274	275	276
60	60	277	278	0	0	279	280
61	61	281	282	0	283	284	0
62	62	285	286	287	288	289	290
63	63	291	292	293	294	295	296
64	64	297	298	299	300	301	302
65	65	303	304	305	306	307	308
66	66	309	310	311	312	313	314
67	67	315	316	317	318	319	320
68	68	321	322	323	324	325	326
69	69	327	328	329	330	331	332
70	70	333	334	0	0	335	336
71	71	337	338	0	339	340	0
72	72	341	342	343	344	345	346
73	73	347	348	349	350	351	352
74	74	353	354	355	356	357	358
75	75	359	360	361	362	363	364
76	76	365	366	367	368	369	370
77	77	371	372	373	374	375	376
78	78	377	378	379	380	381	382
79	79	383	384	385	386	387	388
80	80	389	390	0	0	391	392

COOLING TOWER VIBRATION: INCREASED NODE MODEL

```
IMAGES-3D   s/n:800189                                      02-14-87
                                                           PAGE    4
=============== I M A G E S   3 D ===============
= Copyright (c) 1984   Celestial Software Inc. =
================================================

ASSEMBLE STIFFNESS MATRIX   Version 1.4   12/01/86

COOLING TOWER VIBRATION - NODE INCREASE
```

NODE		TRANSLATION			ROTATION		
WAS	IS	X	Y	Z	X	Y	Z
81	81	393	394	0	395	396	0
82	82	397	398	399	400	401	402
83	83	403	404	405	406	407	408
84	84	409	410	411	412	413	414
85	85	415	416	417	418	419	420
86	86	421	422	423	424	425	426
87	87	427	428	429	430	431	432
88	88	433	434	435	436	437	438
89	89	439	440	441	442	443	444
90	90	445	446	0	0	447	448
91	91	449	450	0	451	452	0
92	92	453	454	455	456	457	458
93	93	459	460	461	462	463	464
94	94	465	466	467	468	469	470
95	95	471	472	473	474	475	476
96	96	477	478	479	480	481	482
97	97	483	484	485	486	487	488
98	98	489	490	491	492	493	494
99	99	495	496	497	498	499	500
100	100	501	502	0	0	503	504
101	101	505	506	0	507	508	0
102	102	509	510	511	512	513	514
103	103	515	516	517	518	519	520
104	104	521	522	523	524	525	526
105	105	527	528	529	530	531	532
106	106	533	534	535	536	537	538
107	107	539	540	541	542	543	544
108	108	545	546	547	548	549	550
109	109	551	552	553	554	555	556
110	110	557	558	0	0	559	560
111	111	561	562	0	563	564	0
112	112	565	566	567	568	569	570
113	113	571	572	573	574	575	576
114	114	577	578	579	580	581	582
115	115	583	584	585	586	587	588
116	116	589	590	591	592	593	594
117	117	595	596	597	598	599	600
118	118	601	602	603	604	605	606
119	119	607	608	609	610	611	612
120	120	613	614	0	0	615	616

================ I M A G E S   3 D ===============
= Copyright (c) 1984   Celestial Software Inc. =
==

ASSEMBLE STIFFNESS MATRIX   Version 1.4   12/01/86

COOLING TOWER VIBRATION - NODE INCREASE

STIFFNESS SUMMARY IN 3 BLOCKS

Minimum Diagonal Stiffness.....   .1904D+08
Eq No of Minimum Diagonal......   562
Maximum Diagonal Stiffness.....   .1376D+12
Eq No of Maximum Diagonal......   46

================ I M A G E S   3 D ===============
= Copyright (c) 1984   Celestial Software Inc. =
==

WEIGHTS   Version 1.4   12/01/86

COOLING TOWER VIBRATION - NODE INCREASE

Weight Matrix

Node	W e i g h t s X	Y	Z	/	R o t a r y I n e r t i a s X	Y	Z
1	.6904E+04	.6904E+04	.6904E+04	/	.0000E+00	.0000E+00	.0000E+00
2	.6904E+04	.6904E+04	.6904E+04	/	.0000E+00	.0000E+00	.0000E+00
3	.6904E+04	.6904E+04	.6904E+04	/	.0000E+00	.0000E+00	.0000E+00
4	.6904E+04	.6904E+04	.6904E+04	/	.0000E+00	.0000E+00	.0000E+00
5	.6904E+04	.6904E+04	.6904E+04	/	.0000E+00	.0000E+00	.0000E+00
6	.6904E+04	.6904E+04	.6904E+04	/	.0000E+00	.0000E+00	.0000E+00
7	.6904E+04	.6904E+04	.6904E+04	/	.0000E+00	.0000E+00	.0000E+00
8	.6904E+04	.6904E+04	.6904E+04	/	.0000E+00	.0000E+00	.0000E+00
9	.6904E+04	.6904E+04	.6904E+04	/	.0000E+00	.0000E+00	.0000E+00
10	.3063E+04	.3063E+04	.3063E+04	/	.0000E+00	.0000E+00	.0000E+00
11	.3324E+04	.3324E+04	.3324E+04	/	.0000E+00	.0000E+00	.0000E+00
12	.7427E+04	.7427E+04	.7427E+04	/	.0000E+00	.0000E+00	.0000E+00
13	.7427E+04	.7427E+04	.7427E+04	/	.0000E+00	.0000E+00	.0000E+00
14	.7427E+04	.7427E+04	.7427E+04	/	.0000E+00	.0000E+00	.0000E+00
15	.7427E+04	.7427E+04	.7427E+04	/	.0000E+00	.0000E+00	.0000E+00
16	.7427E+04	.7427E+04	.7427E+04	/	.0000E+00	.0000E+00	.0000E+00
17	.7427E+04	.7427E+04	.7427E+04	/	.0000E+00	.0000E+00	.0000E+00
18	.7427E+04	.7427E+04	.7427E+04	/	.0000E+00	.0000E+00	.0000E+00
19	.7427E+04	.7427E+04	.7427E+04	/	.0000E+00	.0000E+00	.0000E+00
20	.7165E+04	.7165E+04	.7165E+04	/	.0000E+00	.0000E+00	.0000E+00
21	.4914E+03	.4914E+03	.4914E+03	/	.0000E+00	.0000E+00	.0000E+00
22	.9827E+03	.9827E+03	.9827E+03	/	.0000E+00	.0000E+00	.0000E+00
23	.9827E+03	.9827E+03	.9827E+03	/	.0000E+00	.0000E+00	.0000E+00
24	.9827E+03	.9827E+03	.9827E+03	/	.0000E+00	.0000E+00	.0000E+00
25	.9827E+03	.9827E+03	.9827E+03	/	.0000E+00	.0000E+00	.0000E+00
26	.9827E+03	.9827E+03	.9827E+03	/	.0000E+00	.0000E+00	.0000E+00
27	.9827E+03	.9827E+03	.9827E+03	/	.0000E+00	.0000E+00	.0000E+00
28	.9827E+03	.9827E+03	.9827E+03	/	.0000E+00	.0000E+00	.0000E+00
29	.9827E+03	.9827E+03	.9827E+03	/	.0000E+00	.0000E+00	.0000E+00
30	.4914E+03	.4914E+03	.4914E+03	/	.0000E+00	.0000E+00	.0000E+00
31	.4509E+03	.4509E+03	.4509E+03	/	.0000E+00	.0000E+00	.0000E+00
32	.9018E+03	.9018E+03	.9018E+03	/	.0000E+00	.0000E+00	.0000E+00
33	.9018E+03	.9018E+03	.9018E+03	/	.0000E+00	.0000E+00	.0000E+00
34	.9018E+03	.9018E+03	.9018E+03	/	.0000E+00	.0000E+00	.0000E+00
35	.9018E+03	.9018E+03	.9018E+03	/	.0000E+00	.0000E+00	.0000E+00
36	.9018E+03	.9018E+03	.9018E+03	/	.0000E+00	.0000E+00	.0000E+00
37	.9018E+03	.9018E+03	.9018E+03	/	.0000E+00	.0000E+00	.0000E+00
38	.9018E+03	.9018E+03	.9018E+03	/	.0000E+00	.0000E+00	.0000E+00
39	.9018E+03	.9018E+03	.9018E+03	/	.0000E+00	.0000E+00	.0000E+00

COOLING TOWER VIBRATION: INCREASED NODE MODEL

```
================ I M A G E S   3 D ================
= Copyright (c) 1984   Celestial Software Inc. =
==================================================
```

WEIGHTS   Version 1.4 12/01/86

COOLING TOWER VIBRATION - NODE INCREASE

	W e i g h t s		/	R o t a r y I n e r t i a s	
Node	X	Y	Z	/	X	Y	Z
40	.4509E+03	.4509E+03	.4509E+03	/	.0000E+00	.0000E+00	.0000E+00
41	.4174E+03	.4174E+03	.4174E+03	/	.0000E+00	.0000E+00	.0000E+00
42	.8347E+03	.8347E+03	.8347E+03	/	.0000E+00	.0000E+00	.0000E+00
43	.8347E+03	.8347E+03	.8347E+03	/	.0000E+00	.0000E+00	.0000E+00
44	.8347E+03	.8347E+03	.8347E+03	/	.0000E+00	.0000E+00	.0000E+00
45	.8347E+03	.8347E+03	.8347E+03	/	.0000E+00	.0000E+00	.0000E+00
46	.8347E+03	.8347E+03	.8347E+03	/	.0000E+00	.0000E+00	.0000E+00
47	.8347E+03	.8347E+03	.8347E+03	/	.0000E+00	.0000E+00	.0000E+00
48	.8347E+03	.8347E+03	.8347E+03	/	.0000E+00	.0000E+00	.0000E+00
49	.8347E+03	.8347E+03	.8347E+03	/	.0000E+00	.0000E+00	.0000E+00
50	.4174E+03	.4174E+03	.4174E+03	/	.0000E+00	.0000E+00	.0000E+00
51	.3852E+03	.3852E+03	.3852E+03	/	.0000E+00	.0000E+00	.0000E+00
52	.7704E+03	.7704E+03	.7704E+03	/	.0000E+00	.0000E+00	.0000E+00
53	.7704E+03	.7704E+03	.7704E+03	/	.0000E+00	.0000E+00	.0000E+00
54	.7704E+03	.7704E+03	.7704E+03	/	.0000E+00	.0000E+00	.0000E+00
55	.7704E+03	.7704E+03	.7704E+03	/	.0000E+00	.0000E+00	.0000E+00
56	.7704E+03	.7704E+03	.7704E+03	/	.0000E+00	.0000E+00	.0000E+00
57	.7704E+03	.7704E+03	.7704E+03	/	.0000E+00	.0000E+00	.0000E+00
58	.7704E+03	.7704E+03	.7704E+03	/	.0000E+00	.0000E+00	.0000E+00
59	.7704E+03	.7704E+03	.7704E+03	/	.0000E+00	.0000E+00	.0000E+00
60	.3852E+03	.3852E+03	.3852E+03	/	.0000E+00	.0000E+00	.0000E+00
61	.3532E+03	.3532E+03	.3532E+03	/	.0000E+00	.0000E+00	.0000E+00
62	.7064E+03	.7064E+03	.7064E+03	/	.0000E+00	.0000E+00	.0000E+00
63	.7064E+03	.7064E+03	.7064E+03	/	.0000E+00	.0000E+00	.0000E+00
64	.7064E+03	.7064E+03	.7064E+03	/	.0000E+00	.0000E+00	.0000E+00
65	.7064E+03	.7064E+03	.7064E+03	/	.0000E+00	.0000E+00	.0000E+00
66	.7064E+03	.7064E+03	.7064E+03	/	.0000E+00	.0000E+00	.0000E+00
67	.7064E+03	.7064E+03	.7064E+03	/	.0000E+00	.0000E+00	.0000E+00
68	.7064E+03	.7064E+03	.7064E+03	/	.0000E+00	.0000E+00	.0000E+00
69	.7064E+03	.7064E+03	.7064E+03	/	.0000E+00	.0000E+00	.0000E+00
70	.3532E+03	.3532E+03	.3532E+03	/	.0000E+00	.0000E+00	.0000E+00
71	.3226E+03	.3226E+03	.3226E+03	/	.0000E+00	.0000E+00	.0000E+00
72	.6452E+03	.6452E+03	.6452E+03	/	.0000E+00	.0000E+00	.0000E+00
73	.6452E+03	.6452E+03	.6452E+03	/	.0000E+00	.0000E+00	.0000E+00
74	.6452E+03	.6452E+03	.6452E+03	/	.0000E+00	.0000E+00	.0000E+00
75	.6452E+03	.6452E+03	.6452E+03	/	.0000E+00	.0000E+00	.0000E+00
76	.6452E+03	.6452E+03	.6452E+03	/	.0000E+00	.0000E+00	.0000E+00
77	.6452E+03	.6452E+03	.6452E+03	/	.0000E+00	.0000E+00	.0000E+00
78	.6452E+03	.6452E+03	.6452E+03	/	.0000E+00	.0000E+00	.0000E+00
79	.6452E+03	.6452E+03	.6452E+03	/	.0000E+00	.0000E+00	.0000E+00
80	.3226E+03	.3226E+03	.3226E+03	/	.0000E+00	.0000E+00	.0000E+00

IMAGES-3D   s/n:800189   02-14-87
                                                          PAGE   3
          =============== I M A G E S   3 D ===============
          = Copyright (c) 1984   Celestial Software Inc. =
          ==

          WEIGHTS   Version 1.4 12/01/86

          COOLING TOWER VIBRATION - NODE INCREASE

Node	W e i g h t s X	Y	Z	/	R o t a r y X	I n e r t i a s Y	Z
81	.3027E+03	.3027E+03	.3027E+03	/	.0000E+00	.0000E+00	.0000E+00
82	.6055E+03	.6055E+03	.6055E+03	/	.0000E+00	.0000E+00	.0000E+00
83	.6055E+03	.6055E+03	.6055E+03	/	.0000E+00	.0000E+00	.0000E+00
84	.6055E+03	.6055E+03	.6055E+03	/	.0000E+00	.0000E+00	.0000E+00
85	.6055E+03	.6055E+03	.6055E+03	/	.0000E+00	.0000E+00	.0000E+00
86	.6055E+03	.6055E+03	.6055E+03	/	.0000E+00	.0000E+00	.0000E+00
87	.6055E+03	.6055E+03	.6055E+03	/	.0000E+00	.0000E+00	.0000E+00
88	.6055E+03	.6055E+03	.6055E+03	/	.0000E+00	.0000E+00	.0000E+00
89	.6055E+03	.6055E+03	.6055E+03	/	.0000E+00	.0000E+00	.0000E+00
90	.3027E+03	.3027E+03	.3027E+03	/	.0000E+00	.0000E+00	.0000E+00
91	.2990E+03	.2990E+03	.2990E+03	/	.0000E+00	.0000E+00	.0000E+00
92	.5980E+03	.5980E+03	.5980E+03	/	.0000E+00	.0000E+00	.0000E+00
93	.5980E+03	.5980E+03	.5980E+03	/	.0000E+00	.0000E+00	.0000E+00
94	.5980E+03	.5980E+03	.5980E+03	/	.0000E+00	.0000E+00	.0000E+00
95	.5980E+03	.5980E+03	.5980E+03	/	.0000E+00	.0000E+00	.0000E+00
96	.5980E+03	.5980E+03	.5980E+03	/	.0000E+00	.0000E+00	.0000E+00
97	.5980E+03	.5980E+03	.5980E+03	/	.0000E+00	.0000E+00	.0000E+00
98	.5980E+03	.5980E+03	.5980E+03	/	.0000E+00	.0000E+00	.0000E+00
99	.5980E+03	.5980E+03	.5980E+03	/	.0000E+00	.0000E+00	.0000E+00
100	.2990E+03	.2990E+03	.2990E+03	/	.0000E+00	.0000E+00	.0000E+00
101	.2973E+03	.2973E+03	.2973E+03	/	.0000E+00	.0000E+00	.0000E+00
102	.5947E+03	.5947E+03	.5947E+03	/	.0000E+00	.0000E+00	.0000E+00
103	.5947E+03	.5947E+03	.5947E+03	/	.0000E+00	.0000E+00	.0000E+00
104	.5947E+03	.5947E+03	.5947E+03	/	.0000E+00	.0000E+00	.0000E+00
105	.5947E+03	.5947E+03	.5947E+03	/	.0000E+00	.0000E+00	.0000E+00
106	.5947E+03	.5947E+03	.5947E+03	/	.0000E+00	.0000E+00	.0000E+00
107	.5947E+03	.5947E+03	.5947E+03	/	.0000E+00	.0000E+00	.0000E+00
108	.5947E+03	.5947E+03	.5947E+03	/	.0000E+00	.0000E+00	.0000E+00
109	.5947E+03	.5947E+03	.5947E+03	/	.0000E+00	.0000E+00	.0000E+00
110	.2973E+03	.2973E+03	.2973E+03	/	.0000E+00	.0000E+00	.0000E+00
111	.1488E+03	.1488E+03	.1488E+03	/	.0000E+00	.0000E+00	.0000E+00
112	.2976E+03	.2976E+03	.2976E+03	/	.0000E+00	.0000E+00	.0000E+00
113	.2976E+03	.2976E+03	.2976E+03	/	.0000E+00	.0000E+00	.0000E+00
114	.2976E+03	.2976E+03	.2976E+03	/	.0000E+00	.0000E+00	.0000E+00
115	.2976E+03	.2976E+03	.2976E+03	/	.0000E+00	.0000E+00	.0000E+00
116	.2976E+03	.2976E+03	.2976E+03	/	.0000E+00	.0000E+00	.0000E+00
117	.2976E+03	.2976E+03	.2976E+03	/	.0000E+00	.0000E+00	.0000E+00
118	.2976E+03	.2976E+03	.2976E+03	/	.0000E+00	.0000E+00	.0000E+00
119	.2976E+03	.2976E+03	.2976E+03	/	.0000E+00	.0000E+00	.0000E+00
120	.1488E+03	.1488E+03	.1488E+03	/	.0000E+00	.0000E+00	.0000E+00

COOLING TOWER VIBRATION: INCREASED NODE MODEL

```
IMAGES-3D   s/n:800189                                    02-14-87
                                                         PAGE    4
           =============== I M A G E S  3 D ===============
           = Copyright (c) 1984   Celestial Software Inc. =
           ================================================

           WEIGHTS                    Version 1.4  12/01/86

       COOLING TOWER VIBRATION - NODE INCREASE

                    W e i g h t s         /  R o t a r y   I n e r t i a s
       Node     X          Y          Z    /     X          Y          Z
       ----  ---------  ---------  --------- / --------- --------- ---------

Total:    .1975E+06  .1975E+06  .1975E+06 /  .0000E+00  .0000E+00  .0000E+00

Total:    .1323E+06  .1323E+06  .1149E+06 /  .0000E+00  .0000E+00  .0000E+00
(used)

                    Center of Gravity Based on X-Weights

         X =  .921194E+02    Y =  .779517E+02   Z = -.919312E+02

                    Center of Gravity Based on Y-Weights

         X =  .921194E+02    Y =  .779517E+02   Z = -.919312E+02

                    Center of Gravity Based on Z-Weights

         X =  .921194E+02    Y =  .779517E+02   Z = -.919312E+02




IMAGES-3D   s/n:800189                                    02-14-87
                                                         PAGE    1
           =============== I M A G E S  3 D ===============
           = Copyright (c) 1984   Celestial Software Inc. =
           ================================================

           SOLVE FREQUENCIES             Version 1.4  12/01/86

       COOLING TOWER VIBRATION - NODE INCREASE

                 Number of frequencies requested    1
                 Number of frequencies printed      1
                 Acceleration of gravity        386.40

            Mode   Eigenvalue    Frequency      Period
            ----  ------------  ------------  ------------
              1    .218460E+05   .235237E+02   .425103E-01
```

```
IMAGES-3D   s/n:800189                           02-14-87
                                                 PAGE    1
        =============== I M A G E S   3 D ===============
        = Copyright (c) 1984   Celestial Software Inc. =
        ================================================

        SOLVE MODE SHAPES              Version 1.4  12/01/86

        COOLING TOWER VIBRATION - NODE INCREASE


                        Number of modes          1

                ***Mode  1***      Eigenvalue= .218460E+05
```

	T r a n s l a t i o n s		/	R o t a t i o n s		
Node	X	Y	Z	/	X	Y	Z
----	---------	---------	---------	/	---------	---------	---------
1	.000000	.000000	.000000	/	.000000	.000000	.000000
2	.000000	.000000	.000000	/	.000000	.000000	.000000
3	.000000	.000000	.000000	/	.000000	.000000	.000000
4	.000000	.000000	.000000	/	.000000	.000000	.000000
5	.000000	.000000	.000000	/	.000000	.000000	.000000
6	.000000	.000000	.000000	/	.000000	.000000	.000000
7	.000000	.000000	.000000	/	.000000	.000000	.000000
8	.000000	.000000	.000000	/	.000000	.000000	.000000
9	.000000	.000000	.000000	/	.000000	.000000	.000000
10	.000000	.000000	.000000	/	.000000	.000000	.000000
11	.000361	-.000733	.000000	/	.000001	-.000003	.000000
12	.000976	-.000957	-.001033	/	-.000042	-.000013	-.000027
13	.000834	-.000834	-.001125	/	-.000045	-.000010	-.000023
14	.000633	-.000634	-.000993	/	-.000039	-.000006	-.000018
15	.000509	-.000341	-.000706	/	-.000030	-.000001	-.000017
16	.000547	.000057	-.000307	/	-.000018	.000005	-.000023
17	.000810	.000572	.000149	/	-.000006	.000012	-.000036
18	.001347	.001214	.000620	/	.000004	.000022	-.000059
19	.002125	.002041	.001091	/	.000012	.000030	-.000091
20	.002528	.001904	.000000	/	.000000	.000008	-.000103
21	.012237	-.019675	.000000	/	-.003683	-.001349	.000000
22	.015715	-.017913	-.000852	/	.000075	-.000077	-.000558
23	.017784	-.015139	-.002000	/	-.000174	-.000100	-.000593
24	.019212	-.010862	-.003185	/	-.000229	-.000079	-.000683
25	.020487	-.004797	-.004500	/	-.000364	-.000097	-.000725
26	.021841	.003292	-.006060	/	-.000526	-.000125	-.000731
27	.023252	.013521	-.007685	/	-.000671	-.000127	-.000718
28	.024618	.025980	-.008455	/	-.000732	-.000035	-.000769
29	.026602	.040351	-.006658	/	-.000730	.000134	-.000790
30	.032463	.062214	.000000	/	.000000	.002001	-.006463
31	.047388	-.032068	.000000	/	-.012263	-.003715	.000000
32	.054243	-.028980	-.002285	/	.000050	-.000255	-.001257
33	.059689	-.023392	-.005475	/	.000006	-.000192	-.001373
34	.064002	-.014940	-.009240	/	-.000074	-.000153	-.001501
35	.067526	-.003199	-.013368	/	-.000157	-.000123	-.001625
36	.070338	.012118	-.017465	/	-.000236	-.000071	-.001730

COOLING TOWER VIBRATION: INCREASED NODE MODEL

```
IMAGES-3D   s/n:800189                                    02-14-87
                                                          PAGE    2
          =============== I M A G E S   3 D ===============
          = Copyright (c) 1984   Celestial Software Inc. =
          ===============================================

          SOLVE MODE SHAPES              Version 1.4  12/01/86

       COOLING TOWER VIBRATION - NODE INCREASE
```

Node	Translations X	Y	Z	/	Rotations X	Y	Z
37	.072257	.031125	-.020394	/	-.000287	.000046	-.001805
38	.073384	.053621	-.019914	/	-.000312	.000258	-.001802
39	.074945	.080267	-.013000	/	-.000326	.000515	-.001611
40	.078558	.110598	.000000	/	.000000	.001452	-.004410
41	.099987	-.039824	.000000	/	-.016265	-.004872	.000000
42	.110534	-.035214	-.003454	/	.000060	-.000474	-.001656
43	.119980	-.026841	-.008500	/	.000038	-.000425	-.001797
44	.128184	-.014253	-.014753	/	-.000022	-.000375	-.001969
45	.134804	.002876	-.021565	/	-.000113	-.000295	-.002123
46	.139281	.024726	-.027563	/	-.000215	-.000149	-.002224
47	.141269	.051170	-.030440	/	-.000310	.000086	-.002226
48	.141332	.081991	-.027347	/	-.000375	.000388	-.002095
49	.141313	.116599	-.016648	/	-.000378	.000674	-.001893
50	.144080	.155327	.000000	/	.000000	.001691	-.005156
51	.167110	-.042713	.000000	/	-.021332	-.006390	.000000
52	.181981	-.036589	-.003795	/	.000079	-.000739	-.002124
53	.196085	-.025564	-.010055	/	.000184	-.000699	-.002312
54	.209176	-.009216	-.018583	/	.000077	-.000665	-.002487
55	.219912	.012772	-.028237	/	-.000069	-.000544	-.002635
56	.226827	.040297	-.036612	/	-.000227	-.000288	-.002718
57	.229247	.072938	-.040166	/	-.000359	.000106	-.002702
58	.228241	.109988	-.035455	/	-.000416	.000572	-.002597
59	.226811	.150962	-.021136	/	-.000371	.000977	-.002488
60	.228449	.195787	.000000	/	.000000	.001863	-.005197
61	.250838	-.038313	.000000	/	-.024757	-.007230	.000000
62	.269464	-.030968	-.003739	/	.001640	-.000608	-.002976
63	.287901	-.018338	-.010886	/	.000255	-.001034	-.002774
64	.306140	.000866	-.021506	/	.000056	-.001079	-.002913
65	.321748	.026677	-.034279	/	-.000135	-.000945	-.003026
66	.332032	.058701	-.045855	/	-.000324	-.000580	-.003068
67	.335689	.096095	-.051310	/	-.000456	.000010	-.003037
68	.334041	.137867	-.045875	/	-.000482	.000730	-.002973
69	.331012	.183193	-.027606	/	-.000413	.001356	-.002854
70	.331393	.231801	.000000	/	.000000	.002238	-.005290
71	.345545	-.027568	.000000	/	-.018959	-.005002	.000000
72	.370324	-.020422	-.003793	/	.001045	-.001286	-.003263
73	.394890	-.006698	-.012000	/	-.000189	-.001616	-.003117
74	.418959	.014028	-.024752	/	-.000027	-.001570	-.003403
75	.439361	.041970	-.040342	/	-.000167	-.001361	-.003475
76	.452825	.076670	-.054603	/	-.000274	-.000852	-.003484
77	.457810	.117185	-.061547	/	-.000346	-.000062	-.003407

```
IMAGES-3D  s/n:800189                              02-14-87
                                                  PAGE   3
          =============== I M A G E S  3 D ===============
          = Copyright (c) 1984   Celestial Software Inc. =
          ================================================

          SOLVE MODE SHAPES            Version 1.4  12/01/86

     COOLING TOWER VIBRATION - NODE INCREASE

          T r a n s l a t i o n s   /   R o t a t i o n s
    Node     X         Y         Z    /     X         Y         Z
    ----  --------- --------- --------- / --------- --------- ---------
     78   .455881   .162295  -.055384  / -.000339   .000893  -.003291
     79   .451702   .210844  -.033520  / -.000267   .001759  -.003141
     80   .451029   .262473   .000000  /  .000000   .002303  -.004036
     81   .458423  -.017770   .000000  / -.026346  -.004259   .000000
     82   .489879  -.011874  -.003989  /  .002422  -.001759  -.004239
     83   .520870   .000911  -.013825  / -.000114  -.002034  -.003677
     84   .549654   .021235  -.028919  / -.000055  -.001935  -.003868
     85   .572767   .049561  -.046609  / -.000199  -.001592  -.003856
     86   .586983   .085577  -.061877  / -.000239  -.000919  -.003824
     87   .591276   .128411  -.068275  / -.000234   .000029  -.003740
     88   .588010   .176883  -.060175  / -.000187   .001094  -.003640
     89   .582567   .229807  -.035775  / -.000127   .001997  -.003529
     90   .581034   .286013   .000000  /  .000000   .002400  -.003928
     91   .575652  -.010381   .000000  / -.015802  -.001920   .000000
     92   .616139  -.008139  -.005687  /  .000439  -.002601  -.003587
     93   .654626   .001681  -.018336  / -.000503  -.002503  -.003569
     94   .688352   .020618  -.035873  / -.000395  -.002267  -.003800
     95   .713924   .048871  -.054728  / -.000414  -.001733  -.003836
     96   .728714   .085962  -.069487  / -.000337  -.000901  -.003851
     97   .732627   .131071  -.074017  / -.000225   .000146  -.003851
     98   .728809   .183161  -.063555  / -.000108   .001241  -.003876
     99   .722908   .240993  -.037168  / -.000017   .002135  -.003959
    100   .720863   .303295   .000000  /  .000000   .002496  -.003956
    101   .680536  -.015568   .000000  / -.014192  -.000646   .000000
    102   .737590  -.022949  -.007275  / -.000235  -.003494  -.003431
    103   .789139  -.016793  -.022553  / -.000780  -.003012  -.003573
    104   .830464   .001539  -.042366  / -.000490  -.002480  -.003855
    105   .858747   .031268  -.061990  / -.000435  -.001688  -.003881
    106   .873291   .071346  -.075738  / -.000310  -.000739  -.003892
    107   .875897   .120754  -.078051  / -.000170   .000307  -.003914
    108   .870876   .178402  -.065230  / -.000047   .001317  -.003982
    109   .864198   .243053  -.037399  /  .000030   .002084  -.004120
    110   .861517   .313008   .000000  /  .000000   .002407  -.004069
    111   .780957  -.041647   .000000  / -.040860   .007852   .000000
    112   .854355  -.057117  -.003464  /  .003128  -.005121  -.004880
    113   .916386  -.053686  -.019454  / -.000059  -.003395  -.004253
    114   .961757  -.034287  -.042461  / -.000323  -.002502  -.004217
    115   .989217  -.001210  -.065526  / -.000395  -.001571  -.004137
    116  1.000000   .043919  -.081590  / -.000386  -.000576  -.004042
    117   .997773   .099628  -.084483  / -.000306   .000442  -.004001
    118   .988462   .164576  -.070452  / -.000190   .001383  -.004034
```

COOLING TOWER VIBRATION: INCREASED NODE MODEL

```
IMAGES-3D   s/n:800189                               02-14-87
                                                     PAGE    4
        ================ I M A G E S  3 D ================
        = Copyright (c) 1984   Celestial Software Inc. =
        =================================================

        SOLVE MODE SHAPES            Version 1.4  12/01/86

     COOLING TOWER VIBRATION - NODE INCREASE
```

| | T r a n s l a t i o n s | | / | R o t a t i o n s | | |
Node	X	Y	Z	/	X	Y	Z
119	.979144	.237257	-.040172	/	-.000074	.002083	-.004101
120	.975447	.315494	.000000	/	.000000	.002442	-.003827

```
IMAGES-3D   s/n:800189                               02-14-87
                                                     PAGE    1
        ================ I M A G E S  3 D ================
        = Copyright (c) 1984   Celestial Software Inc. =
        =================================================

        SOLVE PARTICIPATION          Version 1.4  12/01/86

     COOLING TOWER VIBRATION - NODE INCREASE
```

| | | PARTICIPATION FACTORS | | |
Mode	Generalized Weight	X	Y	Z
1	.1326E+05	.1620E+01	.2401E+00	-.1160E+00

| | EFFECTIVE MODAL WEIGHTS | | | % TOTAL SYSTEM WEIGHTS | | |
Mode	X	Y	Z	X	Y	Z
1	.3479E+05	.7646E+03	.1785E+03	26.29	.58	.16
			Summation:	26.29	.58	.16

?
```
IMAGES-3D   s/n:800189                              03-02-87
                                                   PAGE   1
        =============== I M A G E S  3 D ===============
        = Copyright (c) 1984   Celestial Software Inc. =
        ================================================

        SOLVE SEISMIC RESPONSE       Version 1.4  12/01/86

     COOLING TOWER VIBRATION - NODE INCREASE


                        Spectrum Multipliers

                    X-Direction =  .1000E+01
                    Y-Direction =  .1000E+01
                    Z-Direction =  .1000E+01

                    INPUT RESPONSE SPECTRA

              X-Direction    /    Y-Direction    /    Z-Direction
      Point Frequency Acceleration Frequency Acceleration Frequency Acceleration
      ----- --------- ------------ --------- ------------ --------- ------------
        1  .235E+02  .100000E+01

                    Interpolated Accelerations

         Mode   Frequency  X-Direction  Y-Direction  Z-Direction
         ----   ---------  -----------  -----------  -----------
           1   .2352E+02   .10000E+01   .00000E+00   .00000E+00

                    Generalized Displacements

                                                              A B S
      Mode  Frequency  X-Direction  Y-Direction  Z-Direction  Combination
      ----  ---------  -----------  -----------  -----------  -----------
        1  .2352E+02   .28651E-01   .00000E+00   .00000E+00   .28651E-01

                    Generalized Accelerations

                                                              A B S
      Mode  Frequency  X-Direction  Y-Direction  Z-Direction  Combination
      ----  ---------  -----------  -----------  -----------  -----------
        1  .2352E+02   .62590E+03   .00000E+00   .00000E+00   .62590E+03

                    DISPLACEMENTS for MODE  1

              T r a n s l a t i o n s   /     R o t a t i o n s
      Node    X          Y          Z    /     X          Y          Z
      ----  ---------- ---------- ---------- / ---------- ---------- ----------
        1  .0000E+00  .0000E+00  .0000E+00 /  .0000E+00  .0000E+00  .0000E+00
        2  .0000E+00  .0000E+00  .0000E+00 /  .0000E+00  .0000E+00  .0000E+00
        3  .0000E+00  .0000E+00  .0000E+00 /  .0000E+00  .0000E+00  .0000E+00
```

COOLING TOWER VIBRATION: INCREASED NODE MODEL

```
IMAGES-3D   s/n:800189                            03-02-87
                                                  PAGE    2
       ================ I M A G E S   3 D ================
       = Copyright (c) 1984   Celestial Software Inc. =
       ================================================

       SOLVE SEISMIC RESPONSE      Version 1.4  12/01/86

    COOLING TOWER VIBRATION - NODE INCREASE
```

Node	Translations X	Y	Z	/	Rotations X	Y	Z
4	.0000E+00	.0000E+00	.0000E+00	/	.0000E+00	.0000E+00	.0000E+00
5	.0000E+00	.0000E+00	.0000E+00	/	.0000E+00	.0000E+00	.0000E+00
6	.0000E+00	.0000E+00	.0000E+00	/	.0000E+00	.0000E+00	.0000E+00
7	.0000E+00	.0000E+00	.0000E+00	/	.0000E+00	.0000E+00	.0000E+00
8	.0000E+00	.0000E+00	.0000E+00	/	.0000E+00	.0000E+00	.0000E+00
9	.0000E+00	.0000E+00	.0000E+00	/	.0000E+00	.0000E+00	.0000E+00
10	.0000E+00	.0000E+00	.0000E+00	/	.0000E+00	.0000E+00	.0000E+00
11	.1035E-04	-.2099E-04	.0000E+00	/	.2422E-07	-.8532E-07	.0000E+00
12	.2796E-04	-.2741E-04	-.2960E-04	/	-.1205E-05	-.3867E-06	-.7615E-06
13	.2388E-04	-.2389E-04	-.3224E-04	/	-.1277E-05	-.2947E-06	-.6677E-06
14	.1813E-04	-.1816E-04	-.2846E-04	/	-.1127E-05	-.1849E-06	-.5249E-06
15	.1458E-04	-.9768E-05	-.2022E-04	/	-.8474E-06	-.4127E-07	-.4907E-06
16	.1569E-04	.1619E-05	-.8806E-05	/	-.5052E-06	.1388E-06	-.6477E-06
17	.2320E-04	.1639E-04	.4260E-05	/	-.1712E-06	.3575E-06	-.1044E-05
18	.3859E-04	.3477E-04	.1777E-04	/	.1075E-06	.6253E-06	-.1704E-05
19	.6087E-04	.5848E-04	.3125E-04	/	.3420E-06	.8510E-06	-.2608E-05
20	.7242E-04	.5455E-04	.0000E+00	/	.0000E+00	.2338E-06	-.2953E-05
21	.3506E-03	-.5637E-03	.0000E+00	/	-.1055E-03	-.3864E-04	.0000E+00
22	.4503E-03	-.5132E-03	-.2442E-04	/	.2140E-05	-.2219E-05	-.1599E-04
23	.5095E-03	-.4337E-03	-.5729E-04	/	-.4999E-05	-.2868E-05	-.1699E-04
24	.5504E-03	-.3112E-03	-.9124E-04	/	-.6569E-05	-.2263E-05	-.1956E-04
25	.5870E-03	-.1374E-03	-.1289E-03	/	-.1042E-04	-.2781E-05	-.2077E-04
26	.6258E-03	.9431E-04	-.1736E-03	/	-.1506E-04	-.3590E-05	-.2094E-04
27	.6662E-03	.3874E-03	-.2202E-03	/	-.1923E-04	-.3630E-05	-.2057E-04
28	.7053E-03	.7443E-03	-.2422E-03	/	-.2097E-04	-.1001E-05	-.2203E-04
29	.7622E-03	.1156E-02	-.1908E-03	/	-.2090E-04	.3835E-05	-.2265E-04
30	.9301E-03	.1782E-02	.0000E+00	/	.0000E+00	.5732E-04	-.1852E-03
31	.1358E-02	-.9188E-03	.0000E+00	/	-.3514E-03	-.1064E-03	.0000E+00
32	.1554E-02	-.8303E-03	-.6546E-04	/	.1423E-05	-.7300E-05	-.3602E-04
33	.1710E-02	-.6702E-03	-.1569E-03	/	.1659E-06	-.5487E-05	-.3934E-04
34	.1834E-02	-.4280E-03	-.2647E-03	/	-.2117E-05	-.4393E-05	-.4300E-04
35	.1935E-02	-.9167E-04	-.3830E-03	/	-.4512E-05	-.3513E-05	-.4655E-04
36	.2015E-02	.3472E-03	-.5004E-03	/	-.6753E-05	-.2030E-05	-.4956E-04
37	.2070E-02	.8918E-03	-.5843E-03	/	-.8232E-05	.1330E-05	-.5171E-04
38	.2102E-02	.1536E-02	-.5705E-03	/	-.8934E-05	.7400E-05	-.5164E-04

```
IMAGES-3D   s/n:800189                            03-02-87
                                                 PAGE    3
         ================ I M A G E S   3 D ================
         = Copyright (c) 1984   Celestial Software Inc. =
         =================================================

         SOLVE SEISMIC RESPONSE      Version 1.4  12/01/86

     COOLING TOWER VIBRATION - NODE INCREASE
```

Node	T r a n s l a t i o n s X	Y	Z	/	R o t a t i o n s X	Y	Z
39	.2147E-02	.2300E-02	-.3725E-03	/	-.9344E-05	.1477E-04	-.4615E-04
40	.2251E-02	.3169E-02	.0000E+00	/	.0000E+00	.4161E-04	-.1263E-03
41	.2865E-02	-.1141E-02	.0000E+00	/	-.4660E-03	-.1396E-03	.0000E+00
42	.3167E-02	-.1009E-02	-.9897E-04	/	.1710E-05	-.1357E-04	-.4745E-04
43	.3438E-02	-.7690E-03	-.2435E-03	/	.1098E-05	-.1218E-04	-.5149E-04
44	.3673E-02	-.4083E-03	-.4227E-03	/	-.6260E-06	-.1073E-04	-.5642E-04
45	.3862E-02	.8241E-04	-.6178E-03	/	-.3241E-05	-.8446E-05	-.6082E-04
46	.3990E-02	.7084E-03	-.7897E-03	/	-.6160E-05	-.4264E-05	-.6371E-04
47	.4047E-02	.1466E-02	-.8721E-03	/	-.8871E-05	.2455E-05	-.6379E-04
48	.4049E-02	.2349E-02	-.7835E-03	/	-.1074E-04	.1112E-04	-.6001E-04
49	.4049E-02	.3341E-02	-.4770E-03	/	-.1082E-04	.1930E-04	-.5422E-04
50	.4128E-02	.4450E-02	.0000E+00	/	.0000E+00	.4844E-04	-.1477E-03
51	.4788E-02	-.1224E-02	.0000E+00	/	-.6112E-03	-.1831E-03	.0000E+00
52	.5214E-02	-.1048E-02	-.1087E-03	/	.2276E-05	-.2119E-04	-.6085E-04
53	.5618E-02	-.7324E-03	-.2881E-03	/	.5261E-05	-.2002E-04	-.6623E-04
54	.5993E-02	-.2641E-03	-.5324E-03	/	.2217E-05	-.1905E-04	-.7127E-04
55	.6301E-02	.3659E-03	-.8090E-03	/	-.1990E-05	-.1558E-04	-.7550E-04
56	.6499E-02	.1155E-02	-.1049E-02	/	-.6507E-05	-.8242E-05	-.7786E-04
57	.6568E-02	.2090E-02	-.1151E-02	/	-.1027E-04	.3028E-05	-.7740E-04
58	.6539E-02	.3151E-02	-.1016E-02	/	-.1193E-04	.1638E-04	-.7441E-04
59	.6498E-02	.4325E-02	-.6055E-03	/	-.1063E-04	.2800E-04	-.7129E-04
60	.6545E-02	.5609E-02	.0000E+00	/	.0000E+00	.5339E-04	-.1489E-03
61	.7187E-02	-.1098E-02	.0000E+00	/	-.7093E-03	-.2072E-03	.0000E+00
62	.7720E-02	-.8872E-03	-.1071E-03	/	.4700E-04	-.1742E-04	-.8526E-04
63	.8249E-02	-.5254E-03	-.3119E-03	/	.7318E-05	-.2962E-04	-.7947E-04
64	.8771E-02	.2481E-04	-.6162E-03	/	.1596E-05	-.3092E-04	-.8345E-04
65	.9218E-02	.7643E-03	-.9821E-03	/	-.3874E-05	-.2708E-04	-.8670E-04
66	.9513E-02	.1682E-02	-.1314E-02	/	-.9273E-05	-.1662E-04	-.8790E-04
67	.9618E-02	.2753E-02	-.1470E-02	/	-.1306E-04	.2917E-06	-.8703E-04
68	.9570E-02	.3950E-02	-.1314E-02	/	-.1382E-04	.2090E-04	-.8518E-04
69	.9484E-02	.5249E-02	-.7909E-03	/	-.1183E-04	.3886E-04	-.8176E-04
70	.9495E-02	.6641E-02	.0000E+00	/	.0000E+00	.6412E-04	-.1516E-03
71	.9900E-02	-.7898E-03	.0000E+00	/	-.5432E-03	-.1433E-03	.0000E+00
72	.1061E-01	-.5851E-03	-.1087E-03	/	.2993E-04	-.3684E-04	-.9348E-04
73	.1131E-01	-.1919E-03	-.3438E-03	/	-.5408E-05	-.4630E-04	-.8931E-04

COOLING TOWER VIBRATION: INCREASED NODE MODEL

```
IMAGES-3D   s/n:800189                                03-02-87
                                                      PAGE    4
        ================ I M A G E S   3 D ================
        = Copyright (c) 1984   Celestial Software Inc. =
        ==================================================

        SOLVE SEISMIC RESPONSE      Version 1.4  12/01/86

    COOLING TOWER VIBRATION - NODE INCREASE
```

| | T r a n s l a t i o n s | | / | R o t a t i o n s | | |
Node	X	Y	Z	/	X	Y	Z
74	.1200E-01	.4019E-03	-.7092E-03	/	-.7837E-06	-.4497E-04	-.9751E-04
75	.1259E-01	.1202E-02	-.1156E-02	/	-.4791E-05	-.3899E-04	-.9955E-04
76	.1297E-01	.2197E-02	-.1564E-02	/	-.7864E-05	-.2440E-04	-.9982E-04
77	.1312E-01	.3357E-02	-.1763E-02	/	-.9917E-05	-.1784E-05	-.9760E-04
78	.1306E-01	.4650E-02	-.1587E-02	/	-.9723E-05	.2560E-04	-.9430E-04
79	.1294E-01	.6041E-02	-.9604E-03	/	-.7645E-05	.5040E-04	-.8999E-04
80	.1292E-01	.7520E-02	.0000E+00	/	.0000E+00	.6598E-04	-.1156E-03
81	.1313E-01	-.5091E-03	.0000E+00	/	-.7548E-03	-.1220E-03	.0000E+00
82	.1404E-01	-.3402E-03	-.1143E-03	/	.6939E-04	-.5040E-04	-.1214E-03
83	.1492E-01	.2609E-04	-.3961E-03	/	-.3276E-05	-.5828E-04	-.1054E-03
84	.1575E-01	.6084E-03	-.8286E-03	/	-.1577E-05	-.5545E-04	-.1108E-03
85	.1641E-01	.1420E-02	-.1335E-02	/	-.5698E-05	-.4561E-04	-.1105E-03
86	.1682E-01	.2452E-02	-.1773E-02	/	-.6861E-05	-.2634E-04	-.1095E-03
87	.1694E-01	.3679E-02	-.1956E-02	/	-.6705E-05	.8406E-06	-.1072E-03
88	.1685E-01	.5068E-02	-.1724E-02	/	-.5359E-05	.3135E-04	-.1043E-03
89	.1669E-01	.6584E-02	-.1025E-02	/	-.3641E-05	.5723E-04	-.1011E-03
90	.1665E-01	.8194E-02	.0000E+00	/	.0000E+00	.6876E-04	-.1125E-03
91	.1649E-01	-.2974E-03	.0000E+00	/	-.4527E-03	-.5502E-04	.0000E+00
92	.1765E-01	-.2332E-03	-.1629E-03	/	.1256E-04	-.7451E-04	-.1028E-03
93	.1876E-01	.4817E-04	-.5253E-03	/	-.1440E-04	-.7172E-04	-.1022E-03
94	.1972E-01	.5907E-03	-.1028E-02	/	-.1133E-04	-.6496E-04	-.1089E-03
95	.2045E-01	.1400E-02	-.1568E-02	/	-.1187E-04	-.4965E-04	-.1099E-03
96	.2088E-01	.2463E-02	-.1991E-02	/	-.9656E-05	-.2581E-04	-.1103E-03
97	.2099E-01	.3755E-02	-.2121E-02	/	-.6445E-05	.4190E-05	-.1103E-03
98	.2088E-01	.5248E-02	-.1821E-02	/	-.3091E-05	.3555E-04	-.1111E-03
99	.2071E-01	.6905E-02	-.1065E-02	/	-.4957E-06	.6118E-04	-.1134E-03
100	.2065E-01	.8690E-02	.0000E+00	/	.0000E+00	.7150E-04	-.1133E-03
101	.1950E-01	-.4460E-03	.0000E+00	/	-.4066E-03	-.1851E-04	.0000E+00
102	.2113E-01	-.6575E-03	-.2084E-03	/	-.6725E-05	-.1001E-03	-.9829E-04
103	.2261E-01	-.4811E-03	-.6462E-03	/	-.2236E-04	-.8630E-04	-.1024E-03
104	.2379E-01	.4409E-04	-.1214E-02	/	-.1404E-04	-.7105E-04	-.1104E-03
105	.2460E-01	.8959E-03	-.1776E-02	/	-.1247E-04	-.4837E-04	-.1112E-03
106	.2502E-01	.2044E-02	-.2170E-02	/	-.8876E-05	-.2119E-04	-.1115E-03
107	.2509E-01	.3460E-02	-.2236E-02	/	-.4861E-05	.8808E-05	-.1121E-03
108	.2495E-01	.5111E-02	-.1869E-02	/	-.1350E-05	.3774E-04	-.1141E-03

```
IMAGES-3D  s/n:800189                                    03-02-87
                                                        PAGE   5
            =============== I M A G E S   3 D ===============
            = Copyright (c) 1984   Celestial Software Inc. =
            ================================================

            SOLVE SEISMIC RESPONSE        Version 1.4  12/01/86

            COOLING TOWER VIBRATION - NODE INCREASE


         T r a n s l a t i o n s      /      R o t a t i o n s
  Node      X         Y         Z     /      X         Y         Z
  ----   --------- --------- --------- / --------- --------- ---------
  109   .2476E-01 .6964E-02 -.1071E-02 / .8549E-06  .5971E-04 -.1181E-03
  110   .2468E-01 .8968E-02  .0000E+00 / .0000E+00  .6897E-04 -.1166E-03
  111   .2237E-01 -.1193E-02 .0000E+00 / -.1171E-02 .2250E-03  .0000E+00
  112   .2448E-01 -.1636E-02 -.9926E-04 / .8961E-04 -.1467E-03 -.1398E-03
  113   .2625E-01 -.1538E-02 -.5574E-03 / -.1688E-05 -.9727E-04 -.1219E-03
  114   .2755E-01 -.9823E-03 -.1217E-02 / -.9263E-05 -.7169E-04 -.1208E-03
  115   .2834E-01 -.3468E-04 -.1877E-02 / -.1132E-04 -.4501E-04 -.1185E-03
  116   .2865E-01  .1258E-02 -.2338E-02 / -.1106E-04 -.1650E-04 -.1158E-03
  117   .2859E-01  .2854E-02 -.2420E-02 / -.8760E-05  .1265E-04 -.1146E-03
  118   .2832E-01  .4715E-02 -.2018E-02 / -.5430E-05  .3963E-04 -.1156E-03
  119   .2805E-01  .6798E-02 -.1151E-02 / -.2112E-05  .5968E-04 -.1175E-03
  120   .2795E-01  .9039E-02  .0000E+00 / .0000E+00  .6996E-04 -.1097E-03

                 ACCELERATIONS for MODE  1

         T r a n s l a t i o n a l    /      R o t a t i o n a l
  Node      X         Y         Z     /      X         Y         Z
  ----   --------- --------- --------- / --------- --------- ---------
   1    .0000E+00 .0000E+00 .0000E+00 / .0000E+00  .0000E+00  .0000E+00
   2    .0000E+00 .0000E+00 .0000E+00 / .0000E+00  .0000E+00  .0000E+00
   3    .0000E+00 .0000E+00 .0000E+00 / .0000E+00  .0000E+00  .0000E+00
   4    .0000E+00 .0000E+00 .0000E+00 / .0000E+00  .0000E+00  .0000E+00
   5    .0000E+00 .0000E+00 .0000E+00 / .0000E+00  .0000E+00  .0000E+00
   6    .0000E+00 .0000E+00 .0000E+00 / .0000E+00  .0000E+00  .0000E+00
   7    .0000E+00 .0000E+00 .0000E+00 / .0000E+00  .0000E+00  .0000E+00
   8    .0000E+00 .0000E+00 .0000E+00 / .0000E+00  .0000E+00  .0000E+00
   9    .0000E+00 .0000E+00 .0000E+00 / .0000E+00  .0000E+00  .0000E+00
  10    .0000E+00 .0000E+00 .0000E+00 / .0000E+00  .0000E+00  .0000E+00
  11    .5851E-03 -.1187E-02 .0000E+00 / .1369E-05 -.4824E-05  .0000E+00
  12    .1581E-02 -.1550E-02 -.1673E-02 / -.6812E-04 -.2186E-04 -.4305E-04
  13    .1350E-02 -.1351E-02 -.1823E-02 / -.7222E-04 -.1666E-04 -.3775E-04
  14    .1025E-02 -.1027E-02 -.1609E-02 / -.6372E-04 -.1046E-04 -.2968E-04
  15    .8244E-03 -.5523E-03 -.1143E-02 / -.4791E-04 -.2333E-05 -.2774E-04
  16    .8868E-03  .9153E-04 -.4979E-03 / -.2856E-04  .7847E-05 -.3662E-04
  17    .1311E-02  .9266E-03  .2408E-03 / -.9682E-05  .2021E-04 -.5901E-04
```

COOLING TOWER VIBRATION: INCREASED NODE MODEL

```
IMAGES-3D   s/n:800189                                03-02-87
                                                      PAGE    6
            =============== I M A G E S   3 D ===============
            = Copyright (c) 1984   Celestial Software Inc. =
            ================================================

            SOLVE SEISMIC RESPONSE      Version 1.4  12/01/86

      COOLING TOWER VIBRATION - NODE INCREASE
```

Node	Translational X	Y	Z	/	Rotational X	Y	Z
18	.2182E-02	.1966E-02	.1005E-02	/	.6076E-05	.3535E-04	-.9633E-04
19	.3441E-02	.3306E-02	.1767E-02	/	.1934E-04	.4812E-04	-.1474E-03
20	.4094E-02	.3084E-02	.0000E+00	/	.0000E+00	.1322E-04	-.1670E-03
21	.1982E-01	-.3187E-01	.0000E+00	/	-.5965E-02	-.2185E-02	.0000E+00
22	.2546E-01	-.2902E-01	-.1381E-02	/	.1210E-03	-.1254E-03	-.9042E-03
23	.2881E-01	-.2452E-01	-.3239E-02	/	-.2826E-03	-.1622E-03	-.9608E-03
24	.3112E-01	-.1760E-01	-.5159E-02	/	-.3714E-03	-.1280E-03	-.1106E-02
25	.3319E-01	-.7770E-02	-.7290E-02	/	-.5891E-03	-.1572E-03	-.1174E-02
26	.3538E-01	.5332E-02	-.9816E-02	/	-.8515E-03	-.2030E-03	-.1184E-02
27	.3766E-01	.2190E-01	-.1245E-01	/	-.1087E-02	-.2052E-03	-.1163E-02
28	.3988E-01	.4208E-01	-.1370E-01	/	-.1186E-02	-.5660E-04	-.1245E-02
29	.4309E-01	.6536E-01	-.1078E-01	/	-.1182E-02	.2168E-03	-.1280E-02
30	.5258E-01	.1008E+00	.0000E+00	/	.0000E+00	.3241E-02	-.1047E-01
31	.7676E-01	-.5194E-01	.0000E+00	/	-.1986E-01	-.6017E-02	.0000E+00
32	.8786E-01	-.4694E-01	-.3701E-02	/	.8044E-04	-.4127E-03	-.2036E-02
33	.9669E-01	-.3789E-01	-.8868E-02	/	.9380E-05	-.3102E-03	-.2224E-02
34	.1037E+00	-.2420E-01	-.1497E-01	/	-.1197E-03	-.2484E-03	-.2431E-02
35	.1094E+00	-.5183E-02	-.2165E-01	/	-.2551E-03	-.1986E-03	-.2632E-02
36	.1139E+00	.1963E-01	-.2829E-01	/	-.3818E-03	-.1148E-03	-.2802E-02
37	.1170E+00	.5042E-01	-.3303E-01	/	-.4654E-03	.7519E-04	-.2923E-02
38	.1189E+00	.8686E-01	-.3226E-01	/	-.5051E-03	.4184E-03	-.2919E-02
39	.1214E+00	.1300E+00	-.2106E-01	/	-.5283E-03	.8350E-03	-.2609E-02
40	.1273E+00	.1791E+00	.0000E+00	/	.0000E+00	.2353E-02	-.7143E-02
41	.1620E+00	-.6451E-01	.0000E+00	/	-.2635E-01	-.7891E-02	.0000E+00
42	.1790E+00	-.5704E-01	-.5596E-02	/	.9671E-04	-.7674E-03	-.2682E-02
43	.1943E+00	-.4348E-01	-.1377E-01	/	.6208E-04	-.6888E-03	-.2911E-02
44	.2076E+00	-.2309E-01	-.2390E-01	/	-.3539E-04	-.6067E-03	-.3190E-02
45	.2184E+00	.4659E-02	-.3493E-01	/	-.1832E-03	-.4775E-03	-.3439E-02
46	.2256E+00	.4005E-01	-.4465E-01	/	-.3483E-03	-.2411E-03	-.3602E-02
47	.2288E+00	.8289E-01	-.4931E-01	/	-.5015E-03	.1388E-03	-.3606E-02
48	.2289E+00	.1328E+00	-.4430E-01	/	-.6070E-03	.6286E-03	-.3393E-02
49	.2289E+00	.1889E+00	-.2697E-01	/	-.6118E-03	.1091E-02	-.3066E-02
50	.2334E+00	.2516E+00	.0000E+00	/	.0000E+00	.2739E-02	-.8351E-02
51	.2707E+00	-.6919E-01	.0000E+00	/	-.3455E-01	-.1035E-01	.0000E+00
52	.2948E+00	-.5927E-01	-.6146E-02	/	.1287E-03	-.1198E-02	-.3440E-02

```
=============== I M A G E S  3 D ===============
= Copyright (c) 1984   Celestial Software Inc. =
================================================
```

         SOLVE SEISMIC RESPONSE   Version 1.4 12/01/86

      COOLING TOWER VIBRATION - NODE INCREASE

| | T r a n s l a t i o n a l | | / | R o t a t i o n a l | | |
Node	X	Y	Z	/	X	Y	Z
53	.3176E+00	-.4141E-01	-.1629E-01	/	.2974E-03	-.1132E-02	-.3744E-02
54	.3388E+00	-.1493E-01	-.3010E-01	/	.1254E-03	-.1077E-02	-.4029E-02
55	.3562E+00	.2069E-01	-.4574E-01	/	-.1125E-03	-.8806E-03	-.4269E-02
56	.3674E+00	.6527E-01	-.5930E-01	/	-.3679E-03	-.4660E-03	-.4402E-02
57	.3713E+00	.1181E+00	-.6506E-01	/	-.5807E-03	.1712E-03	-.4376E-02
58	.3697E+00	.1782E+00	-.5743E-01	/	-.6746E-03	.9261E-03	-.4207E-02
59	.3674E+00	.2445E+00	-.3424E-01	/	-.6008E-03	.1583E-02	-.4031E-02
60	.3700E+00	.3171E+00	.0000E+00	/	.0000E+00	.3018E-02	-.8419E-02
61	.4063E+00	-.6206E-01	.0000E+00	/	-.4010E-01	-.1171E-01	.0000E+00
62	.4365E+00	-.5016E-01	-.6057E-02	/	.2657E-02	-.9849E-03	-.4820E-02
63	.4663E+00	-.2970E-01	-.1763E-01	/	.4137E-03	-.1675E-02	-.4493E-02
64	.4959E+00	.1403E-02	-.3484E-01	/	.9025E-04	-.1748E-02	-.4718E-02
65	.5212E+00	.4321E-01	-.5553E-01	/	-.2190E-03	-.1531E-02	-.4902E-02
66	.5378E+00	.9508E-01	-.7428E-01	/	-.5243E-03	-.9395E-03	-.4970E-02
67	.5438E+00	.1557E+00	-.8311E-01	/	-.7386E-03	.1649E-04	-.4920E-02
68	.5411E+00	.2233E+00	-.7431E-01	/	-.7813E-03	.1182E-02	-.4816E-02
69	.5362E+00	.2967E+00	-.4472E-01	/	-.6688E-03	.2197E-02	-.4623E-02
70	.5368E+00	.3755E+00	.0000E+00	/	.0000E+00	.3625E-02	-.8568E-02
71	.5597E+00	-.4466E-01	.0000E+00	/	-.3071E-01	-.8103E-02	.0000E+00
72	.5999E+00	-.3308E-01	-.6144E-02	/	.1692E-02	-.2083E-02	-.5285E-02
73	.6397E+00	-.1085E-01	-.1944E-01	/	-.3058E-03	-.2618E-02	-.5049E-02
74	.6786E+00	.2272E-01	-.4009E-01	/	-.4431E-04	-.2543E-02	-.5513E-02
75	.7117E+00	.6798E-01	-.6535E-01	/	-.2709E-03	-.2205E-02	-.5628E-02
76	.7335E+00	.1242E+00	-.8845E-01	/	-.4446E-03	-.1380E-02	-.5644E-02
77	.7416E+00	.1898E+00	-.9970E-01	/	-.5607E-03	-.1009E-03	-.5518E-02
78	.7384E+00	.2629E+00	-.8971E-01	/	-.5497E-03	.1447E-02	-.5332E-02
79	.7317E+00	.3415E+00	-.5430E-01	/	-.4322E-03	.2849E-02	-.5088E-02
80	.7306E+00	.4252E+00	.0000E+00	/	.0000E+00	.3730E-02	-.6538E-02
81	.7426E+00	-.2878E-01	.0000E+00	/	-.4268E-01	-.6899E-02	.0000E+00
82	.7935E+00	-.1923E-01	-.6462E-02	/	.3923E-02	-.2849E-02	-.6866E-02
83	.8437E+00	.1475E-02	-.2239E-01	/	-.1852E-03	-.3295E-02	-.5957E-02
84	.8903E+00	.3440E-01	-.4684E-01	/	-.8914E-04	-.3135E-02	-.6265E-02
85	.9278E+00	.8028E-01	-.7550E-01	/	-.3221E-03	-.2579E-02	-.6247E-02
86	.9508E+00	.1386E+00	-.1002E+00	/	-.3879E-03	-.1489E-02	-.6194E-02
87	.9578E+00	.2080E+00	-.1106E+00	/	-.3791E-03	.4753E-04	-.6059E-02

COOLING TOWER VIBRATION: INCREASED NODE MODEL

```
IMAGES-3D   s/n:800189                                      03-02-87
                                                           PAGE    8
           =============== I M A G E S  3 D ===============
           = Copyright (c) 1984   Celestial Software Inc. =
           ================================================

              SOLVE SEISMIC RESPONSE      Version 1.4  12/01/86

        COOLING TOWER VIBRATION - NODE INCREASE
```

Node	T r a n s l a t i o n a l			/	R o t a t i o n a l		
	X	Y	Z	/	X	Y	Z
88	.9525E+00	.2865E+00	-.9747E-01	/	-.3030E-03	.1773E-02	-.5896E-02
89	.9437E+00	.3722E+00	-.5795E-01	/	-.2059E-03	.3236E-02	-.5716E-02
90	.9412E+00	.4633E+00	.0000E+00	/	.0000E+00	.3887E-02	-.6362E-02
91	.9325E+00	-.1682E-01	.0000E+00	/	-.2560E-01	-.3111E-02	.0000E+00
92	.9980E+00	-.1318E-01	-.9212E-02	/	.7104E-03	-.4213E-02	-.5811E-02
93	.1060E+01	.2723E-02	-.2970E-01	/	-.8144E-03	-.4055E-02	-.5781E-02
94	.1115E+01	.3340E-01	-.5811E-01	/	-.6404E-03	-.3673E-02	-.6156E-02
95	.1156E+01	.7916E-01	-.8865E-01	/	-.6709E-03	-.2807E-02	-.6214E-02
96	.1180E+01	.1392E+00	-.1126E+00	/	-.5459E-03	-.1459E-02	-.6237E-02
97	.1187E+01	.2123E+00	-.1199E+00	/	-.3644E-03	.2369E-03	-.6237E-02
98	.1181E+01	.2967E+00	-.1029E+00	/	-.1747E-03	.2010E-02	-.6279E-02
99	.1171E+01	.3904E+00	-.6020E-01	/	-.2802E-04	.3459E-02	-.6413E-02
100	.1168E+01	.4913E+00	.0000E+00	/	.0000E+00	.4042E-02	-.6408E-02
101	.1102E+01	-.2522E-01	.0000E+00	/	-.2299E-01	-.1047E-02	.0000E+00
102	.1195E+01	-.3717E-01	-.1178E-01	/	-.3802E-03	-.5660E-02	-.5557E-02
103	.1278E+01	-.2720E-01	-.3653E-01	/	-.1264E-02	-.4879E-02	-.5788E-02
104	.1345E+01	.2493E-02	-.6863E-01	/	-.7941E-03	-.4017E-02	-.6245E-02
105	.1391E+01	.5065E-01	-.1004E+00	/	-.7048E-03	-.2735E-02	-.6287E-02
106	.1415E+01	.1156E+00	-.1227E+00	/	-.5018E-03	-.1198E-02	-.6305E-02
107	.1419E+01	.1956E+00	-.1264E+00	/	-.2748E-03	.4980E-03	-.6341E-02
108	.1411E+01	.2890E+00	-.1057E+00	/	-.7633E-04	.2134E-02	-.6449E-02
109	.1400E+01	.3937E+00	-.6058E-01	/	.4833E-04	.3376E-02	-.6674E-02
110	.1396E+01	.5070E+00	.0000E+00	/	.0000E+00	.3899E-02	-.6591E-02
111	.1265E+01	-.6746E-01	.0000E+00	/	-.6619E-01	.1272E-01	.0000E+00
112	.1384E+01	-.9252E-01	-.5612E-02	/	.5066E-02	-.8295E-02	-.7905E-02
113	.1484E+01	-.8696E-01	-.3151E-01	/	-.9544E-03	-.5499E-02	-.6889E-02
114	.1558E+01	-.5554E-01	-.6878E-01	/	-.5237E-03	-.4053E-02	-.6831E-02
115	.1602E+01	-.1960E-02	-.1061E+00	/	-.6398E-03	-.2545E-02	-.6702E-02
116	.1620E+01	.7114E-01	-.1322E+00	/	-.6253E-03	-.9328E-03	-.6547E-02
117	.1616E+01	.1614E+00	-.1368E+00	/	-.4953E-03	.7153E-03	-.6481E-02
118	.1601E+01	.2666E+00	-.1141E+00	/	-.3070E-03	.2240E-02	-.6534E-02
119	.1586E+01	.3843E+00	-.6507E-01	/	-.1194E-03	.3374E-02	-.6643E-02
120	.1580E+01	.5110E+00	.0000E+00	/	.0000E+00	.3955E-02	-.6200E-02

IMAGES-3D s/n:800189   03-02-87
                                                     PAGE   9
        =============== I M A G E S 3 D ===============
        = Copyright (c) 1984   Celestial Software Inc. =
        ==

        SOLVE SEISMIC RESPONSE   Version 1.4 12/01/86

    COOLING TOWER VIBRATION - NODE INCREASE

                         ABS DISPLACEMENTS

Node		Translations		/	Rotations		
	X	Y	Z	/	X	Y	Z
1	.0000E+00	.0000E+00	.0000E+00	/	.0000E+00	.0000E+00	.0000E+00
2	.0000E+00	.0000E+00	.0000E+00	/	.0000E+00	.0000E+00	.0000E+00
3	.0000E+00	.0000E+00	.0000E+00	/	.0000E+00	.0000E+00	.0000E+00
4	.0000E+00	.0000E+00	.0000E+00	/	.0000E+00	.0000E+00	.0000E+00
5	.0000E+00	.0000E+00	.0000E+00	/	.0000E+00	.0000E+00	.0000E+00
6	.0000E+00	.0000E+00	.0000E+00	/	.0000E+00	.0000E+00	.0000E+00
7	.0000E+00	.0000E+00	.0000E+00	/	.0000E+00	.0000E+00	.0000E+00
8	.0000E+00	.0000E+00	.0000E+00	/	.0000E+00	.0000E+00	.0000E+00
9	.0000E+00	.0000E+00	.0000E+00	/	.0000E+00	.0000E+00	.0000E+00
10	.0000E+00	.0000E+00	.0000E+00	/	.0000E+00	.0000E+00	.0000E+00
11	.1035E-04	.2099E-04	.0000E+00	/	.2422E-07	.8532E-07	.0000E+00
12	.2796E-04	.2741E-04	.2960E-04	/	.1205E-05	.3867E-06	.7615E-06
13	.2388E-04	.2389E-04	.3224E-04	/	.1277E-05	.2947E-06	.6677E-06
14	.1813E-04	.1816E-04	.2846E-04	/	.1127E-05	.1849E-06	.5249E-06
15	.1458E-04	.9768E-05	.2022E-04	/	.8474E-06	.4127E-07	.4907E-06
16	.1569E-04	.1619E-05	.8806E-05	/	.5052E-06	.1388E-06	.6477E-06
17	.2320E-04	.1639E-04	.4260E-05	/	.1712E-06	.3575E-06	.1044E-05
18	.3859E-04	.3477E-04	.1777E-04	/	.1075E-06	.6253E-06	.1704E-05
19	.6087E-04	.5848E-04	.3125E-04	/	.3420E-06	.8510E-06	.2608E-05
20	.7242E-04	.5455E-04	.0000E+00	/	.0000E+00	.2338E-06	.2953E-05
21	.3506E-03	.5637E-03	.0000E+00	/	.1055E-03	.3864E-04	.0000E+00
22	.4503E-03	.5132E-03	.2442E-04	/	.2140E-05	.2219E-05	.1599E-04
23	.5095E-03	.4337E-03	.5729E-04	/	.4999E-05	.2868E-05	.1699E-04
24	.5504E-03	.3112E-03	.9124E-04	/	.6569E-05	.2263E-05	.1956E-04
25	.5870E-03	.1374E-03	.1289E-03	/	.1042E-04	.2781E-05	.2077E-04
26	.6258E-03	.9431E-04	.1736E-03	/	.1506E-04	.3590E-05	.2094E-04
27	.6662E-03	.3874E-04	.2202E-03	/	.1923E-04	.3630E-05	.2057E-04
28	.7053E-03	.7443E-03	.2422E-03	/	.2097E-04	.1001E-05	.2203E-04
29	.7622E-03	.1156E-02	.1908E-03	/	.2090E-04	.3835E-05	.2265E-04
30	.9301E-03	.1782E-02	.0000E+00	/	.0000E+00	.5732E-04	.1852E-03
31	.1358E-02	.9188E-03	.0000E+00	/	.3514E-03	.1064E-03	.0000E+00
32	.1554E-02	.8303E-03	.6546E-04	/	.1423E-05	.7300E-05	.3602E-04
33	.1710E-02	.6702E-03	.1569E-03	/	.1659E-06	.5487E-05	.3934E-04
34	.1834E-02	.4280E-03	.2647E-03	/	.2117E-05	.4393E-05	.4300E-04

COOLING TOWER VIBRATION: INCREASED NODE MODEL

```
    ================ I M A G E S   3 D ================
    = Copyright (c) 1984   Celestial Software Inc. =
    =================================================
```

    SOLVE SEISMIC RESPONSE   Version 1.4 12/01/86

    COOLING TOWER VIBRATION - NODE INCREASE

Node	Translations X	 Y	 Z	/	Rotations X	 Y	 Z
35	.1935E-02	.9167E-04	.3830E-03	/	.4512E-05	.3513E-05	.4655E-04
36	.2015E-02	.3472E-03	.5004E-03	/	.6753E-05	.2030E-05	.4956E-04
37	.2070E-02	.8918E-03	.5843E-03	/	.8232E-05	.1330E-05	.5171E-04
38	.2102E-02	.1536E-02	.5705E-03	/	.8934E-05	.7400E-05	.5164E-04
39	.2147E-02	.2300E-02	.3725E-03	/	.9344E-05	.1477E-04	.4615E-04
40	.2251E-02	.3169E-02	.0000E+00	/	.0000E+00	.4161E-04	.1263E-03
41	.2865E-02	.1141E-02	.0000E+00	/	.4660E-03	.1396E-03	.0000E+00
42	.3167E-02	.1009E-02	.9897E-04	/	.1710E-05	.1357E-04	.4745E-04
43	.3438E-02	.7690E-03	.2435E-03	/	.1098E-05	.1218E-04	.5149E-04
44	.3673E-02	.4083E-03	.4227E-03	/	.6260E-06	.1073E-04	.5642E-04
45	.3862E-02	.8241E-04	.6178E-03	/	.3241E-05	.8446E-05	.6082E-04
46	.3990E-02	.7084E-03	.7897E-03	/	.6160E-05	.4264E-05	.6371E-04
47	.4047E-02	.1466E-02	.8721E-03	/	.8871E-05	.2455E-05	.6379E-04
48	.4049E-02	.2349E-02	.7835E-03	/	.1074E-04	.1112E-04	.6001E-04
49	.4049E-02	.3341E-02	.4770E-03	/	.1082E-04	.1930E-04	.5422E-04
50	.4128E-02	.4450E-02	.0000E+00	/	.0000E+00	.4844E-04	.1477E-03
51	.4788E-02	.1224E-02	.0000E+00	/	.6112E-03	.1831E-03	.0000E+00
52	.5214E-02	.1048E-02	.1087E-02	/	.2276E-05	.2119E-04	.6085E-04
53	.5618E-02	.7324E-03	.2881E-03	/	.5261E-05	.2002E-04	.6623E-04
54	.5993E-02	.2641E-03	.5324E-03	/	.2217E-05	.1905E-04	.7127E-04
55	.6301E-02	.3659E-03	.8090E-03	/	.1990E-05	.1558E-04	.7550E-04
56	.6499E-02	.1155E-02	.1049E-02	/	.6507E-05	.8242E-05	.7786E-04
57	.6568E-02	.2090E-02	.1151E-02	/	.1027E-04	.3028E-05	.7740E-04
58	.6539E-02	.3151E-02	.1016E-02	/	.1193E-04	.1638E-04	.7441E-04
59	.6498E-02	.4325E-02	.6055E-03	/	.1063E-04	.2800E-04	.7129E-04
60	.6545E-02	.5609E-02	.0000E+00	/	.0000E+00	.5339E-04	.1489E-03
61	.7187E-02	.1098E-02	.0000E+00	/	.7093E-03	.2072E-03	.0000E+00
62	.7720E-02	.8872E-03	.1071E-03	/	.4700E-04	.1742E-04	.8526E-04
63	.8249E-02	.5254E-03	.3119E-03	/	.7318E-05	.2962E-04	.7947E-04
64	.8771E-02	.2481E-04	.6162E-03	/	.1596E-05	.3092E-04	.8345E-04
65	.9218E-02	.7643E-03	.9821E-03	/	.3874E-05	.2708E-04	.8670E-04
66	.9513E-02	.1682E-02	.1314E-02	/	.9273E-05	.1662E-04	.8790E-04
67	.9618E-02	.2753E-02	.1470E-02	/	.1306E-04	.2917E-06	.8703E-04
68	.9570E-02	.3950E-02	.1314E-02	/	.1382E-04	.2090E-04	.8518E-04
69	.9484E-02	.5249E-02	.7909E-03	/	.1183E-04	.3886E-04	.8176E-04

IMAGES-3D s/n:800189   03-02-87
                                              PAGE   11
=============== I M A G E S   3 D ===============
= Copyright (c) 1984   Celestial Software Inc. =
==

SOLVE SEISMIC RESPONSE   Version 1.4 12/01/86

COOLING TOWER VIBRATION - NODE INCREASE

| Node | Translations | | / | Rotations | | |
	X	Y	Z /	X	Y	Z
70	.9495E-02	.6641E-02	.0000E+00 /	.0000E+00	.6412E-04	.1516E-03
71	.9900E-02	.7898E-03	.0000E+00 /	.5432E-03	.1433E-03	.0000E+00
72	.1061E-01	.5851E-03	.1087E-03 /	.2993E-04	.3684E-04	.9348E-04
73	.1131E-01	.1919E-03	.3438E-03 /	.5408E-05	.4630E-04	.8931E-04
74	.1200E-01	.4019E-03	.7092E-03 /	.7837E-06	.4497E-04	.9751E-04
75	.1259E-01	.1202E-02	.1156E-02 /	.4791E-05	.3899E-04	.9955E-04
76	.1297E-01	.2197E-02	.1564E-02 /	.7864E-05	.2440E-04	.9982E-04
77	.1312E-01	.3357E-02	.1763E-02 /	.9917E-05	.1784E-05	.9760E-04
78	.1306E-01	.4650E-02	.1587E-02 /	.9723E-05	.2560E-04	.9430E-04
79	.1294E-01	.6041E-02	.9604E-03 /	.7645E-05	.5040E-04	.8999E-04
80	.1292E-01	.7520E-02	.0000E+00 /	.0000E+00	.6598E-04	.1156E-03
81	.1313E-01	.5091E-03	.0000E+00 /	.7548E-03	.1220E-03	.0000E+00
82	.1404E-01	.3402E-03	.1143E-02 /	.6939E-04	.5040E-04	.1214E-03
83	.1492E-01	.2609E-04	.3961E-03 /	.3276E-05	.5828E-04	.1054E-03
84	.1575E-01	.6084E-03	.8286E-03 /	.1577E-05	.5545E-04	.1108E-03
85	.1641E-01	.1420E-02	.1335E-02 /	.5698E-05	.4561E-04	.1105E-03
86	.1682E-01	.2452E-02	.1773E-02 /	.6861E-05	.2634E-04	.1095E-03
87	.1694E-01	.3679E-02	.1956E-02 /	.6705E-05	.8406E-06	.1072E-03
88	.1685E-01	.5068E-02	.1724E-02 /	.5359E-05	.3135E-04	.1043E-03
89	.1669E-01	.6584E-02	.1025E-02 /	.3641E-05	.5723E-04	.1011E-03
90	.1665E-01	.8194E-02	.0000E+00 /	.0000E+00	.6876E-04	.1125E-03
91	.1649E-01	.2974E-03	.0000E+00 /	.4527E-03	.5502E-04	.0000E+00
92	.1765E-01	.2332E-03	.1629E-03 /	.1256E-04	.7451E-04	.1028E-03
93	.1876E-01	.4817E-04	.5253E-03 /	.1440E-04	.7172E-04	.1022E-03
94	.1972E-01	.5907E-03	.1028E-02 /	.1133E-04	.6496E-04	.1089E-03
95	.2045E-01	.1400E-02	.1568E-02 /	.1187E-04	.4965E-04	.1099E-03
96	.2088E-01	.2463E-02	.1991E-02 /	.9656E-05	.2581E-04	.1103E-03
97	.2099E-01	.3755E-02	.2121E-02 /	.6445E-05	.4190E-05	.1103E-03
98	.2088E-01	.5248E-02	.1821E-02 /	.3091E-05	.3555E-04	.1111E-03
99	.2071E-01	.6905E-02	.1065E-02 /	.4957E-06	.6118E-04	.1134E-03
100	.2065E-01	.8690E-02	.0000E+00 /	.0000E+00	.7150E-04	.1133E-03
101	.1950E-01	.4460E-03	.0000E+00 /	.4066E-03	.1851E-04	.0000E+00
102	.2113E-01	.6575E-03	.2084E-03 /	.6725E-05	.1001E-03	.9829E-04
103	.2261E-01	.4811E-03	.6462E-03 /	.2236E-04	.8630E-04	.1024E-03
104	.2379E-01	.4409E-04	.1214E-02 /	.1404E-04	.7105E-04	.1104E-03

COOLING TOWER VIBRATION: INCREASED NODE MODEL

```
IMAGES-3D   s/n:800189                                      03-02-87
                                                           PAGE   12
            =============== I M A G E S   3 D ===============
            = Copyright (c) 1984   Celestial Software Inc. =
            ================================================

            SOLVE SEISMIC RESPONSE      Version 1.4  12/01/86

     COOLING TOWER VIBRATION - NODE INCREASE
```

```
           T r a n s l a t i o n s     /        R o t a t i o n s
    Node     X           Y           Z    /     X           Y           Z
    ----  ----------  ----------  ---------- / ----------  ----------  ----------
    105   .2460E-01   .8959E-03   .1776E-02 /  .1247E-04   .4837E-04   .1112E-03
    106   .2502E-01   .2044E-02   .2170E-02 /  .8876E-05   .2119E-04   .1115E-03
    107   .2509E-01   .3460E-02   .2236E-02 /  .4861E-05   .8808E-05   .1121E-03
    108   .2495E-01   .5111E-02   .1869E-02 /  .1350E-05   .3774E-04   .1141E-03
    109   .2476E-01   .6964E-02   .1071E-02 /  .8549E-06   .5971E-04   .1181E-03
    110   .2468E-01   .8968E-02   .0000E+00 /  .0000E+00   .6897E-04   .1166E-03
    111   .2237E-01   .1193E-02   .0000E+00 /  .1171E-02   .2250E-03   .0000E+00
    112   .2448E-01   .1636E-02   .9926E-04 /  .8961E-04   .1467E-03   .1398E-03
    113   .2625E-01   .1538E-02   .5574E-03 /  .1688E-05   .9727E-04   .1219E-03
    114   .2755E-01   .9823E-03   .1217E-02 /  .9263E-05   .7169E-04   .1208E-03
    115   .2834E-01   .3468E-04   .1877E-02 /  .1132E-04   .4501E-04   .1185E-03
    116   .2865E-01   .1258E-02   .2338E-02 /  .1106E-04   .1650E-04   .1158E-03
    117   .2859E-01   .2854E-02   .2420E-02 /  .8760E-05   .1265E-04   .1146E-03
    118   .2832E-01   .4715E-02   .2018E-02 /  .5430E-05   .3963E-04   .1156E-03
    119   .2805E-01   .6798E-02   .1151E-02 /  .2112E-05   .5968E-04   .1175E-03
    120   .2795E-01   .9039E-02   .0000E+00 /  .0000E+00   .6996E-04   .1097E-03
```

                         ABS ACCELERATIONS

```
           T r a n s l a t i o n a l   /        R o t a t i o n a l
    Node     X           Y           Z    /     X           Y           Z
    ----  ----------  ----------  ---------- / ----------  ----------  ----------
     1    .0000E+00   .0000E+00   .0000E+00 /  .0000E+00   .0000E+00   .0000E+00
     2    .0000E+00   .0000E+00   .0000E+00 /  .0000E+00   .0000E+00   .0000E+00
     3    .0000E+00   .0000E+00   .0000E+00 /  .0000E+00   .0000E+00   .0000E+00
     4    .0000E+00   .0000E+00   .0000E+00 /  .0000E+00   .0000E+00   .0000E+00
     5    .0000E+00   .0000E+00   .0000E+00 /  .0000E+00   .0000E+00   .0000E+00
     6    .0000E+00   .0000E+00   .0000E+00 /  .0000E+00   .0000E+00   .0000E+00
     7    .0000E+00   .0000E+00   .0000E+00 /  .0000E+00   .0000E+00   .0000E+00
     8    .0000E+00   .0000E+00   .0000E+00 /  .0000E+00   .0000E+00   .0000E+00
     9    .0000E+00   .0000E+00   .0000E+00 /  .0000E+00   .0000E+00   .0000E+00
    10    .0000E+00   .0000E+00   .0000E+00 /  .0000E+00   .0000E+00   .0000E+00
    11    .5851E-03   .1187E-02   .0000E+00 /  .1369E-05   .4824E-05   .0000E+00
    12    .1581E-02   .1550E-02   .1673E-02 /  .6812E-04   .2186E-04   .4305E-04
    13    .1350E-02   .1351E-02   .1823E-02 /  .7222E-04   .1666E-04   .3775E-04
```

```
                                                    03-02-87
                                                    PAGE   13
     =============== I M A G E S   3 D ===============
     = Copyright (c) 1984   Celestial Software Inc. =
     ================================================

        SOLVE SEISMIC RESPONSE      Version 1.4  12/01/86

   COOLING TOWER VIBRATION - NODE INCREASE
```

Node	T r a n s l a t i o n a l			/	R o t a t i o n a l		
	X	Y	Z	/	X	Y	Z
----	----------	----------	----------	/	----------	----------	----------
14	.1025E-02	.1027E-02	.1609E-02	/	.6372E-04	.1046E-04	.2968E-04
15	.8244E-03	.5523E-03	.1143E-02	/	.4791E-04	.2333E-05	.2774E-04
16	.8868E-03	.9153E-04	.4979E-03	/	.2856E-04	.7847E-05	.3662E-04
17	.1311E-02	.9266E-03	.2408E-03	/	.9682E-05	.2021E-04	.5901E-04
18	.2182E-02	.1966E-02	.1005E-02	/	.6076E-05	.3535E-04	.9633E-04
19	.3441E-02	.3306E-02	.1767E-02	/	.1934E-04	.4812E-04	.1474E-03
20	.4094E-02	.3084E-02	.0000E+00	/	.0000E+00	.1322E-04	.1670E-03
21	.1982E-01	.3187E-01	.0000E+00	/	.5965E-02	.2185E-02	.0000E+00
22	.2546E-01	.2902E-01	.1381E-02	/	.1210E-03	.1254E-03	.9042E-03
23	.2881E-01	.2452E-01	.3239E-02	/	.2826E-03	.1622E-03	.9608E-03
24	.3112E-01	.1760E-01	.5159E-02	/	.3714E-03	.1280E-03	.1106E-02
25	.3319E-01	.7770E-02	.7290E-02	/	.5891E-03	.1572E-03	.1174E-02
26	.3538E-01	.5332E-02	.9816E-02	/	.8515E-03	.2030E-03	.1184E-02
27	.3766E-01	.2190E-01	.1245E-01	/	.1087E-02	.2052E-03	.1163E-02
28	.3988E-01	.4208E-01	.1370E-01	/	.1186E-02	.5660E-04	.1245E-02
29	.4309E-01	.6536E-01	.1078E-01	/	.1182E-02	.2168E-03	.1280E-02
30	.5258E-01	.1008E+00	.0000E+00	/	.0000E+00	.3241E-02	.1047E-01
31	.7676E-01	.5194E-01	.0000E+00	/	.1986E-01	.6017E-02	.0000E+00
32	.8786E-01	.4694E-01	.3701E-02	/	.8044E-04	.4127E-03	.2036E-02
33	.9669E-01	.3789E-01	.8868E-02	/	.9380E-05	.3102E-03	.2224E-02
34	.1037E+00	.2420E-01	.1497E-01	/	.1197E-03	.2484E-03	.2431E-02
35	.1094E+00	.5183E-02	.2165E-01	/	.2551E-03	.1986E-03	.2632E-02
36	.1139E+00	.1963E-01	.2829E-01	/	.3818E-03	.1148E-03	.2802E-02
37	.1170E+00	.5042E-01	.3303E-01	/	.4654E-03	.7519E-04	.2923E-02
38	.1189E+00	.8686E-01	.3226E-01	/	.5051E-03	.4184E-03	.2919E-02
39	.1214E+00	.1300E+00	.2106E-01	/	.5283E-03	.8350E-03	.2609E-02
40	.1273E+00	.1791E+00	.0000E+00	/	.0000E+00	.2353E-02	.7143E-02
41	.1620E+00	.6451E-01	.0000E+00	/	.2635E-01	.7891E-02	.0000E+00
42	.1790E+00	.5704E-01	.5596E-02	/	.9671E-04	.7674E-03	.2682E-02
43	.1943E+00	.4348E-01	.1377E-01	/	.6208E-04	.6888E-03	.2911E-02
44	.2076E+00	.2309E-01	.2390E-01	/	.3539E-04	.6067E-03	.3190E-02
45	.2184E+00	.4659E-02	.3493E-01	/	.1832E-03	.4775E-03	.3439E-02
46	.2256E+00	.4005E-01	.4465E-01	/	.3483E-03	.2411E-03	.3602E-02
47	.2288E+00	.8289E-01	.4931E-01	/	.5015E-03	.1388E-03	.3606E-02
48	.2289E+00	.1328E+00	.4430E-01	/	.6070E-03	.6286E-03	.3393E-02

COOLING TOWER VIBRATION: INCREASED NODE MODEL

```
IMAGES-3D   s/n:800189                              03-02-87
                                                    PAGE    14
            =============== I M A G E S  3 D ===============
            = Copyright (c) 1984   Celestial Software Inc. =
            ===============================================

            SOLVE SEISMIC RESPONSE      Version 1.4  12/01/86

      COOLING TOWER VIBRATION - NODE INCREASE
```

Node	Translational X	Y	Z	/	Rotational X	Y	Z
49	.2289E+00	.1889E+00	.2697E-01	/	.6118E-03	.1091E-02	.3066E-02
50	.2334E+00	.2516E+00	.0000E+00	/	.0000E+00	.2739E-02	.8351E-02
51	.2707E+00	.6919E-01	.0000E+00	/	.3455E-01	.1035E-01	.0000E+00
52	.2948E+00	.5927E-01	.6146E-02	/	.1287E-03	.1198E-02	.3440E-02
53	.3176E+00	.4141E-01	.1629E-01	/	.2974E-03	.1132E-02	.3744E-02
54	.3388E+00	.1493E-01	.3010E-01	/	.1254E-03	.1077E-02	.4029E-02
55	.3562E+00	.2069E-01	.4574E-01	/	.1125E-03	.8806E-03	.4269E-02
56	.3674E+00	.6527E-01	.5930E-01	/	.3679E-03	.4660E-03	.4402E-02
57	.3713E+00	.1181E+00	.6506E-01	/	.5807E-03	.1712E-03	.4376E-02
58	.3697E+00	.1782E+00	.5743E-01	/	.6746E-03	.9261E-03	.4207E-02
59	.3674E+00	.2445E+00	.3424E-01	/	.6008E-03	.1583E-02	.4031E-02
60	.3700E+00	.3171E+00	.0000E+00	/	.0000E+00	.3018E-02	.8419E-02
61	.4063E+00	.6206E-01	.0000E+00	/	.4010E-01	.1171E-01	.0000E+00
62	.4365E+00	.5016E-01	.6057E-02	/	.2657E-02	.9849E-03	.4820E-02
63	.4663E+00	.2970E-01	.1763E-01	/	.4137E-03	.1675E-02	.4493E-02
64	.4959E+00	.1403E-02	.3484E-01	/	.9025E-04	.1748E-02	.4718E-02
65	.5212E+00	.4321E-01	.5553E-01	/	.2190E-03	.1531E-02	.4902E-02
66	.5378E+00	.9508E-01	.7428E-01	/	.5243E-03	.9395E-03	.4970E-02
67	.5438E+00	.1557E+00	.8311E-01	/	.7386E-03	.1649E-04	.4920E-02
68	.5411E+00	.2233E+00	.7431E-01	/	.7813E-03	.1182E-02	.4816E-02
69	.5362E+00	.2967E+00	.4472E-01	/	.6688E-03	.2197E-02	.4623E-02
70	.5368E+00	.3755E+00	.0000E+00	/	.0000E+00	.3625E-02	.8568E-02
71	.5597E+00	.4466E-01	.0000E+00	/	.3071E-01	.8103E-02	.0000E+00
72	.5999E+00	.3308E-01	.6144E-02	/	.1692E-02	.2083E-02	.5285E-02
73	.6397E+00	.1085E-01	.1944E-01	/	.3058E-03	.2618E-02	.5049E-02
74	.6786E+00	.2272E-01	.4009E-01	/	.4431E-04	.2543E-02	.5513E-02
75	.7117E+00	.6798E-01	.6535E-01	/	.2709E-03	.2205E-02	.5628E-02
76	.7335E+00	.1242E+00	.8845E-01	/	.4446E-03	.1380E-02	.5644E-02
77	.7416E+00	.1898E+00	.9970E-01	/	.5607E-03	.1009E-03	.5518E-02
78	.7384E+00	.2629E+00	.8971E-01	/	.5497E-03	.1447E-02	.5332E-02
79	.7317E+00	.3415E+00	.5430E-01	/	.4322E-03	.2849E-02	.5088E-02
80	.7306E+00	.4252E+00	.0000E+00	/	.0000E+00	.3730E-02	.6538E-02
81	.7426E+00	.2878E-01	.0000E+00	/	.4268E-01	.6899E-02	.0000E+00
82	.7935E+00	.1923E-01	.6462E-02	/	.3923E-02	.2849E-02	.6866E-02
83	.8437E+00	.1475E-02	.2239E-01	/	.1852E-03	.3295E-02	.5957E-02

```
IMAGES-3D   s/n:800189                              03-02-87
                                                   PAGE   15
            ================ I M A G E S  3 D ================
            = Copyright (c) 1984   Celestial Software Inc. =
            ================================================

            SOLVE SEISMIC RESPONSE      Version 1.4  12/01/86

       COOLING TOWER VIBRATION - NODE INCREASE
```

Node	T r a n s l a t i o n a l			/	R o t a t i o n a l		
	X	Y	Z	/	X	Y	Z
84	.8903E+00	.3440E-01	.4684E-01	/	.8914E-04	.3135E-02	.6265E-02
85	.9278E+00	.8028E-01	.7550E-01	/	.3221E-03	.2579E-02	.6247E-02
86	.9508E+00	.1386E+00	.1002E+00	/	.3879E-03	.1489E-02	.6194E-02
87	.9578E+00	.2080E+00	.1106E+00	/	.3791E-03	.4753E-04	.6059E-02
88	.9525E+00	.2865E+00	.9747E-01	/	.3030E-03	.1773E-02	.5896E-02
89	.9437E+00	.3722E+00	.5795E-01	/	.2059E-03	.3236E-02	.5716E-02
90	.9412E+00	.4633E+00	.0000E+00	/	.0000E+00	.3887E-02	.6362E-02
91	.9325E+00	.1682E-01	.0000E+00	/	.2560E-01	.3111E-02	.0000E+00
92	.9980E+00	.1318E-01	.9212E-02	/	.7104E-03	.4213E-02	.5811E-02
93	.1060E+01	.2723E-02	.2970E-01	/	.8144E-03	.4055E-02	.5781E-02
94	.1115E+01	.3340E-01	.5811E-01	/	.6404E-03	.3673E-02	.6156E-02
95	.1156E+01	.7916E-01	.8865E-01	/	.6709E-03	.2807E-02	.6214E-02
96	.1180E+01	.1392E+00	.1126E+00	/	.5459E-03	.1459E-02	.6237E-02
97	.1187E+01	.2123E+00	.1199E+00	/	.3644E-03	.2369E-03	.6237E-02
98	.1181E+01	.2967E+00	.1029E+00	/	.1747E-03	.2010E-02	.6279E-02
99	.1171E+01	.3904E+00	.6020E-01	/	.2802E-04	.3459E-02	.6413E-02
100	.1168E+01	.4913E+00	.0000E+00	/	.0000E+00	.4042E-02	.6408E-02
101	.1102E+01	.2522E-01	.0000E+00	/	.2299E-01	.1047E-02	.0000E+00
102	.1195E+01	.3717E-01	.1178E-01	/	.3802E-03	.5660E-02	.5557E-02
103	.1278E+01	.2720E-01	.3653E-01	/	.1264E-02	.4879E-02	.5788E-02
104	.1345E+01	.2493E-02	.6863E-01	/	.7941E-03	.4017E-02	.6245E-02
105	.1391E+01	.5065E-01	.1004E+00	/	.7048E-03	.2735E-02	.6287E-02
106	.1415E+01	.1156E+00	.1227E+00	/	.5018E-03	.1198E-02	.6305E-02
107	.1419E+01	.1956E+00	.1264E+00	/	.2748E-03	.4980E-03	.6341E-02
108	.1411E+01	.2890E+00	.1057E+00	/	.7633E-04	.2134E-02	.6449E-02
109	.1400E+01	.3937E+00	.6058E-01	/	.4833E-04	.3376E-02	.6674E-02
110	.1396E+01	.5070E+00	.0000E+00	/	.0000E+00	.3899E-02	.6591E-02
111	.1265E+01	.6746E-01	.0000E+00	/	.6619E-01	.1272E-01	.0000E+00
112	.1384E+01	.9252E-01	.5612E-02	/	.5066E-02	.8295E-02	.7905E-02
113	.1484E+01	.8696E-01	.3151E-01	/	.9544E-04	.5499E-02	.6889E-02
114	.1558E+01	.5554E-01	.6878E-01	/	.5237E-03	.4053E-02	.6831E-02
115	.1602E+01	.1960E-02	.1061E+00	/	.6398E-03	.2545E-02	.6702E-02
116	.1620E+01	.7114E-01	.1322E+00	/	.6253E-03	.9328E-03	.6547E-02
117	.1616E+01	.1614E+00	.1368E+00	/	.4953E-03	.7153E-03	.6481E-02
118	.1601E+01	.2666E+00	.1141E+00	/	.3070E-03	.2240E-02	.6534E-02

```
IMAGES-3D   s/n:800189                              03-02-87
                                                   PAGE   16
            ================ I M A G E S  3 D ================
            = Copyright (c) 1984   Celestial Software Inc. =
            ================================================

            SOLVE SEISMIC RESPONSE      Version 1.4  12/01/86

       COOLING TOWER VIBRATION - NODE INCREASE
```

Node	T r a n s l a t i o n a l			/	R o t a t i o n a l		
	X	Y	Z	/	X	Y	Z
119	.1586E+01	.3843E+00	.6507E-01	/	.1194E-03	.3374E-02	.6643E-02
120	.1580E+01	.5110E+00	.0000E+00	/	.0000E+00	.3955E-02	.6200E-02

COOLING TOWER VIBRATION: INCREASED NODE MODEL

```
IMAGES-3D  s/n:800189                              03-02-87
                                                   PAGE    1
      =============== I M A G E S   3 D ===============
      = Copyright (c) 1984   Celestial Software Inc. =
      ================================================

        SOLVE BEAM LOADS/STRESSES    Version 1.4  12/01/86

     COOLING TOWER VIBRATION - NODE INCREASE

Mode  1 -
```

BEAM LOADS AND/OR STRESSES

GLoads LLoads /Stress	Node Node	Fx Axial	Fy Y-Shear	Fz Z-Shear	Mx Torsion	My Y-Bending	Mz Z-Bending
				BEAM NO. 1			
GLoads	1	-.3588E+04	.1030E+05	.3526E+01	-.1331E+03	.6594E+03	.1437E+05
GLoads	11	.3588E+04	-.1030E+05	-.3526E+01	.2594E+03	-.6253E+03	.1437E+05
LLoads	1	.1088E+05	-.7750E+03	-.3526E+01	.6713E+03	.4359E+02	-.1437E+05
LLoads	11	-.1088E+05	.7750E+03	.3526E+01	-.6713E+03	.8719E+02	.1437E+05
Stress	1	-.1857E+02	-.1323E+01	-.6021E-02	-.2759E-01	.5906E+00	-.1792E-02
Stress	11	-.1857E+02	-.1323E+01	-.6021E-02	-.2759E-01	-.5906E+00	.3584E-02
				BEAM NO. 2			
GLoads	1	-.1625E+04	.3841E+04	-.1383E+04	.4513E+05	.2320E+05	.1531E+05
GLoads	12	.1625E+04	-.3841E+04	.1383E+04	.9611E+04	.4239E+04	-.3424E+04
LLoads	1	.4185E+04	.9719E+03	-.9240E+03	-.2770E+04	.3968E+04	.3503E+05
LLoads	12	-.4185E+04	-.9719E+03	.9240E+03	.2770E+04	.3301E+04	.1017E+05
Stress	1	-.7145E+01	.1659E+01	-.1578E+01	.1138E+00	-.1440E+01	-.1631E+01
Stress	12	-.7145E+01	.1659E+01	-.1578E+01	.1138E+00	.4181E+00	.1357E+00
				BEAM NO. 3			
GLoads	2	-.5233E+04	.1571E+05	.1542E+04	.3841E+05	.9413E+04	.3374E+05
GLoads	12	.5233E+04	-.1571E+05	-.1542E+04	-.9624E+04	-.3514E+04	.3875E+04
LLoads	2	.1658E+05	-.1134E+04	-.6094E+03	.7425E+03	.3332E+05	-.3990E+05
LLoads	12	-.1658E+05	.1134E+04	.6094E+03	-.7425E+03	-.1072E+05	-.2145E+04
Stress	2	-.2830E+02	-.1936E+01	-.1041E+01	-.3052E-01	.1640E+01	-.1369E+01
Stress	12	-.2830E+02	-.1936E+01	-.1041E+01	-.3052E-01	-.8817E-01	-.4405E+00
				BEAM NO. 4			
GLoads	2	-.1742E+04	.3393E+04	-.9105E+03	.4353E+05	.2286E+05	.1055E+05
GLoads	13	.1742E+04	-.3393E+04	.9105E+03	.7443E+04	.4951E+04	-.4450E+04
LLoads	2	.3715E+04	.8358E+03	-.9366E+03	-.3514E+04	.4004E+05	.3021E+05
LLoads	13	-.3715E+04	-.8358E+03	.9366E+03	.3514E+04	.3519E+04	.8659E+04
Stress	2	-.6343E+01	.1427E+01	-.1599E+01	.1444E+00	-.1242E+01	-.1646E+01
Stress	13	-.6343E+01	.1427E+01	-.1599E+01	.1444E+00	.3559E+00	.1446E+00
				BEAM NO. 5			
GLoads	3	-.4489E+04	.1426E+05	.2237E+04	.4167E+05	.8185E+04	.2860E+05
GLoads	13	.4489E+04	-.1426E+05	-.2237E+04	-.8793E+04	-.2696E+04	.2386E+04
LLoads	3	.1507E+05	-.1088E+04	-.5670E+03	.2331E+03	.3050E+05	-.4113E+05
LLoads	13	-.1507E+05	.1088E+04	.5670E+03	-.2331E+03	-.9468E+04	.7653E+03
Stress	3	-.2573E+02	-.1858E+01	-.9681E+00	-.9583E-02	.1691E+01	-.1253E+01
Stress	13	-.2573E+02	-.1858E+01	-.9681E+00	-.9583E-02	.3146E-01	-.3892E+00
				BEAM NO. 6			
GLoads	3	-.1738E+04	.3002E+04	-.6380E+03	.3579E+05	.1935E+05	.5767E+04

```
IMAGES-3D   s/n:800189                                    03-02-87
                                                         PAGE    2
        =============== I M A G E S   3 D ===============
        = Copyright (c) 1984   Celestial Software Inc. =
        ================================================

           SOLVE BEAM LOADS/STRESSES      Version 1.4  12/01/86

        COOLING TOWER VIBRATION - NODE INCREASE

Mode  1 -

GLoads Node      Fx          Fy          Fz          Mx          My          Mz
LLoads Node     Axial      Y-Shear     Z-Shear     Torsion     Y-Bending   Z-Bending
/Stress
------ ----  ----------  ----------  ----------  ----------  ----------  ----------
GLoads  14  .1738E+04  -.3002E+04   .6380E+03   .5535E+04   .4654E+04  -.5398E+04
LLoads   3  .3374E+04   .6341E+03  -.8086E+03  -.3611E+04   .3467E+05   .2176E+05
LLoads  14 -.3374E+04  -.6341E+03   .8086E+03   .3611E+04   .2937E+04   .7731E+04
Stress   3 -.5761E+01   .1083E+01  -.1381E+01   .1484E+00  -.8945E+00  -.1425E+01
Stress  14 -.5761E+01   .1083E+01  -.1381E+01   .1484E+00   .3178E+00   .1207E+00
                           ***BEAM  NO.   7***
GLoads   4 -.3306E+04   .1136E+05   .2289E+04   .3565E+05   .5322E+04   .2193E+05
GLoads  14  .3306E+04  -.1136E+05  -.2289E+04  -.8675E+04  -.2139E+04   .1244E+04
LLoads   4  .1201E+05  -.9048E+03  -.3288E+03  -.5854E+02   .2061E+05  -.3682E+05
LLoads  14 -.1201E+05   .9048E+03   .3288E+03   .5854E+02  -.8411E+04   .3261E+04
Stress   4 -.2050E+02  -.1545E+01  -.5614E+00   .2406E-02   .1513E+01  -.8469E+00
Stress  14 -.2050E+02  -.1545E+01  -.5614E+00   .2406E-02  -.1340E+00  -.3457E+00
                           ***BEAM  NO.   8***
GLoads   4 -.1591E+04   .2386E+04  -.4169E+03   .2374E+05   .1393E+05   .5409E+04
GLoads  15  .1591E+04  -.2386E+04   .4169E+03   .3019E+04   .3863E+04  -.5711E+04
LLoads   4  .2814E+04   .3438E+03  -.5993E+03  -.3561E+04   .2611E+05   .9595E+04
LLoads  15 -.2814E+04  -.3438E+03   .5993E+03   .3561E+04   .1758E+04   .6394E+04
Stress   4 -.4805E+01   .5869E+00  -.1023E+01   .1464E+00  -.3944E+00  -.1073E+01
Stress  15 -.4805E+01   .5869E+00  -.1023E+01   .1464E+00   .2628E+00   .7227E-01
                           ***BEAM  NO.   9***
GLoads   5 -.1968E+04   .7032E+04   .1639E+04   .2402E+05   .1317E+04   .1892E+05
GLoads  15  .1968E+04  -.7032E+04  -.1639E+04  -.9106E+04  -.1413E+04  -.5909E+03
LLoads   5  .7456E+04  -.6371E+03   .9834E+01  -.3579E+03   .6371E+04  -.2993E+05
LLoads  15 -.7456E+04   .6371E+03  -.9834E+01   .3579E+03  -.6736E+04   .6306E+04
Stress   5 -.1273E+02  -.1088E+01   .1679E-01   .1471E-01   .1230E+01   .2619E+00
Stress  15 -.1273E+02  -.1088E+01   .1679E-01   .1471E-01   .2592E+00  -.2769E+00
                           ***BEAM  NO.  10***
GLoads   5 -.1316E+04   .1506E+04  -.2899E+03   .9356E+04   .6829E+04   .1206E+05
GLoads  16  .1316E+04  -.1506E+04   .2899E+03  -.9638E+02   .2590E+04  -.5141E+04
LLoads   5  .1995E+04  -.4620E+02  -.3172E+03  -.3490E+04   .1491E+05  -.6725E+04
LLoads  16 -.1995E+04   .4620E+02   .3172E+03   .3490E+04  -.1509E+03   .4576E+04
Stress   5 -.3407E+01  -.7889E-01  -.5416E+00   .1434E+00   .2764E+00  -.6126E+00
Stress  16 -.3407E+01  -.7889E-01  -.5416E+00   .1434E+00   .1881E+00  -.6204E-02
                           ***BEAM  NO.  11***
GLoads   6 -.7577E+03   .1215E+04   .2043E+03   .8915E+04  -.3862E+04   .2275E+05
GLoads  16  .7577E+03  -.1215E+04  -.2043E+03  -.1061E+05  -.4851E+03  -.3182E+04
LLoads   6  .1341E+04  -.3041E+03   .4491E+03  -.6755E+03  -.1230E+05  -.2145E+05
LLoads  16 -.1341E+04   .3041E+03  -.4491E+03   .6755E+03  -.4357E+04   .1017E+05
Stress   6 -.2289E+01  -.5193E+00   .7667E+00   .2777E-01   .8817E+00   .5054E+00
```

COOLING TOWER VIBRATION: INCREASED NODE MODEL

```
IMAGES-3D   s/n:800189                                03-02-87
                                                      PAGE   3
        =============== I M A G E S   3 D ===============
        = Copyright (c) 1984   Celestial Software Inc. =
        ================================================

           SOLVE BEAM LOADS/STRESSES     Version 1.4  12/01/86

        COOLING TOWER VIBRATION - NODE INCREASE

Mode  1 -

GLoads Node     Fx          Fy          Fz          Mx          My          Mz
LLoads Node     Axial       Y-Shear     Z-Shear     Torsion     Y-Bending   Z-Bending
/Stress
------ ----  ----------  ----------  ----------  ----------  ----------  ----------
Stress  16 -.2289E+01  -.5193E+00   .7667E+00   .2777E-01   .4181E+00  -.1791E+00
                        ***BEAM   NO.   12***
GLoads   6 -.9430E+03   .3382E+03  -.3152E+03  -.4911E+04  -.2041E+04   .2747E+05
GLoads  17  .9430E+03  -.3382E+03   .3152E+03  -.3597E+04   .7874E+03  -.3364E+04
LLoads   6  .8944E+03  -.5487E+03   .4223E+02  -.3407E+04   .9099E+03  -.2776E+05
LLoads  17 -.8944E+03   .5487E+03  -.4223E+02   .3407E+04  -.2874E+04   .2239E+04
Stress   6 -.1527E+01  -.9368E+00   .7210E-01   .1400E+00   .1141E+01  -.3740E-01
Stress  17 -.1527E+01  -.9368E+00   .7210E-01   .1400E+00   .9201E-01  -.1181E+00
                        ***BEAM   NO.   13***
GLoads   7  .2453E+02  -.6269E+04  -.2049E+04  -.7233E+04  -.1035E+05   .3563E+05
GLoads  17 -.2453E+02   .6269E+04   .2049E+04  -.1356E+05   .6431E+03  -.6170E+04
LLoads   7 -.6518E+04   .8824E+02   .1003E+04  -.9953E+03  -.3598E+05  -.1155E+05
LLoads  17  .6518E+04  -.8824E+02  -.1003E+04   .9953E+03  -.1217E+04   .1483E+05
Stress   7  .1113E+02   .1507E+00   .1713E+01   .4091E-01   .4749E+00   .1479E+01
Stress  17  .1113E+02   .1507E+00   .1713E+01   .4091E-01   .6094E+00  -.5004E-01
                        ***BEAM   NO.   14***
GLoads   7 -.5910E+03  -.1063E+04  -.5553E+03  -.1652E+05  -.1316E+05   .5267E+05
GLoads  18  .5910E+03   .1063E+04   .5553E+03  -.6681E+04  -.1383E+04  -.1178E+03
LLoads   7 -.4060E+03  -.1176E+04   .4899E+03  -.3185E+04  -.1678E+05  -.5412E+05
LLoads  18  .4060E+03   .1176E+04  -.4899E+03   .3185E+04  -.6006E+04  -.5868E+03
Stress   7  .6932E+00  -.2008E+01   .8365E+00   .1309E+00   .2224E+01   .6897E+00
Stress  18  .6932E+00  -.2008E+01   .8365E+00   .1309E+00  -.2412E-01  -.2469E+00
                        ***BEAM   NO.   15***
GLoads   8  .2919E+02  -.1563E+05  -.5099E+04  -.2251E+05  -.1856E+05   .5940E+05
GLoads  18 -.2919E+02   .1563E+05   .5099E+04  -.1790E+05   .1939E+04  -.8711E+04
LLoads   8 -.1634E+05   .5563E+03   .1717E+04  -.1333E+04  -.6616E+05   .8346E+03
LLoads  18  .1634E+05  -.5563E+03  -.1717E+04   .1333E+04   .2499E+04   .1980E+05
Stress   8  .2789E+02   .9499E+00   .2931E+01   .5479E-01  -.3430E-01   .2719E+01
Stress  18  .2789E+02   .9499E+00   .2931E+01   .5479E-01   .8137E+00   .1027E+00
                        ***BEAM   NO.   16***
GLoads   8  .2515E+03  -.3464E+04  -.1187E+04  -.2506E+05  -.2829E+05   .8787E+05
GLoads  19 -.2515E+03   .3464E+04   .1187E+04  -.1036E+05  -.6370E+04   .5738E+04
LLoads   8 -.2879E+04  -.1955E+04   .1167E+04  -.1947E+04  -.4196E+05  -.8593E+05
LLoads  19  .2879E+04   .1955E+04  -.1167E+04   .1947E+04  -.1233E+05  -.5011E+04
Stress   8  .4915E+01  -.3338E+01   .1993E+01   .8001E-01   .3532E+01   .1725E+01
Stress  19  .4915E+01  -.3338E+01   .1993E+01   .8001E-01  -.2059E+00  -.5066E+00
                        ***BEAM   NO.   17***
GLoads   9 -.8948E+03  -.2794E+05  -.8984E+04  -.3443E+05  -.2698E+05   .9125E+05
GLoads  19  .8948E+03   .2794E+05   .8984E+04  -.2082E+05   .3347E+04  -.1225E+05
```

```
IMAGES-3D  s/n:800189                              03-02-87
                                                   PAGE    4
         =============== I M A G E S   3 D ===============
         = Copyright (c) 1984   Celestial Software Inc. =
         ================================================

         SOLVE BEAM LOADS/STRESSES    Version 1.4  12/01/86

      COOLING TOWER VIBRATION - NODE INCREASE

Mode  1 -

GLoads Node    Fx          Fy          Fz          Mx          My          Mz
LLoads Node    Axial       Y-Shear     Z-Shear     Torsion     Y-Bending   Z-Bending
/Stress
------ ----  ----------  ----------  ----------  ----------  ----------  ----------
LLoads   9  -.2924E+05   .1097E+04   .2441E+04  -.1027E+04  -.9957E+05   .1806E+05
LLoads  19   .2924E+05  -.1097E+04  -.2441E+04   .1027E+04   .9026E+04   .2263E+05
Stress   9   .4993E+02   .1873E+01   .4168E+01   .4221E-01  -.7423E+00   .4092E+01
Stress  19   .4993E+02   .1873E+01   .4168E+01   .4221E-01   .9301E+00   .3710E+00
                         ***BEAM  NO.    18***
GLoads   9  -.2148E+04  -.4011E+03   .4220E+03   .1063E+05  -.7318E+04   .9016E+05
GLoads  20   .2148E+04   .4011E+03  -.4220E+03   .7346E+04   .4104E+04  -.1723E+04
LLoads   9   .1087E+04  -.1939E+04   .1083E+03  -.1658E+04  -.1346E+05  -.9006E+05
LLoads  20  -.1087E+04   .1939E+04  -.1083E+03  -.1658E+04   .8427E+04  -.9956E+02
Stress   9  -.1855E+01  -.3310E+01   .1848E+00  -.6813E-01   .3702E+01   .5533E+00
Stress  20  -.1855E+01  -.3310E+01   .1848E+00  -.6813E-01  -.4092E-02   .3464E+00
                         ***BEAM  NO.    19***
GLoads  10  -.3163E+04  -.2472E+05  -.4152E+04   .4534E+05  -.2629E+05   .1130E+06
GLoads  20   .3163E+04   .2472E+05   .4152E+04   .4534E+05  -.4335E+04   .2206E+03
LLoads  10  -.2495E+05  -.2445E+04   .3163E+04   .4127E+04  -.1160E+06  -.4534E+05
LLoads  20   .2495E+05   .2445E+04  -.3163E+04  -.4127E+04  -.1344E+04  -.4534E+05
Stress  10   .4260E+02  -.4175E+01   .5401E+01  -.1696E+00   .1864E+01   .4767E+01
Stress  20   .4260E+02  -.4175E+01   .5401E+01  -.1696E+00  -.1864E+01  -.5526E-01
```

```
IMAGES-3D  s/n:800189                              03-02-87
                                                   PAGE    5
         =============== I M A G E S   3 D ===============
         = Copyright (c) 1984   Celestial Software Inc. =
         ================================================

         SOLVE BEAM LOADS/STRESSES    Version 1.4  12/01/86

      COOLING TOWER VIBRATION - NODE INCREASE

  Mode  1 -


              MAXIMUM STRESS SUMMARY FOR BEAMS/TRUSSES
              WITHIN SPECIFIED RANGE     1-   19


        Maximum (absolute) Stress = .4993E+02 at BEAM   17

  Beam    Axial      Y-Shear     Z-Shear    Torsion    Y-Bending   Z-Bending
  ----  ----------  ----------  ----------  ----------  ----------  ----------
   17   .4993E+02   .1873E+01   .4168E+01   .4221E-01  -.7423E+00   .4092E+01
```

COOLING TOWER VIBRATION: INCREASED NODE MODEL

```
IMAGES-3D   s/n:800189                                      03-02-87
                                                           PAGE    1
           ================ I M A G E S   3 D ================
           = Copyright (c) 1984   Celestial Software Inc. =
           ==================================================

              SOLVE PLATE LOADS/STRESSES   Version 1.4  12/01/86

        COOLING TOWER VIBRATION - NODE INCREASE

  Mode  1 -

                        PLATE LOADS AND/OR STRESSES
```

GLoads Stress Stress	Node Surf	Fx Sigma X Shear XZ	Fy Sigma Y Shear YZ	Fz Tau XY	Mx Sigma 1	My Sigma 2	Mz Angle
			***PLATE	1***			
Loads	11	-.3586E+04	.1030E+05	-.2138E+04	-.2594E+03	.6253E+03	-.6823E+03
Loads	12	-.3031E+04	.9471E+04	.2238E+04	.1251E+03	-.1092E+03	.7903E+03
Loads	22	.3954E+04	-.1001E+05	.2228E+04	-.1259E+04	.2059E+04	-.6404E+04
Loads	21	.2663E+04	-.9754E+04	-.2328E+04	-.2069E+04	.4942E+04	-.3278E+04
Stress	TOP	-.1211E+02	-.7446E+02	-.3617E+01	-.1190E+02	-.7467E+02	-3.3
Stress	MID	-.9947E+01	-.6641E+02	-.1521E+01	-.9906E+01	-.6645E+02	-1.5
Stress	BOT	-.7788E+01	-.5835E+02	.5746E+00	-.7782E+01	-.5836E+02	.7
			***PLATE	2***			
Loads	12	-.3815E+04	.1007E+05	-.2091E+04	-.1119E+03	-.6157E+03	-.1241E+04
Loads	13	-.2364E+04	.8427E+04	.2805E+04	.1187E+04	-.1665E+04	.2196E+04
Loads	23	.3960E+04	-.9698E+04	.1889E+04	-.1373E+04	-.8000E+03	-.6194E+04
Loads	22	.2220E+04	-.8795E+04	-.2602E+04	-.1272E+04	.7355E+02	-.4660E+04
Stress	TOP	-.1210E+02	-.7203E+02	-.4312E+01	-.1179E+02	-.7234E+02	-4.1
Stress	MID	-.1089E+02	-.6211E+02	-.2900E+01	-.1073E+02	-.6227E+02	-3.2
Stress	BOT	-.9679E+01	-.5218E+02	-.1489E+01	-.9627E+01	-.5223E+02	-2.0
			***PLATE	3***			
Loads	13	-.3857E+04	.9217E+04	-.1492E+04	.1639E+03	-.5896E+03	-.1318E+03
Loads	14	-.1565E+04	.6735E+04	.2526E+04	.2308E+04	-.2139E+04	.3191E+04
Loads	24	.3931E+04	-.8601E+04	.1265E+04	-.2329E+04	-.9688E+03	-.5794E+04
Loads	23	.1491E+04	-.7351E+04	-.2300E+04	-.2231E+04	.8301E+03	-.4103E+04
Stress	TOP	-.1177E+02	-.6539E+02	-.5716E+01	-.1117E+02	-.6599E+02	-6.0
Stress	MID	-.9769E+01	-.5362E+02	-.4314E+01	-.9348E+01	-.5404E+02	-5.6
Stress	BOT	-.7764E+01	-.4185E+02	-.2913E+01	-.7517E+01	-.4209E+02	-4.9
			***PLATE	4***			
Loads	14	-.3472E+04	.7614E+04	-.8876E+03	.8327E+03	-.3764E+03	.9635E+03
Loads	15	-.8831E+03	.4229E+04	.1708E+04	.3963E+04	-.2448E+04	.4177E+04
Loads	25	.3481E+04	-.6730E+04	.6707E+03	-.2895E+04	-.1146E+04	-.4830E+04
Loads	24	.8740E+03	-.5112E+04	-.1491E+04	-.2732E+04	.1058E+04	-.3264E+04

IMAGES-3D   s/n:800189   03-02-87
                                                      PAGE   2
          =============== I M A G E S 3 D ===============
          = Copyright (c) 1984   Celestial Software Inc. =
          ===

          SOLVE PLATE LOADS/STRESSES   Version 1.4 12/01/86

          COOLING TOWER VIBRATION - NODE INCREASE

Mode 1 -

GLoads	Node	Fx	Fy	Fz	Mx	My	Mz
Stress	Surf	Sigma X	Sigma Y	Tau XY	Sigma 1	Sigma 2	Angle
Stress		Shear XZ	Shear YZ				
Stress	TOP	-.1008E+02	-.5330E+02	-.7440E+01	-.8839E+01	-.5454E+02	-9.5
Stress	MID	-.7577E+01	-.3987E+02	-.5818E+01	-.6560E+01	-.4089E+02	-9.9
Stress	BOT	-.5069E+01	-.2645E+02	-.4197E+01	-.4274E+01	-.2725E+02	-10.7
			***PLATE	5***			
Loads	15	-.2670E+04	.5185E+04	-.4950E+03	.2124E+04	-.3111E+01	.2125E+04
Loads	16	-.5091E+03	.8583E+03	.4113E+03	.6329E+04	-.2699E+04	.5003E+04
Loads	26	.2651E+04	-.4001E+04	.3041E+03	-.2828E+04	-.1026E+04	-.3333E+04
Loads	25	.5287E+03	-.2042E+04	-.2204E+03	-.2807E+04	.1233E+04	-.2204E+04
Stress	TOP	-.7315E+01	.3554E+02	-.9288E+01	-.4533E+01	-.3832E+02	-16.7
Stress	MID	-.4455E+01	-.2048E+02	-.7353E+01	-.1592E+01	-.2334E+02	-21.3
Stress	BOT	-.1595E+01	-.5425E+01	-.5419E+01	.2237E+01	-.9257E+01	-35.3
			***PLATE	6***			
Loads	16	-.1558E+04	.1864E+04	-.5006E+03	.4377E+04	.5944E+03	.3320E+04
Loads	17	-.6175E+03	-.3452E+04	-.1298E+04	.9386E+04	-.2850E+04	.5364E+04
Loads	27	.1548E+04	-.3592E+04	.3845E+03	-.1692E+04	-.5034E+03	-.1399E+04
Loads	26	.6270E+03	.1947E+04	.1415E+04	-.2185E+04	.1158E+04	-.1041E+04
Stress	TOP	-.3429E+01	-.1151E+02	-.1111E+02	.4351E+01	-.1929E+02	-35.0
Stress	MID	-.5302E+00	.5016E+01	-.8967E+01	.1163E+02	-.7143E+01	-53.6
Stress	BOT	.2369E+01	.2154E+02	-.6824E+01	.2372E+02	.1882E+00	-72.3
			***PLATE	7***			
Loads	17	-.2912E+03	-.2473E+04	-.1064E+04	.7769E+04	.1419E+04	.4171E+04
Loads	18	-.1336E+04	-.8820E+04	-.3332E+04	.1261E+05	-.2801E+04	.4809E+04
Loads	28	.3620E+03	.4330E+04	.1143E+04	.7946E+03	.4643E+03	.5452E+03
Loads	27	.1265E+04	.6962E+04	.3253E+04	-.6363E+03	.5877E+03	-.7363E+02
Stress	TOP	.1577E+01	.2011E+02	-.1258E+02	.2647E+02	-.4777E+01	-63.2
Stress	MID	.3984E+01	.3744E+02	-.1062E+02	.4053E+02	.8974E+00	-73.8
Stress	BOT	.6391E+01	.5477E+02	-.8666E+01	.5628E+02	.4886E+01	-80.1
			***PLATE	8***			
Loads	18	.7905E+03	-.7855E+04	-.2315E+04	.1197E+05	.2245E+04	.4020E+04
Loads	19	-.2814E+04	-.1542E+05	-.5602E+04	.1504E+05	-.2626E+04	.3089E+04
Loads	29	-.3453E+03	.1004E+05	.2766E+04	.4593E+04	.1706E+04	.1840E+04
Loads	28	.2369E+04	.1324E+05	.5151E+04	.1995E+04	-.5740E+03	.3241E+03

COOLING TOWER VIBRATION: INCREASED NODE MODEL

```
IMAGES-3D   s/n:800189                                    03-02-87
                                                         PAGE    3
         =============== I M A G E S   3 D ===============
         = Copyright (c) 1984   Celestial Software Inc. =
         ================================================

         SOLVE PLATE LOADS/STRESSES   Version 1.4  12/01/86

    COOLING TOWER VIBRATION - NODE INCREASE

Mode   1 -

GLoads Node    Fx          Fy          Fz          Mx          My          Mz
Stress Surf  Sigma  X    Sigma  Y     Tau XY     Sigma  1    Sigma  2     Angle
Stress       Shear XZ    Shear YZ
------ ----  ----------  ----------  ----------  ----------  ----------  ----------
Stress TOP   .7030E+01   .6101E+02  -.1382E+02   .6434E+02   .3696E+01     -76.4
Stress MID   .8374E+01   .7753E+02  -.1276E+02   .7981E+02   .6094E+01     -79.9
Stress BOT   .9717E+01   .9405E+02  -.1170E+02   .9564E+02   .8125E+01     -82.2
                         ***PLATE    9***
Loads    19  .2196E+04  -.1596E+05  -.4556E+04   .1614E+05   .5648E+04   .3419E+04
Loads    20 -.5282E+04  -.2510E+05  -.8715E+04   .1584E+05   .2311E+03   .1502E+04
Loads    30 -.2374E+03   .1936E+05   .6557E+04   .1090E+05  -.7899E+04  -.1898E+04
Loads    29  .3323E+04   .2170E+05   .6714E+04   .4813E+04  -.5040E+04  -.1316E+04
Stress TOP   .8245E+01   .1240E+03  -.1345E+02   .1255E+03   .6703E+01     -83.5
Stress MID   .1270E+02   .1372E+03  -.1348E+02   .1386E+03   .1125E+02     -83.9
Stress BOT   .1715E+02   .1503E+03  -.1352E+02   .1517E+03   .1579E+02     -84.3
                         ***PLATE   10***
Loads    21 -.2654E+04   .9738E+04  -.1991E+04   .2069E+04  -.4942E+04   .7829E+04
Loads    22 -.2895E+04   .8867E+04   .2070E+04   .1427E+04  -.2948E+04   .6068E+04
Loads    32  .3247E+04  -.9498E+04   .1510E+04  -.4775E+03   .1480E+04  -.1558E+04
Loads    31  .2301E+04  -.9106E+04  -.1589E+04  -.8942E+03   .2769E+04   .3172E+03
Stress TOP  -.7575E+01  -.8014E+02  -.4458E+01  -.7302E+01  -.8041E+02      -3.5
Stress MID  -.8102E+01  -.6686E+02  -.1393E+01  -.8069E+01  -.6690E+02      -1.4
Stress BOT  -.8629E+01  -.5359E+02   .1671E+01  -.8567E+01  -.5365E+02       2.1
                         ***PLATE   11***
Loads    22 -.3253E+04   .9912E+04  -.1698E+04   .1104E+04   .8156E+03   .4997E+04
Loads    23 -.1998E+04   .7592E+04   .1724E+04   .1827E+04  -.8062E+03   .5967E+04
Loads    33  .3364E+04  -.9554E+04   .1237E+04  -.4567E+03  -.2438E+03  -.1974E+04
Loads    32  .1887E+04  -.7950E+04  -.1264E+04  -.5111E+03   .1276E+03  -.1778E+04
Stress TOP  -.8369E+01  -.7586E+02  -.7457E+01  -.7555E+01  -.7667E+02      -6.2
Stress MID  -.7241E+01  -.6279E+02  -.4595E+01  -.6863E+01  -.6317E+02      -4.7
Stress BOT  -.6112E+01  -.4972E+02  -.1733E+01  -.6044E+01  -.4979E+02      -2.3
                         ***PLATE   12***
Loads    23 -.3424E+04   .9433E+04  -.1316E+04   .1776E+04   .7760E+03   .4330E+04
Loads    24 -.1484E+04   .5512E+04   .1102E+04   .2860E+04  -.1113E+04   .5410E+04
Loads    34  .3405E+04  -.8857E+04   .8613E+03  -.8216E+03  -.6076E+03  -.2171E+04
Loads    33  .1504E+04  -.6088E+04  -.6471E+03  -.6213E+03   .2285E+03  -.1191E+04
```

```
IMAGES-3D  s/n:800189                                03-02-87
                                                     PAGE    4
             =============== I M A G E S   3 D ===============
             = Copyright (c) 1984  Celestial Software Inc. =
             =================================================

             SOLVE PLATE LOADS/STRESSES    Version 1.4  12/01/86

             COOLING TOWER VIBRATION - NODE INCREASE

Mode  1 -

GLoads Node      Fx          Fy          Fz          Mx          My          Mz
Stress Surf   Sigma  X    Sigma  Y     Tau XY     Sigma  1    Sigma  2      Angle
Stress        Shear XZ    Shear YZ
------ ----  ----------  ----------  ----------  ----------  ----------  ----------
Stress TOP  -.7398E+01  -.6603E+02  -.1095E+02  -.5421E+01  -.6801E+02     -10.2
Stress MID  -.5845E+01  -.5363E+02  -.7816E+01  -.4599E+01  -.5488E+02      -9.1
Stress BOT  -.4291E+01  -.4123E+02  -.4685E+01  -.3706E+01  -.4182E+02      -7.1
                         ***PLATE   13***
Loads   24  -.3290E+04   .8184E+04  -.8807E+03   .2201E+04   .1023E+04   .3648E+04
Loads   25  -.1148E+04   .2698E+04   .1185E+03   .3379E+04  -.1079E+04   .4312E+04
Loads   35   .3172E+04  -.7362E+04   .5012E+03  -.1089E+04  -.5777E+03  -.1845E+04
Loads   34   .1266E+04  -.3520E+04   .2611E+03  -.9560E+03   .4595E+03  -.1151E+04
Stress TOP  -.5575E+01  -.5054E+02  -.1416E+02  -.1485E+01  -.5463E+02     -16.1
Stress MID  -.3773E+01  -.3907E+02  -.1092E+02  -.6669E+00  -.4218E+02     -15.9
Stress BOT  -.1970E+01  -.2761E+02  -.7678E+01   .1536E+00  -.2973E+02     -15.5
                         ***PLATE   14***
Loads   25  -.2830E+04   .6067E+04  -.5759E+03   .2323E+04   .9921E+03   .2722E+04
Loads   26  -.1104E+04  -.8951E+03  -.1144E+04   .3097E+04  -.7549E+03   .2811E+04
Loads   36   .2668E+04  -.4975E+04   .3385E+03  -.1072E+04   .3082E+02  -.1070E+04
Loads   35   .1266E+04  -.1968E+03   .1381E+04  -.1275E+04   .3798E+03  -.1135E+04
Stress TOP  -.2439E+01  -.2803E+02  -.1689E+02   .5951E+01  -.3642E+02     -26.4
Stress MID  -.1226E+01  -.1861E+02  -.1377E+02   .6368E+01  -.2620E+02     -28.9
Stress BOT  -.1229E-01  -.9191E+01  -.1066E+02   .7003E+01  -.1621E+02     -33.4
                         ***PLATE   15***
Loads   26  -.2139E+04   .2954E+04  -.5846E+03   .1915E+04   .6221E+03   .1562E+04
Loads   27  -.1358E+04  -.5279E+04  -.2577E+04   .1652E+04   .1185E+03   .1209E+04
Loads   37   .1998E+04  -.1574E+04   .5842E+03  -.6307E+03   .1390E+04   .1898E+02
Loads   36   .1499E+04   .3899E+04   .2578E+04  -.1395E+04  -.2902E+03  -.1085E+04
Stress TOP   .2037E+01   .2425E+01  -.1861E+02   .2084E+02  -.1637E+02     -45.3
Stress MID   .1379E+01   .8257E+01  -.1612E+02   .2130E+02  -.1166E+02     -51.0
Stress BOT   .7200E+00   .1409E+02  -.1363E+02   .2259E+02  -.7779E+01     -58.1
                         ***PLATE   16***
Loads   27  -.1418E+04  -.1302E+04  -.1072E+04   .6764E+03  -.2027E+03   .2644E+03
Loads   28  -.1794E+04  -.1036E+05  -.4044E+04  -.1041E+04   .1739E+04   .5865E+02
Loads   38   .1403E+04   .2946E+04   .1428E+04  -.1168E+02   .3297E+04   .9956E+03
Loads   37   .1808E+04   .8716E+04   .3689E+04  -.1121E+04  -.1714E+04  -.1060E+04
```

COOLING TOWER VIBRATION: INCREASED NODE MODEL

```
IMAGES-3D   s/n:800189                                    03-02-87
                                                         PAGE    5
        =============== I M A G E S   3 D ===============
        = Copyright (c) 1984   Celestial Software Inc. =
        ================================================

        SOLVE PLATE LOADS/STRESSES   Version 1.4  12/01/86

     COOLING TOWER VIBRATION - NODE INCREASE

Mode  1 -
```

GLoads	Node	Fx	Fy	Fz	Mx	My	Mz
Stress	Surf	Sigma X	Sigma Y	Tau XY	Sigma 1	Sigma 2	Angle
Stress		Shear XZ	Shear YZ				
Stress	TOP	.7261E+01	.4098E+02	-.1866E+02	.4927E+02	-.1030E+01	-66.0
Stress	MID	.3284E+01	.4174E+02	-.1748E+02	.4849E+02	-.3474E+01	-68.9
Stress	BOT	-.6933E+00	.4249E+02	-.1629E+02	.4795E+02	-.6151E+01	-71.5
				PLATE 17			
Loads	28	-.8982E+03	-.7165E+04	-.2263E+04	-.1749E+04	-.1629E+04	-.9279E+03
Loads	29	-.1951E+04	-.1614E+05	-.5401E+04	-.3964E+04	.3191E+04	-.1344E+04
Loads	39	.1087E+04	.9232E+04	.3083E+04	-.3062E+03	.4317E+04	.1153E+04
Loads	38	.1762E+04	.1407E+05	.4580E+04	-.3571E+03	-.3676E+04	-.1165E+04
Stress	TOP	.1060E+02	.8791E+04	-.1569E+02	.9097E+02	.7534E+01	-79.0
Stress	MID	.3248E+01	.8351E+02	-.1632E+02	.8670E+02	.5795E-01	-78.9
Stress	BOT	-.4101E+01	.7912E+02	-.1694E+02	.8244E+02	-.7418E+01	-78.9
				PLATE 18			
Loads	29	-.9839E+03	-.1554E+05	-.4090E+04	-.5442E+04	.1432E+03	-.3897E+03
Loads	30	.2632E+03	-.1931E+05	-.5965E+04	-.2883E+04	.7899E+04	.1898E+04
Loads	40	.4212E+03	.1703E+05	.5328E+04	-.1536E+04	-.7637E+03	-.3681E+03
Loads	39	.2995E+03	.1782E+05	.4726E+04	.7077E+03	-.5268E+04	-.1379E+04
Stress	TOP	.5671E+01	.1310E+03	-.3022E+01	.1311E+03	.5598E+01	-88.6
Stress	MID	-.1441E+01	.1249E+03	-.5493E+01	.1251E+03	-.1680E+01	-87.5
Stress	BOT	-.8554E+01	.1187E+03	-.7963E+01	.1192E+03	-.9050E+01	-86.4
				PLATE 19			
Loads	31	-.2267E+04	.9083E+04	-.8209E+03	.8942E+03	-.2769E+04	.2631E+04
Loads	32	-.2418E+04	.7823E+04	.5819E+03	.6571E+03	-.1162E+04	.2560E+04
Loads	42	.2837E+04	-.8834E+04	.3905E+03	-.4959E+03	.1554E+04	-.2213E+04
Loads	41	.1848E+04	-.8071E+04	-.1516E+03	-.9762E+03	.3103E+04	.1734E+03
Stress	TOP	-.2239E+01	-.7192E+02	-.6154E+01	-.1700E+01	-.7246E+02	-5.0
Stress	MID	-.2291E+01	-.6502E+02	-.2397E+01	-.2200E+01	-.6511E+02	-2.2
Stress	BOT	-.2343E+01	-.5812E+02	.1360E+01	-.2310E+01	-.5815E+02	1.4
				PLATE 20			
Loads	32	-.2637E+04	.9583E+04	-.8316E+03	.3315E+03	-.4456E+03	.7753E+03
Loads	33	-.1909E+04	.6188E+04	.7661E+02	.5272E+03	.4902E+01	.1975E+04
Loads	43	.2895E+04	-.9212E+04	.4056E+03	-.6590E+03	.1029E+03	-.2355E+04
Loads	42	.1651E+04	-.6559E+04	.3494E+03	-.6948E+03	.2272E+03	-.2357E+04

```
IMAGES-3D   s/n:800189                              03-02-87
                                                   PAGE    6
       ================ I M A G E S   3 D ================
       = Copyright (c) 1984   Celestial Software Inc. =
       ================================================

       SOLVE PLATE LOADS/STRESSES    Version 1.4  12/01/86

   COOLING TOWER VIBRATION - NODE INCREASE

Mode   1 -
```

GLoads Stress Stress	Node Surf	Fx Sigma X Shear XZ	Fy Sigma Y Shear YZ	Fz Tau XY	Mx Sigma 1	My Sigma 2	Mz Angle
Stress	TOP	-.1457E+01	-.6775E+02	-.1070E+02	.2277E+00	-.6944E+02	-8.9
Stress	MID	-.1649E+01	-.6052E+02	-.7068E+01	-.8127E+00	-.6136E+02	-6.8
Stress	BOT	-.1841E+01	-.5329E+02	-.3433E+01	-.1613E+01	-.5352E+02	-3.8
			PLATE 21				
Loads	33	-.2872E+04	.9419E+04	-.6748E+03	.5508E+03	.1043E+02	.1191E+04
Loads	34	-.1691E+04	.3880E+04	-.6535E+03	.1020E+04	-.1399E+03	.2101E+04
Loads	44	.3018E+04	-.8849E+04	.2919E+03	-.1046E+04	.2909E+02	-.2227E+04
Loads	43	.1544E+04	-.4451E+04	.1036E+04	-.7164E+03	.9103E+02	-.1481E+04
Stress	TOP	-.6748E+00	-.5826E+02	-.1588E+02	.3413E+01	-.6235E+02	-14.4
Stress	MID	-.5487E+00	-.5107E+02	-.1161E+02	.1993E+01	-.5361E+02	-12.3
Stress	BOT	-.4225E+00	-.4388E+02	-.7350E+01	.7870E+00	-.4509E+02	-9.3
			PLATE 22				
Loads	34	-.2886E+04	.8475E+04	-.4824E+03	.7578E+03	.2880E+03	.1221E+04
Loads	35	-.1630E+04	.9625E+03	-.1537E+04	.1261E+04	-.7771E+02	.1768E+04
Loads	45	.2962E+04	-.7685E+04	.1794E+03	-.1372E+04	.6968E+03	-.1636E+04
Loads	44	.1555E+04	-.1753E+04	.1840E+04	-.8907E+03	-.7099E+02	-.1308E+04
Stress	TOP	.9759E+00	-.4303E+02	-.1993E+02	.8662E+01	-.5072E+02	-21.1
Stress	MID	.7646E+00	-.3625E+02	-.1574E+02	.6554E+01	-.4203E+02	-20.2
Stress	BOT	.5532E+00	-.2946E+02	-.1155E+02	.4483E+01	-.3339E+02	-18.8
			PLATE 23				
Loads	35	-.2709E+04	.6591E+04	-.3652E+03	.1103E+04	.2757E+03	.1212E+04
Loads	36	-.1692E+04	-.2507E+04	-.2467E+04	.1025E+04	.5444E+03	.1236E+04
Loads	46	.2704E+04	-.5578E+04	.2404E+03	-.1423E+04	.1968E+04	-.6762E+03
Loads	45	.1697E+04	.1494E+04	.2591E+04	-.8873E+03	-.7691E+03	-.1184E+04
Stress	TOP	.3622E+01	-.2143E+02	-.2239E+02	.1675E+02	-.3456E+02	-30.4
Stress	MID	.1921E+01	-.1570E+02	-.1897E+02	.1402E+02	-.2780E+02	-32.5
Stress	BOT	.2202E+00	-.9960E+01	-.1554E+02	.1148E+02	-.2122E+02	-35.9
			PLATE 24				
Loads	36	-.2372E+04	.3600E+04	-.4755E+03	.1442E+04	-.2850E+03	.9199E+03
Loads	37	-.1761E+04	-.6400E+04	-.3324E+04	.2678E+03	.1849E+04	.7989E+03
Loads	47	.2303E+04	-.2383E+04	.6277E+03	-.1133E+04	.3580E+04	.3816E+03
Loads	46	.1830E+04	.5183E+04	.3172E+04	-.6558E+03	-.2106E+04	-.1156E+04

COOLING TOWER VIBRATION: INCREASED NODE MODEL

```
IMAGES-3D   s/n:800189                                    03-02-87
                                                         PAGE    7
          =============== I M A G E S   3 D ===============
          = Copyright (c) 1984   Celestial Software Inc. =
          ================================================

          SOLVE PLATE LOADS/STRESSES   Version 1.4  12/01/86

       COOLING TOWER VIBRATION - NODE INCREASE

Mode  1 -
```

GLoads Stress Stress	Node Surf	Fx Sigma X Shear XZ	Fy Sigma Y Shear YZ	Fz Tau XY	Mx Sigma 1	My Sigma 2	Mz Angle
Stress	TOP	.6961E+01	.6757E+01	-.2258E+02	.2944E+02	-.1572E+02	-44.9
Stress	MID	.2480E+01	.1074E+02	-.2064E+02	.2766E+02	-.1443E+02	-50.7
Stress	BOT	-.2001E+01	.1473E+02	-.1869E+02	.2684E+02	-.1411E+02	-57.1
				PLATE 25			
Loads	37	-.1940E+04	-.6959E+03	-.9791E+03	.1484E+04	-.1524E+04	.2416E+03
Loads	38	-.1610E+04	-.1050E+05	-.3990E+04	-.5259E+04	.3559E+04	.8160E+03
Loads	48	.1822E+04	.2114E+04	.1445E+04	-.6233E+03	.4698E+04	.1105E+04
Loads	47	.1728E+04	.9079E+04	.3525E+04	-.1836E+03	-.3810E+04	-.1222E+04
Stress	TOP	.9836E+01	.4103E+02	-.1986E+02	.5069E+02	.1806E+00	-64.1
Stress	MID	.1993E+01	.4298E+02	-.1972E+02	.5092E+02	-.5954E+01	-68.0
Stress	BOT	-.5850E+01	.4492E+02	-.1958E+02	.5160E+02	-.1252E+02	-71.2
				PLATE 26			
Loads	38	-.1448E+04	-.6445E+04	-.2047E+04	.8947E+03	-.3180E+04	-.6465E+03
Loads	39	-.8485E+03	-.1395E+05	-.4239E+04	-.3562E+03	.4310E+04	.1111E+04
Loads	49	.1162E+04	.7871E+04	.2647E+04	-.2789E+03	.3731E+04	.9655E+03
Loads	48	.1135E+04	.1252E+05	.3639E+04	.4536E+03	-.5043E+04	-.1290E+04
Stress	TOP	.9681E+01	.7778E+02	-.1317E+02	.8024E+02	.7222E+01	-79.4
Stress	MID	.2356E+00	.7830E+02	-.1426E+02	.8082E+02	-.2288E+01	-80.0
Stress	BOT	-.9210E+01	.7881E+02	-.1535E+02	.8141E+02	-.1181E+02	-80.4
				PLATE 27			
Loads	39	-.4289E+03	-.1299E+05	-.3589E+04	-.4532E+02	-.3359E+04	-.8853E+03
Loads	40	-.3638E+03	-.1695E+05	-.4487E+04	.2064E+04	.7637E+03	.3681E+03
Loads	50	.3033E+03	.1453E+05	.4206E+04	.3806E+04	-.7873E+03	-.2114E+03
Loads	49	.4894E+03	.1541E+05	.3870E+04	.1234E+04	-.4340E+04	-.1044E+04
Stress	TOP	.4670E+01	.1145E+03	-.3984E+01	.1146E+03	.4526E+01	-87.9
Stress	MID	.1844E+00	.1149E+03	-.5539E+01	.1152E+03	-.8236E-01	-87.2
Stress	BOT	-.4301E+01	.1153E+03	-.7094E+01	.1157E+03	-.4721E+01	-86.6
				PLATE 28			
Loads	41	-.1780E+04	.8044E+04	.1222E+03	.9762E+03	-.3103E+04	.3066E+04
Loads	42	-.2198E+04	.6426E+04	-.5709E+03	.8993E+03	-.1692E+04	.3568E+04
Loads	52	.2202E+04	-.7863E+04	-.2241E+03	.7820E+02	-.7541E+03	-.3006E+04
Loads	51	.1775E+04	-.6608E+04	.6728E+03	.1238E+03	-.6812E+03	.1458E+04

```
IMAGES-3D   s/n:800189                              03-02-87
                                                   PAGE    8
            =============== I M A G E S  3 D ===============
            = Copyright (c) 1984   Celestial Software Inc. =
            ================================================

            SOLVE PLATE LOADS/STRESSES   Version 1.4  12/01/86

        COOLING TOWER VIBRATION - NODE INCREASE

Mode  1 -

GLoads Node    Fx          Fy          Fz          Mx          My          Mz
Stress Surf  Sigma  X    Sigma  Y    Tau XY      Sigma  1    Sigma  2     Angle
Stress       Shear XZ    Shear YZ
------ ----  ----------  ----------  ----------  ----------  ----------  ----------
Stress TOP   .2534E+01  -.6830E+02  -.8267E+01   .3486E+01  -.6926E+02     -6.6
Stress MID   .1777E+01  -.6004E+02  -.3178E+01   .1940E+01  -.6020E+02     -2.9
Stress BOT   .1020E+01  -.5177E+02   .1911E+01   .1089E+01  -.5184E+02      2.1
                            ***PLATE    29***
Loads   42  -.2141E+04   .8920E+04  -.1737E+03   .2914E+03  -.8895E+02   .1002E+04
Loads   43  -.1812E+04   .4538E+04  -.1189E+04   .7319E+03  -.4187E+03   .2312E+04
Loads   53   .2614E+04  -.8543E+04  -.1961E+02  -.5318E+03  -.2631E+03  -.2258E+04
Loads   52   .1340E+04  -.4915E+04   .1382E+04  -.1010E+04   .1126E+04  -.2635E+04
Stress TOP   .1173E+01  -.6427E+02  -.1383E+02   .3976E+01  -.6707E+02    -11.5
Stress MID   .2155E+01  -.5565E+02  -.9367E+01   .3635E+01  -.5713E+02     -9.0
Stress BOT   .3138E+01  -.4704E+02  -.4904E+01   .3613E+01  -.4751E+02     -5.5
                            ***PLATE    30***
Loads   43  -.2465E+04   .9089E+04  -.2649E+03   .6436E+03   .2247E+03   .1524E+04
Loads   44  -.1788E+04   .2157E+04  -.1877E+04   .1018E+04   .8852E+02   .2242E+04
Loads   54   .2860E+04  -.8550E+04   .4328E+01  -.9096E+03   .4360E+03  -.1685E+04
Loads   53   .1392E+04  -.2696E+04   .2138E+04  -.8833E+03   .6351E+03  -.1506E+04
Stress TOP   .2607E+01  -.5433E+02  -.2038E+02   .9151E+01  -.6088E+02    -17.8
Stress MID   .2796E+01  -.4656E+02  -.1497E+02   .6982E+01  -.5074E+02    -15.6
Stress BOT   .2985E+01  -.3879E+02  -.9560E+01   .5069E+01  -.4087E+02    -12.3
                            ***PLATE    31***
Loads   44  -.2612E+04   .8425E+04  -.2743E+03   .9182E+03  -.4669E+02   .1294E+04
Loads   45  -.1860E+04  -.5647E+03  -.2495E+04   .1046E+04   .9187E+03   .1916E+04
Loads   55   .2947E+04  -.7685E+04  -.1342E+02  -.1274E+04   .2270E+04  -.7898E+03
Loads   54   .1525E+04  -.1753E+03   .2783E+04  -.8469E+03  -.1906E+03  -.1300E+04
Stress TOP   .5186E+01  -.3924E+02  -.2471E+02   .1620E+02  -.5025E+02    -24.0
Stress MID   .3220E+01  -.3255E+02  -.1936E+02   .1169E+02  -.4102E+02    -23.6
Stress BOT   .1253E+01  -.2586E+02  -.1400E+02   .7187E+01  -.3180E+02    -23.0
                            ***PLATE    32***
Loads   45  -.2617E+04   .6760E+04  -.3048E+03   .1213E+04  -.8463E+03   .9037E+03
Loads   46  -.1891E+04  -.3551E+04  -.2982E+04   .6937E+03   .2352E+04   .1565E+04
Loads   56   .2803E+04  -.5844E+04   .9616E+02  -.1454E+04   .4589E+04   .2360E+03
Loads   55   .1705E+04   .2634E+04   .3191E+04  -.5782E+03  -.1940E+04  -.1297E+04
```

COOLING TOWER VIBRATION: INCREASED NODE MODEL

```
IMAGES-3D  s/n:800189                                    03-02-87
                                                         PAGE    9
            ================ I M A G E S  3 D ================
            = Copyright (c) 1984  Celestial Software Inc. =
            ================================================

            SOLVE PLATE LOADS/STRESSES   Version 1.4  12/01/86

        COOLING TOWER VIBRATION - NODE INCREASE

Mode  1 -
```

GLoads	Node	Fx	Fy	Fz	Mx	My	Mz
Stress	Surf	Sigma X	Sigma Y	Tau XY	Sigma 1	Sigma 2	Angle
Stress		Shear XZ	Shear YZ				
Stress	TOP	.8737E+01	-.1859E+02	-.2640E+02	.2480E+02	-.3465E+02	-31.3
Stress	MID	.3143E+01	-.1331E+02	-.2207E+02	.1847E+02	-.2864E+02	-34.8
Stress	BOT	-.2451E+01	-.8023E+01	-.1775E+02	.1273E+02	-.2320E+02	-40.5
PLATE 33							
Loads	46	-.2455E+04	.3979E+04	-.4669E+03	.1385E+04	-.2214E+04	.2673E+03
Loads	47	-.1759E+04	-.6653E+04	-.3310E+04	.1284E+03	.4068E+04	.1389E+04
Loads	57	.2438E+04	-.2920E+04	.4644E+03	-.1375E+04	.6575E+04	.1129E+04
Loads	56	.1776E+04	.5593E+04	.3313E+04	-.1532E+03	-.4307E+04	-.1483E+04
Stress	TOP	.1232E+02	.7395E+01	-.2491E+02	.3489E+02	-.1517E+02	-42.2
Stress	MID	.2423E+01	.1104E+02	-.2250E+02	.2964E+02	-.1617E+02	-50.4
Stress	BOT	-.7471E+01	.1469E+02	-.2009E+02	.2656E+02	-.1933E+02	-59.4
PLATE 34							
Loads	47	-.2081E+04	.2563E+02	-.8829E+03	.1188E+04	-.3838E+04	-.5484E+03
Loads	48	-.1373E+04	-.9591E+04	-.3471E+04	-.2696E+03	.5176E+04	.1369E+04
Loads	58	.1869E+04	.1109E+04	.1154E+04	-.1060E+04	.7014E+04	.1522E+04
Loads	57	.1585E+04	.8457E+04	.3200E+04	.2597E+04	-.6458E+04	-.1742E+04
Stress	TOP	.1418E+02	.3758E+02	-.2005E+02	.4909E+02	.2664E+01	-60.1
Stress	MID	.1186E+02	.3958E+02	-.1990E+02	.4804E+02	-.7270E+01	-67.0
Stress	BOT	-.1181E+02	.4158E+02	-.1976E+02	.4810E+02	-.1832E+02	-71.7
PLATE 35							
Loads	48	-.1393E+04	-.4931E+04	-.1649E+04	.4393E+03	-.4831E+04	-.1184E+04
Loads	49	-.8994E+03	-.1229E+05	-.3631E+04	-.3620E+03	.4228E+04	.1051E+04
Loads	59	.1147E+04	.6140E+04	.2210E+04	-.7644E+03	.4734E+04	.1070E+04
Loads	58	.1145E+04	.1108E+05	.3071E+04	.4727E+03	-.7187E+04	-.1820E+04
Stress	TOP	.1243E+02	.7076E+02	-.1271E+02	.7341E+02	.9782E+01	-78.2
Stress	MID	.2814E+00	.7129E+02	-.1434E+02	.7408E+02	-.2505E+01	-79.0
Stress	BOT	-.1187E+02	.7182E+02	-.1597E+02	.7476E+02	-.1481E+02	-79.6
PLATE 36							
Loads	49	-.5609E+03	-.1083E+05	-.2908E+04	-.5930E+03	-.3619E+04	-.9722E+03
Loads	50	-.2059E+03	-.1443E+05	-.3778E+04	.3152E+03	.7873E+03	.2114E+03
Loads	60	.3225E+03	.1205E+05	.3609E+04	-.2668E+03	-.2180E+03	-.1041E+03
Loads	59	.4442E+03	.1320E+05	.3077E+04	.6870E+03	-.5233E+04	-.1293E+04

```
IMAGES-3D   s/n:800189                            03-02-87
                                                 PAGE   10
         =============== I M A G E S   3 D ===============
         = Copyright (c) 1984   Celestial Software Inc. =
         ================================================

         SOLVE PLATE LOADS/STRESSES    Version 1.4  12/01/86

      COOLING TOWER VIBRATION - NODE INCREASE

Mode  1 -

GLoads Node     Fx          Fy          Fz          Mx          My          Mz
Stress Surf  Sigma  X    Sigma  Y    Tau XY      Sigma  1    Sigma  2    Angle
Stress       Shear XZ    Shear YZ
------ ----  ----------  ----------  ----------  ----------  ----------  ----------
Stress TOP   .5213E+01   .1052E+03  -.2679E+01   .1052E+03   .5141E+01    -88.5
Stress MID  -.2441E+00   .1045E+03  -.5392E+01   .1048E+03  -.5209E+00    -87.1
Stress BOT  -.5701E+01   .1039E+03  -.8106E+01   .1045E+03  -.6298E+01    -85.8
                          ***PLATE     37***
Loads   51  -.1671E+04   .6581E+04   .4934E+03  -.1238E+03   .6812E+03   .2265E+04
Loads   52  -.1455E+04   .4732E+04  -.1173E+04   .8032E+03  -.9503E+03   .4549E+04
Loads   62   .2547E+04  -.6301E+04  -.4868E+03  -.1164E+04   .2144E+04  -.8933E+04
Loads   61   .5794E+03  -.5012E+04   .1166E+04  -.2836E+04   .1218E+05   .6205E+04
Stress TOP  -.3149E+01  -.6166E+02  -.1186E+02  -.8358E+00  -.6397E+02    -11.0
Stress MID   .3518E+01  -.5092E+02  -.4126E+01   .3829E+01  -.5123E+02     -4.3
Stress BOT   .1018E+02  -.4019E+02   .3610E+01   .1044E+02  -.4045E+02      4.1
                          ***PLATE     38***
Loads   52  -.1860E+04   .8000E+04   .1042E+02   .1290E+03   .5783E+03   .1091E+04
Loads   53  -.1500E+04   .2628E+04  -.1952E+04   .8310E+03  -.9752E+03   .2138E+04
Loads   63   .2272E+04  -.7763E+04  -.8769E+02  -.1133E+03  -.2204E+04  -.2629E+04
Loads   62   .1088E+04  -.2865E+04   .2029E+04  -.7145E+03   .1045E+04  -.1675E+04
Stress TOP   .1239E+01  -.5612E+02  -.1694E+02   .5867E+01  -.6075E+02    -15.3
Stress MID   .3990E+01  -.4770E+02  -.1193E+02   .6610E+01  -.5032E+02    -12.4
Stress BOT   .6742E+01  -.3928E+02  -.6918E+01   .7760E+01  -.4030E+02     -8.4
                          ***PLATE     39***
Loads   53  -.2261E+04   .8579E+04  -.1791E+04   .5841E+03   .6032E+03   .1626E+04
Loads   54  -.1634E+04   .4813E+03  -.2554E+04   .7493E+03   .2990E+03   .1793E+04
Loads   64   .2828E+04  -.8122E+04  -.3426E+02  -.5807E+03   .1486E+03  -.1153E+04
Loads   63   .1067E+04  -.9381E+03   .2768E+04  -.1219E+04   .1697E+04  -.1570E+04
Stress TOP   .3108E+01  -.4819E+02  -.2424E+02   .1275E+02  -.5783E+02    -21.7
Stress MID   .4152E+01  -.4071E+02  -.1792E+02   .1043E+02  -.4698E+02    -19.3
Stress BOT   .5196E+01  -.3322E+02  -.1160E+02   .8427E+01  -.3645E+02    -15.6
                          ***PLATE     40***
Loads   54  -.2490E+04   .8233E+04  -.2561E+04   .1007E+04  -.5444E+03   .1192E+04
Loads   55  -.1812E+04  -.1802E+04  -.2933E+04   .3732E+03   .2268E+04   .1565E+04
Loads   65   .3079E+04  -.7582E+04  -.1017E+03  -.8088E+03   .3681E+04   .5176E+03
Loads   64   .1223E+04   .1151E+04   .3290E+04  -.1329E+04  -.9817E+02  -.1944E+04
```

COOLING TOWER VIBRATION: INCREASED NODE MODEL

```
IMAGES-3D   s/n:800189                                    03-02-87
                                                          PAGE    11
          =============== I M A G E S  3 D ===============
          = Copyright (c) 1984   Celestial Software Inc. =
          =================================================

          SOLVE PLATE LOADS/STRESSES   Version 1.4  12/01/86

     COOLING TOWER VIBRATION - NODE INCREASE

Mode  1 -
```

GLoads	Node	Fx	Fy	Fz	Mx	My	Mz
Stress	Surf	Sigma X	Sigma Y	Tau XY	Sigma 1	Sigma 2	Angle
Stress		Shear XZ	Shear YZ				
Stress	TOP	.7783E+01	-.3481E+02	-.2857E+02	.2212E+02	-.4915E+02	-26.6
Stress	MID	.3965E+01	-.2890E+02	-.2208E+02	.1505E+02	-.3999E+02	-26.7
Stress	BOT	.1475E+00	-.2299E+02	.1558E+02	.7988E+01	-.3083E+02	-26.7

```
                       ***PLATE    41***
```

Loads	55	-.2566E+04	.6868E+04	-.2801E+03	.1479E+04	-.2597E+04	.5223E+03
Loads	56	-.1876E+04	-.4131E+04	-.3091E+04	-.1223E+03	.4854E+04	.1668E+04
Loads	66	.2984E+04	-.6061E+04	-.1154E+03	-.6874E+03	.7668E+04	.2139E+04
Loads	65	.1457E+04	.3323E+04	.3487E+04	-.7866E+03	-.3455E+04	-.2062E+04
Stress	TOP	.1407E+02	-.1613E+02	-.2949E+02	.3211E+02	-.3416E+02	-31.4
Stress	MID	.3344E+01	-.1234E+02	-.2401E+02	.2076E+02	-.2976E+02	-36.0
Stress	BOT	-.7386E+01	-.8551E+01	-.1852E+02	.1056E+02	-.2650E+02	-44.1

```
                       ***PLATE    42***
```

Loads	56	-.2420E+04	.4431E+04	-.3633E+03	.1730E+04	-.5135E+04	-.4217E+03
Loads	57	-.1729E+04	-.6405E+04	-.3086E+04	-.4039E+03	.7118E+04	.1984E+04
Loads	67	.2551E+04	-.3523E+04	.1108E+03	-.3149E+03	.1079E+05	.3211E+04
Loads	66	.1598E+04	.5497E+04	.3339E+04	.2088E+04	-.7423E+04	-.2218E+04
Stress	TOP	.1992E+02	.7172E+01	-.2668E+02	.4098E+02	-.1388E+02	-38.3
Stress	MID	.2324E+01	.8788E+01	-.2337E+02	.2915E+02	-.1803E+02	-48.9
Stress	BOT	-.1527E+02	.1040E+02	-.2006E+02	.2138E+02	-.2625E+02	-61.3

```
                       ***PLATE    43***
```

Loads	57	-.2008E+04	.9586E+03	-.6287E+03	.1519E+04	-.7235E+04	-.1371E+04
Loads	58	-.1371E+04	-.8558E+04	-.3040E+04	-.2479E+03	.7746E+04	.2115E+04
Loads	68	.1864E+04	.4042E+01	.7313E+03	-.1038E+02	.1141E+05	.3274E+04
Loads	67	.1514E+04	.7595E+04	.2937E+04	.1351E+04	-.1073E+05	-.2458E+04
Stress	TOP	.2263E+02	.3416E+02	-.2041E+02	.4960E+02	.7182E+01	-52.9
Stress	MID	.1182E+01	.3404E+02	-.2004E+02	.4353E+02	-.8306E+01	-64.7
Stress	BOT	-.2026E+02	.3392E+02	-.1968E+02	.4031E+02	-.2665E+02	-72.0

```
                       ***PLATE    44***
```

Loads	58	-.1358E+04	-.3496E+04	-.1230E+04	.8351E+03	-.7573E+04	-.1818E+04
Loads	59	-.7990E+03	-.1044E+05	-.3035E+04	.8518E+02	.5583E+04	.1532E+04
Loads	69	.1036E+04	.4449E+04	.1839E+04	-.3285E+03	.7915E+04	.2045E+04
Loads	68	.1121E+04	.9486E+04	.2426E+04	.2295E+04	-.1179E+05	-.2570E+04

IMAGES-3D   s/n:800189   03-02-87
                                                          PAGE   12
                ================ I M A G E S   3 D ================
                = Copyright (c) 1984   Celestial Software Inc. =
                ==

                SOLVE PLATE LOADS/STRESSES   Version 1.4 12/01/86

          COOLING TOWER VIBRATION - NODE INCREASE

Mode 1 -

GLoads	Node	Fx	Fy	Fz	Mx	My	Mz
Stress	Surf	Sigma X	Sigma Y	Tau XY	Sigma 1	Sigma 2	Angle
Stress		Shear XZ	Shear YZ				
Stress	TOP	.1914E+02	.6355E+02	-.1124E+02	.6624E+02	.1646E+02	-76.6
Stress	MID	.1668E+00	.6251E+02	-.1385E+02	.6545E+02	-.2773E+01	-78.0
Stress	BOT	-.1881E+02	.6146E+02	-.1646E+02	.6471E+02	-.2206E+02	-78.8
		***PLATE	45***				
Loads	59	-.5094E+03	-.8714E+04	-.2278E+04	-.7776E+01	-.5083E+04	-.1310E+04
Loads	60	-.1799E+03	-.1193E+05	-.3134E+04	.7350E+03	.2180E+03	.1041E+03
Loads	70	.2165E+03	.9584E+04	.3482E+04	.1283E+04	-.2086E+04	-.6780E+03
Loads	69	.4728E+03	.1106E+05	.1929E+04	.2796E+04	-.9457E+04	-.2215E+04
Stress	TOP	.7153E+01	.9338E+02	-.1681E+00	.9338E+02	.7153E+01	-89.9
Stress	MID	-.1684E+00	.9267E+02	-.5034E+01	.9294E+02	-.4406E+00	-86.9
Stress	BOT	-.7490E+01	.9197E+02	-.9900E+01	.9294E+02	-.8466E+01	-84.4
		***PLATE	46***				
Loads	61	-.4359E+03	.4990E+04	.6041E+03	.2836E+04	-.1218E+05	.1682E+04
Loads	62	-.2036E+04	.2310E+04	-.1560E+04	.2440E+04	-.5883E+04	.1030E+05
Loads	72	.3942E+04	-.4327E+04	-.7536E+03	-.2483E+04	.6694E+04	-.1195E+05
Loads	71	-.1469E+04	-.2973E+04	.1710E+04	-.6479E+04	.3596E+05	.2416E+05
Stress	TOP	-.7709E+01	-.3713E+02	-.2193E+02	.3990E+01	-.4883E+02	-28.1
Stress	MID	.4862E+01	-.3618E+02	-.5522E+01	.5592E+01	-.3691E+02	-7.5
Stress	BOT	.1743E+02	-.3523E+02	.1089E+02	.1959E+02	-.3739E+02	11.2
		***PLATE	47***				
Loads	62	-.1290E+04	.6820E+04	.1379E+02	-.5614E+03	.2694E+04	.3041E+03
Loads	63	-.1124E+04	.8312E+03	-.2331E+04	.1048E+04	-.1762E+04	.2333E+04
Loads	73	.1809E+04	-.6674E+04	.4956E+02	-.2876E+04	-.3264E+04	-.3028E+04
Loads	72	.6058E+03	-.9772E+03	.2268E+04	-.5205E+03	-.5468E+02	-.2009E+04
Stress	TOP	.1234E+00	-.4620E+02	-.2035E+02	.7790E+01	-.5387E+02	-20.6
Stress	MID	.4611E+01	-.3710E+02	-.1354E+02	.8622E+01	-.4111E+02	-16.5
Stress	BOT	.9099E+01	-.2799E+02	-.6736E+01	.1028E+02	-.2918E+02	-10.0
		***PLATE	48***				
Loads	63	-.1885E+04	.7849E+04	-.3611E+03	.2836E+03	.2269E+04	.1865E+04
Loads	64	-.1315E+04	-.9970E+03	-.2820E+04	.5232E+03	.1119E+03	.1183E+04
Loads	74	.2707E+04	-.7518E+04	.1854E+03	-.4869E+03	.4895E+03	-.7731E+03
Loads	73	.4930E+03	.6656E+03	.2995E+04	-.2338E+04	.2741E+04	-.3482E+04

COOLING TOWER VIBRATION: INCREASED NODE MODEL

```
IMAGES-3D   s/n:800189                                          03-02-87
                                                               PAGE   13
          =============== I M A G E S  3 D ===============
          = Copyright (c) 1984   Celestial Software Inc. =
          ================================================

          SOLVE PLATE LOADS/STRESSES   Version 1.4  12/01/86

       COOLING TOWER VIBRATION - NODE INCREASE


Mode  1 -

GLoads Node     Fx           Fy          Fz          Mx          My          Mz
Stress Surf  Sigma  X     Sigma  Y     Tau XY     Sigma  1    Sigma  2     Angle
Stress       Shear XZ     Shear YZ
------ ----  ----------   ----------  ----------  ----------  ----------  ----------
Stress TOP   .1532E+01  -.4289E+02  -.2725E+02   .1447E+02  -.5583E+02     -25.4
Stress MID   .4095E+01  -.3323E+02  -.2003E+02   .1281E+02  -.4194E+02     -23.5
Stress BOT   .6658E+01  -.2356E+02  -.1281E+02   .1136E+02  -.2826E+02     -20.1
                            ***PLATE    49***
Loads   64 -.2385E+04   .7969E+04  -.4612E+03   .1386E+04  -.1622E+03   .1914E+04
Loads   65 -.1537E+04  -.2745E+04  -.3026E+04  -.6565E+03   .3892E+04   .6589E+03
Loads   75  .3087E+04  -.7506E+04   .1653E+03  -.1729E+03   .5124E+04   .1851E+04
Loads   74  .8353E+03   .2281E+04   .3322E+04  -.2365E+04  -.1078E+04  -.3821E+04
Stress TOP  .9233E+01  -.3199E+02  -.3075E+02   .2564E+02  -.4840E+02     -28.1
Stress MID  .3312E+01  -.2536E+02  -.2415E+02   .1706E+02  -.3911E+02     -29.7
Stress BOT -.2610E+01  -.1874E+02  -.1754E+02   .8633E+01  -.2998E+02     -32.7
                            ***PLATE    50***
Loads   65 -.2631E+04   .7034E+04  -.3984E+03   .2252E+04  -.4118E+04   .8852E+03
Loads   66 -.1646E+04  -.4380E+04  -.2954E+04  -.1901E+04   .8328E+04   .8693E+03
Loads   76  .3156E+04  -.6436E+04   .6841E+02   .5995E+03   .1077E+05   .4178E+04
Loads   75  .1122E+04   .3782E+04   .3284E+04  -.1656E+04  -.6053E+04  -.3669E+04
Stress TOP  .1916E+02  -.1514E+02  -.3067E+02   .3715E+02  -.3313E+02     -30.4
Stress MID  .2217E+01  -.1293E+02  -.2551E+02   .2126E+02  -.3197E+02     -36.7
Stress BOT -.1472E+02  -.1071E+02  -.2035E+02   .7736E+01  -.3317E+02     -47.8
                            ***PLATE    51***
Loads   66 -.2556E+04   .5011E+04  -.3220E+03   .2379E+04  -.8574E+04  -.7910E+03
Loads   67 -.1536E+04  -.5900E+04  -.2699E+04  -.2528E+04   .1200E+05   .1675E+04
Loads   77  .2822E+04  -.4306E+04   .1360E+03   .1570E+04   .1536E+05   .5502E+04
Loads   76  .1270E+04   .5195E+04   .2885E+04   .8084E+01  -.1179E+05  -.3377E+04
Stress TOP  .2858E+02   .6762E+01  -.2686E+02   .4666E+02  -.1132E+02     -33.9
Stress MID  .9544E+00   .4230E+01  -.2404E+02   .2669E+02  -.2151E+02     -46.9
Stress BOT -.2667E+02   .1698E+01  -.2123E+02   .1305E+02  -.3802E+02     -61.9
                            ***PLATE    52***
Loads   67 -.2145E+04   .1938E+04  -.4075E+03   .1492E+04  -.1206E+05  -.2428E+04
Loads   68 -.1171E+04  -.7290E+04  -.2394E+04  -.2017E+04   .1315E+05   .2471E+04
Loads   78  .2141E+04  -.1171E+04   .5553E+03   .1952E+04   .1678E+05   .5258E+04
Loads   77  .1175E+04   .6523E+04   .2247E+04   .2320E+04  -.1643E+05  -.3179E+04
```

STRUCTURAL MODELS

```
IMAGES-3D   s/n:800189
                                                              03-02-87
          ================ I M A G E S   3 D ================   PAGE   14
          = Copyright (c) 1984   Celestial Software Inc. =
          ================================================

          SOLVE PLATE LOADS/STRESSES   Version 1.4  12/01/86

     COOLING TOWER VIBRATION - NODE INCREASE

Mode  1 -

GLoads Node     Fx          Fy          Fz          Mx          My          Mz
Stress Surf   Sigma  X    Sigma  Y    Tau  XY     Sigma  1    Sigma  2    Angle
Stress        Shear XZ    Shear YZ
------ ----  ----------  ----------  ----------  ----------  ----------  ----------
Stress TOP   .3359E+02   .3206E+02  -.1964E+02   .5248E+02   .1316E+02     -43.9
Stress MID  -.2246E+00   .2587E+02  -.1981E+02   .3654E+02  -.1090E+02     -61.7
Stress BOT  -.3404E+02   .1968E+02  -.1998E+02   .2630E+02  -.4066E+02     -71.7
                      ***PLATE    53***
Loads   68  -.1432E+04  -.2043E+04  -.8157E+03  -.2679E+03  -.1277E+05  -.3175E+04
Loads   69  -.6341E+03  -.8510E+04  -.2176E+04  -.3702E+03   .9908E+04   .2313E+04
Loads   79   .1238E+04   .2810E+04   .1466E+04   .6790E+03   .1272E+05   .3275E+04
Loads   78   .8273E+03   .7743E+04   .1525E+04  -.1783E+05  -.3065E+04
Stress TOP   .2989E+02   .5843E+02  -.9854E+01   .6150E+02   .2682E+02     -72.7
Stress MID  -.9082E+00   .5113E+02  -.1310E+02   .5424E+02  -.4019E+01     -76.6
Stress BOT  -.3171E+02   .4382E+02  -.1634E+02   .4721E+02  -.3509E+02     -78.3
                      ***PLATE    54***
Loads   69  -.4962E+03  -.6791E+04  -.1623E+04  -.2097E+04  -.8366E+04  -.2143E+04
Loads   70  -.2689E+02  -.9451E+04  -.2136E+04   .2335E+04   .2086E+04   .6780E+03
Loads   80   .2127E+03   .7455E+04   .3089E+04  -.2315E+04  -.2172E+04  -.7187E+03
Loads   79   .3104E+03   .8788E+04   .6708E+03   .6715E+04  -.1404E+05  -.2715E+04
Stress TOP   .1233E+02   .8351E+02   .1750E+01   .8356E+02   .1229E+02      88.6
Stress MID  -.6222E+00   .7878E+02  -.4014E+01   .7898E+02  -.8246E+00     -87.1
Stress BOT  -.1358E+02   .7404E+02  -.9778E+01   .7511E+02  -.1465E+02     -83.7
                      ***PLATE    55***
Loads   71   .1650E+04   .2959E+04   .8239E+03   .6479E+04  -.3596E+05   .2137E+05
Loads   72  -.3560E+04   .2629E+03  -.1679E+04   .2547E+04  -.1211E+05   .9512E+04
Loads   82   .3584E+04  -.2483E+04  -.1123E+04  -.9186E+03   .3848E+04  -.6397E+04
Loads   81  -.1674E+04  -.7388E+03   .1978E+04  -.3392E+04   .4176E+05   .2570E+05
Stress TOP  -.8969E+00  -.3306E+02  -.2907E+02   .1624E+02  -.5020E+02     -30.5
Stress MID   .6446E+01  -.1735E+02  -.5149E+01   .7513E+01  -.1842E+02     -11.7
Stress BOT   .1379E+02  -.1637E+01   .1877E+02   .2637E+02  -.1422E+02      33.8
                      ***PLATE    56***
Loads   72  -.6007E+03   .5020E+04   .1607E+04   .4564E+03   .5473E+04   .4451E+04
Loads   73  -.7926E+03  -.7061E+03  -.2566E+04   .9458E+03  -.1423E+04   .2858E+04
Loads   83   .6761E+03  -.5150E+04  -.1739E+03   .3766E+03  -.3727E+04  -.4343E+03
Loads   82   .7172E+03   .8358E+03   .2579E+04   .5545E+02  -.3049E+04  -.1402E+04
```

COOLING TOWER VIBRATION: INCREASED NODE MODEL

```
IMAGES-3D  s/n:800189                                    03-02-87
                                                         PAGE   15
        =============== I M A G E S  3 D ===============
        = Copyright (c) 1984   Celestial Software Inc. =
        ================================================

            SOLVE PLATE LOADS/STRESSES   Version 1.4  12/01/86

        COOLING TOWER VIBRATION - NODE INCREASE

Mode  1 -
```

GLoads	Node	Fx	Fy	Fz	Mx	My	Mz
Stress	Surf	Sigma X	Sigma Y	Tau XY	Sigma 1	Sigma 2	Angle
Stress		Shear XZ	Shear YZ				
Stress	TOP	.1686E+01	-.3356E+02	-.2055E+02	.1113E+02	-.4300E+02	-24.7
Stress	MID	.6179E+01	-.2211E+02	-.1357E+02	.1164E+02	-.2757E+02	-21.9
Stress	BOT	.1067E+02	-.1066E+02	-.6597E+01	.1255E+02	-.1253E+02	-15.9
			***PLATE	57***			
Loads	73	-.1097E+04	.6708E+04	-.4915E+03	.1420E+04	.1947E+04	.3652E+04
Loads	74	-.1245E+04	-.2040E+04	-.2855E+04	.4106E+03	.1145E+04	.1233E+04
Loads	84	.1667E+04	-.6647E+04	.1143E+03	.2381E+03	.2217E+04	.1190E+04
Loads	83	.6750E+03	.1980E+04	.3232E+04	-.1309E+04	.1027E+04	-.2490E+04
Stress	TOP	.5513E+01	-.3248E+02	-.2661E+02	.1921E+02	-.4618E+02	-27.2
Stress	MID	.5315E+01	-.2387E+02	-.2034E+02	.1576E+02	-.3431E+02	-27.2
Stress	BOT	.5118E+01	-.1525E+02	-.1407E+02	.1230E+02	-.2244E+02	-27.1
			***PLATE	58***			
Loads	74	-.1859E+04	.7292E+04	-.6786E+03	.2441E+04	-.5559E+03	.3362E+04
Loads	75	-.1379E+04	-.3070E+04	-.2831E+04	-.1124E+04	.6866E+04	-.4209E+02
Loads	85	.2107E+04	-.7121E+04	.3686E+03	.8523E+03	.8075E+04	.3052E+04
Loads	84	.1131E+04	.2900E+04	.3141E+04	-.1494E+04	-.3525E+04	-.2937E+04
Stress	TOP	.1470E+02	-.2631E+02	-.2835E+02	.2918E+02	-.4079E+02	-27.1
Stress	MID	.3685E+01	-.2159E+02	-.2393E+02	.1811E+02	-.3602E+02	-31.1
Stress	BOT	-.7327E+01	-.1687E+02	-.1952E+02	.7996E+01	-.3219E+02	-38.1
			***PLATE	59***			
Loads	75	-.2370E+04	.6838E+04	-.6601E+03	.2952E+04	-.5937E+04	.1859E+04
Loads	76	-.1395E+04	-.4005E+04	-.2500E+04	-.2886E+04	.1289E+05	-.5075E+03
Loads	86	.2380E+04	-.6532E+04	.4997E+03	.1512E+04	.1442E+05	.4169E+04
Loads	85	.1385E+04	.3699E+04	.2660E+04	-.1190E+04	-.9495E+04	-.2942E+04
Stress	TOP	.2646E+02	-.1436E+02	-.2697E+02	.3986E+02	-.2777E+02	-26.4
Stress	MID	.1650E+01	-.1449E+02	-.2476E+02	.1963E+02	-.3246E+02	-36.0
Stress	BOT	-.2316E+02	-.1461E+02	-.2256E+02	.4075E+01	-.4185E+02	-50.4
			***PLATE	60***			
Loads	76	-.2558E+04	.5326E+04	-.5111E+03	·.2278E+04	-.1188E+05	-.2931E+03
Loads	77	-.1205E+04	-.4894E+04	-.2029E+04	-.4013E+04	.1770E+05	.9109E+01
Loads	87	.2339E+04	-.4887E+04	.6005E+03	.1725E+04	.1895E+05	.4221E+04
Loads	86	.1424E+04	.4455E+04	.1940E+04	-.4155E+02	-.1587E+05	-.2557E+04

STRUCTURAL MODELS   173

```
=============== I M A G E S  3 D ===============
= Copyright (c) 1984   Celestial Software Inc. =
================================================
```

SOLVE PLATE LOADS/STRESSES   Version 1.4   12/01/86

COOLING TOWER VIBRATION - NODE INCREASE

Mode 1 -

GLoads	Node	Fx	Fy	Fz	Mx	My	Mz
Stress	Surf	Sigma X	Sigma Y	Tau XY	Sigma 1	Sigma 2	Angle
Stress		Shear XZ	Shear YZ				
Stress	TOP	.3715E+02	.2737E+01	-.2286E+02	.4855E+02	-.8667E+01	-26.5
Stress	MID	-.2591E+00	-.2215E+01	-.2296E+02	.2174E+02	-.2422E+02	-43.8
Stress	BOT	-.3766E+02	-.7168E+01	-.2306E+02	.5226E+01	-.5006E+02	-61.7
			***PLATE	61***			
Loads	77	-.2314E+04	.2800E+04	-.4175E+03	.1237E+03	-.1663E+05	-.2333E+04
Loads	78	-.8613E+03	-.5766E+04	-.1512E+04	-.3582E+04	.1911E+05	.1081E+04
Loads	88	.1951E+04	-.2230E+04	.7668E+03	.9118E+03	.1952E+05	.3231E+04
Loads	87	.1224E+04	.5196E+04	.1163E+04	.1788E+04	-.2032E+05	-.2092E+04
Stress	TOP	.4237E+02	.2347E+02	-.1669E+02	.5210E+02	.1374E+02	-30.2
Stress	MID	-.1562E+01	.1514E+02	-.1868E+02	.2725E+02	-.1367E+02	-57.0
Stress	BOT	-.4550E+02	.6818E+01	-.2067E+02	.1400E+02	-.5268E+02	-70.8
			***PLATE	62***			
Loads	78	-.1631E+04	-.6367E+03	-.6259E+04	-.3131E+04	-.1806E+05	-.3274E+04
Loads	79	-.4356E+03	-.6624E+04	-.9549E+03	-.8452E+03	.1440E+05	.1719E+04
Loads	89	.1239E+04	.1336E+04	.1124E+04	-.1229E+04	.1394E+05	.1549E+04
Loads	88	.8275E+03	.5925E+04	.4572E+03	.3977E+04	-.2052E+05	-.1706E+04
Stress	TOP	.3701E+02	.4571E+02	-.9136E+01	.5148E+02	.3124E+02	-57.7
Stress	MID	-.1906E+01	.3706E+02	-.1216E+02	.4055E+02	-.5390E+01	-74.0
Stress	BOT	-.4082E+02	.2841E+02	-.1519E+02	.3159E+02	-.4401E+02	-78.2
			***PLATE	63***			
Loads	79	-.6411E+03	-.4753E+04	-.1217E+04	-.6549E+04	-.1308E+05	-.2279E+04
Loads	80	.2303E+02	-.7318E+04	-.4103E+03	.5017E+04	.2172E+04	.7187E+03
Loads	90	.3035E+03	.5485E+04	.1816E+04	-.4424E+04	-.6042E+03	-.4696E+03
Loads	89	.3146E+03	.6586E+04	-.1880E+03	.6464E+04	-.1437E+05	-.1313E+04
Stress	TOP	.1586E+02	.6603E+02	-.2296E+00	.6603E+02	.1586E+02	-89.7
Stress	MID	-.1050E+01	.6156E+02	-.3831E+01	.6179E+02	-.1283E+01	-86.5
Stress	BOT	-.1796E+02	.5708E+02	-.7433E+01	.5781E+02	-.1869E+02	-84.4
			***PLATE	64***			
Loads	81	.1899E+04	.7301E+03	.1829E+04	.3392E+04	-.4176E+05	.1998E+05
Loads	82	-.3798E+04	-.1480E+04	-.2727E+04	.1134E+04	-.7408E+04	.7781E+04
Loads	92	.3372E+04	-.9070E+03	-.2773E+04	-.5878E+03	.5758E+04	-.5095E+04
Loads	91	-.1472E+04	.1657E+04	.3671E+04	.1888E+04	.4190E+05	.4318E+05

COOLING TOWER VIBRATION: INCREASED NODE MODEL

```
IMAGES-3D   s/n:800189                              03-02-87
                                                    PAGE   17
=============== I M A G E S   3 D ===============
= Copyright (c) 1984   Celestial Software Inc. =
================================================

        SOLVE PLATE LOADS/STRESSES   Version 1.4  12/01/86

     COOLING TOWER VIBRATION - NODE INCREASE
```

Mode 1 -

GLoads Stress Stress	Node Surf	Fx Sigma X Shear XZ	Fy Sigma Y Shear YZ	Fz Tau XY	Mx Sigma 1	My Sigma 2	Mz Angle
Stress	TOP	.1210E+02	.1674E+02	-.3585E+02	.5035E+02	-.2151E+02	-46.9
Stress	MID	.1282E+02	.3505E+01	-.5542E+01	.1540E+02	.9224E+00	-25.0
Stress	BOT	.1354E+02	-.9734E+01	.2477E+02	.2927E+02	-.2546E+02	32.4
			***PLATE	65***			
Loads	82	-.2221E+02	.3115E+04	.1266E+04	-.2711E+03	.6610E+04	.1825E+02
Loads	83	-.7258E+03	-.1599E+04	-.3390E+04	.3007E+03	.2259E+04	.1528E+04
Loads	93	.5991E+02	-.3429E+04	.1708E+04	-.8027E+03	.1129E+04	-.6610E+02
Loads	92	.6881E+03	.1913E+04	.3832E+04	.7370E+03	-.2730E+04	.2300E+04
Stress	TOP	.1152E+02	-.6872E+01	-.1960E+02	.2398E+02	-.1933E+02	-32.4
Stress	MID	.1185E+02	-.7961E+01	-.1174E+02	.1730E+02	-.1342E+02	-24.9
Stress	BOT	.1217E+02	-.9050E+01	-.3881E+01	.1286E+02	-.9737E+01	-10.0
			***PLATE	66***			
Loads	83	-.1145E+03	.4770E+04	.3188E+04	.6312E+03	.4408E+03	.1396E+04
Loads	84	-.1317E+04	-.1798E+04	-.2942E+04	-.3594E+03	.4506E+04	-.3142E+03
Loads	94	.6980E+03	-.4875E+04	-.6776E+03	.6505E+03	.6788E+04	.2095E+04
Loads	93	.7339E+03	.1903E+04	.3301E+04	-.4333E+03	-.1035E+04	-.1028E+04
Stress	TOP	.1514E+02	-.1561E+02	-.2147E+02	.2617E+02	-.2665E+02	-27.2
Stress	MID	.8236E+01	-.1557E+02	-.1560E+02	.1595E+02	-.2329E+02	-26.3
Stress	BOT	.1336E+01	-.1553E+02	-.9725E+01	.5776E+01	-.1996E+02	-24.5
			***PLATE	67***			
Loads	84	-.9415E+03	.5566E+04	-.3416E+03	.1615E+04	-.3198E+04	.2061E+04
Loads	85	-.1200E+04	-.2158E+04	-.2356E+04	-.1603E+04	.1079E+05	-.1474E+04
Loads	95	.1253E+04	-.5529E+04	.3996E+04	.1738E+04	.1244E+05	.3426E+04
Loads	94	.8880E+03	.2121E+04	.2298E+04	-.1006E+04	-.6835E+04	-.1951E+04
Stress	TOP	.2299E+02	-.1636E+02	-.2081E+02	.3195E+02	-.2532E+02	-23.3
Stress	MID	.3659E+01	-.1786E+02	-.1798E+02	.1385E+02	-.2805E+02	-29.6
Stress	BOT	-.1567E+02	-.1935E+02	-.1515E+02	-.2250E+01	-.3278E+02	-41.5
			***PLATE	68***			
Loads	85	-.1731E+04	.5629E+04	-.7188E+03	.1940E+04	-.9373E+04	.1364E+04
Loads	86	-.9860E+03	-.2724E+04	-.1740E+04	-.2673E+04	.1699E+05	-.1631E+04
Loads	96	.1924E+04	-.5463E+04	.1079E+04	.2597E+04	.1783E+05	.3692E+04
Loads	95	.7930E+03	.2558E+04	.1379E+04	-.8134E+03	-.1311E+05	-.1617E+04

STRUCTURAL MODELS

IMAGES-3D   s/n:800189   03-02-87
                                                PAGE   18
          =============== I M A G E S   3 D ===============
          = Copyright (c) 1984   Celestial Software Inc. =
          ==

          SOLVE PLATE LOADS/STRESSES   Version 1.4 12/01/86

     COOLING TOWER VIBRATION - NODE INCREASE

Mode 1 -

GLoads	Node	Fx	Fy	Fz	Mx	My	Mz
Stress	Surf	Sigma X	Sigma Y	Tau XY	Sigma 1	Sigma 2	Angle
Stress		Shear XZ	Shear YZ				
Stress	TOP	.3288E+02	-.1073E+02	-.1924E+02	.4015E+02	-.1800E+02	-20.7
Stress	MID	-.4485E+00	-.1522E+02	-.1914E+02	.1268E+02	-.2835E+02	-34.4
Stress	BOT	-.3377E+02	-.1972E+02	-.1904E+02	-.6453E+01	-.4704E+02	-55.1
				PLATE 69			
Loads	86	-.2242E+04	.4884E+04	-.7607E+03	.1203E+04	-.1554E+05	.1887E+02
Loads	87	-.7468E+03	-.3398E+04	-.1184E+04	-.2849E+04	.2120E+05	-.8709E+03
Loads	97	.2348E+04	-.4585E+04	.1295E+04	.2723E+04	.2091E+05	.3013E+04
Loads	96	.6402E+03	.3099E+04	.6497E+03	.3880E+03	-.1886E+05	-.7275E+03
Stress	TOP	.4132E+02	-.3401E+00	-.1681E+02	.4726E+02	-.6279E+01	-19.5
Stress	MID	-.3191E+01	-.7792E+01	-.1864E+02	.1329E+02	-.2427E+02	-41.5
Stress	BOT	-.4770E+02	-.1524E+02	-.2046E+02	-.5356E+01	-.5759E+02	-64.2
				PLATE 70			
Loads	87	-.2237E+04	.3215E+04	-.6462E+03	-.6630E+03	-.1983E+05	-.1258E+04
Loads	88	-.5383E+03	-.4095E+04	-.6318E+03	-.1637E+04	.2114E+05	.2503E+03
Loads	98	.2286E+04	-.2767E+04	.1161E+04	.1744E+04	.2013E+05	.1776E+04
Loads	97	.4889E+03	.3647E+04	.1166E+04	.2275E+04	-.2201E+05	.6317E+01
Stress	TOP	.4416E+02	.1363E+02	-.1337E+02	.4918E+02	.8604E+01	-20.6
Stress	MID	-.4199E+01	.4596E+01	-.1598E+02	.1677E+02	-.1637E+02	-52.7
Stress	BOT	-.5256E+02	-.4437E+01	-.1859E+02	.1908E+01	-.5890E+02	-71.2
				PLATE 71			
Loads	88	-.1663E+04	.5734E+03	-.6512E+03	-.3251E+04	-.2014E+05	-.1775E+04
Loads	89	-.3196E+03	-.4746E+04	-.2639E+02	.9839E+03	.1456E+05	.9201E+03
Loads	99	.1617E+04	.1894E+02	.1002E+04	-.6096E+03	.1371E+05	.4499E+03
Loads	98	.3652E+03	.4154E+04	-.3240E+03	.4398E+04	-.2098E+05	.2347E+03
Stress	TOP	.3688E+02	.3001E+02	-.8546E+01	.4265E+02	.2423E+02	-34.1
Stress	MID	-.3501E+01	.2183E+02	-.1093E+02	.2590E+02	-.7567E+01	-69.6
Stress	BOT	-.4388E+02	.1366E+02	-.1332E+02	.1660E+02	-.4681E+02	-77.6
				PLATE 72			
Loads	89	-.6625E+03	-.2950E+04	-.9443E+03	-.6219E+04	-.1413E+05	-.1156E+04
Loads	90	-.1853E+02	-.5345E+04	.5427E+03	.5077E+04	.6042E+03	.4696E+03
Loads	100	.4564E+03	.3674E+04	.1142E+04	-.4674E+04	.2826E+03	-.3979E+03
Loads	99	.2246E+03	.4621E+04	-.7403E+03·	.6610E+04	-.1397E+05	-.3159E+02

COOLING TOWER VIBRATION: INCREASED NODE MODEL

```
IMAGES-3D  s/n:800189                              03-02-87
                                                   PAGE   19
        =============== I M A G E S  3 D ===============
        = Copyright (c) 1984  Celestial Software Inc. =
        ===============================================

        SOLVE PLATE LOADS/STRESSES   Version 1.4  12/01/86

     COOLING TOWER VIBRATION - NODE INCREASE

Mode   1 -

GLoads Node     Fx          Fy          Fz          Mx          My          Mz
Stress Surf  Sigma  X    Sigma  Y    Tau XY      Sigma  1    Sigma  2    Angle
Stress       Shear XZ    Shear YZ
------ ---- ---------- ---------- ---------- ---------- ---------- ----------
Stress TOP   .1551E+02  .4746E+02 -.1895E+01  .4757E+02  .1540E+02    -86.6
Stress MID  -.1360E+01  .4342E+02 -.3731E+01  .4373E+02 -.1668E+01    -85.3
Stress BOT  -.1823E+02  .3938E+02 -.5567E+01  .3991E+02 -.1876E+02    -84.5
                       ***PLATE   73***
Loads   91   .1751E+04 -.1662E+04  .3710E+04 -.1888E+04 -.4190E+05  .3800E+05
Loads   92  -.4486E+04 -.1045E+04 -.3862E+04 -.2066E+03 -.4896E+04  .3487E+04
Loads  102   .3702E+04  .1971E+03 -.4204E+04  .1986E+04  .1120E+05  .6128E+04
Loads  101  -.9670E+03  .2510E+04  .4357E+04  .7846E+04  .3812E+05  .3730E+05
Stress TOP   .2378E+02  .1880E+02 -.3867E+02  .6004E+02 -.1746E+02    -43.2
Stress MID   .1873E+02  .1553E+02 -.2010E+01  .1970E+02  .1456E+02    -25.7
Stress BOT   .1367E+02  .1225E+02  .3465E+02  .4762E+02 -.2169E+02     44.4
                       ***PLATE   74***
Loads   92   .1023E+04  .3143E+02  .2798E+04  .5734E+02  .1867E+04 -.6914E+03
Loads   93  -.9441E+03  .3935E+03 -.2808E+04  .3828E+03  .3585E+04 -.2467E+03
Loads  103  -.2957E+02 -.4180E+03 -.2573E+04  .1138E+04  .7634E+04  .6761E+03
Loads  102  -.4911E+02 -.6928E+01  .2583E+04 -.9320E+03 -.5474E+04 -.9065E+03
Stress TOP   .2135E+02 -.1412E+01 -.1040E+02  .2538E+02 -.5449E+01    -21.2
Stress MID   .1257E+02 -.2222E+01  .5509E-01  .1257E+02 -.2222E+01      .2
Stress BOT   .3783E+01 -.3031E+01  .1051E+02  .1143E+02 -.1067E+02     36.0
                       ***PLATE   75***
Loads   93   .7842E+03  .1134E+04  .1197E+04  .1307E+03 -.3680E+04  .1341E+04
Loads   94  -.6934E+03  .6129E+03 -.1546E+04 -.7513E+02  .7689E+04 -.2380E+04
Loads  104   .5401E+02 -.1290E+04 -.9117E+03  .2432E+04  .1212E+05  .1722E+04
Loads  103  -.1447E+03 -.4562E+03  .1260E+04 -.1764E+04 -.7249E+04 -.1690E+04
Stress TOP   .2374E+02 -.7488E+01 -.7056E+01  .2526E+02 -.9008E+01    -12.2
Stress MID   .5738E+01 -.9073E+01 -.1430E+01  .5875E+01 -.9210E+01     -5.5
Stress BOT  -.1226E+02 -.1066E+02  .4195E+01 -.7190E+01 -.1573E+02     50.4
                       ***PLATE   76***
Loads   94  -.2257E+03  .2161E+04 -.1088E+04  .4303E+03 -.7642E+04  .2236E+04
Loads   95  -.1113E+03  .3024E+01 -.7826E+03 -.4771E+03  .1409E+05 -.3672E+04
Loads  105   .3872E+03 -.2119E+04  .5817E+03  .3454E+04  .1551E+05  .1641E+04
Loads  104  -.5025E+02 -.4507E+02  .3096E+03 -.1881E+04 -.1247E+05 -.4064E+02
```

```
IMAGES-3D   s/n:800189                              03-02-87
                                                   PAGE   20
          =============== I M A G E S   3 D ===============
          = Copyright (c) 1984   Celestial Software Inc. =
          ================================================

          SOLVE PLATE LOADS/STRESSES   Version 1.4   12/01/86

       COOLING TOWER VIBRATION - NODE INCREASE

Mode  1 -


GLoads Node     Fx          Fy          Fz          Mx          My          Mz
Stress Surf   Sigma  X    Sigma  Y    Tau XY      Sigma  1    Sigma  2      Angle
Stress        Shear XZ    Shear YZ
------ ----   ----------  ----------  ----------  ----------  ----------  ----------
Stress  TOP   .2901E+02  -.6819E+01  -.5704E+01   .2989E+02  -.7705E+01      -8.8
Stress  MID  -.1167E-01  -.1121E+02  -.4756E+01   .1736E+01  -.1296E+02     -20.2
Stress  BOT  -.2903E+02  -.1560E+02  -.3809E+01  -.1460E+02  -.3003E+02     -75.2
                            ***PLATE    77***
Loads    95  -.1244E+04   .3016E+04  -.1049E+04  -.4480E+03  -.1342E+05   .1863E+04
Loads    96   .1986E+03  -.9863E+03  -.3252E+03  -.8789E+03   .1939E+05  -.4217E+04
Loads   106   .1118E+04  -.2865E+04   .1555E+04   .3813E+04   .1862E+05   .6071E+03
Loads   105  -.7355E+02   .8351E+03  -.1809E+03  -.1415E+04  -.1630E+05   .1392E+04
Stress  TOP   .3527E+02  -.3175E+01  -.6857E+01   .3646E+02  -.4361E+01      -9.8
Stress  MID  -.4176E+01  -.1052E+02  -.8811E+01   .2015E+01  -.1671E+02     -35.1
Stress  BOT  -.4363E+02  -.1787E+02  -.1076E+02  -.1396E+02   .4753E+02     -70.1
                            ***PLATE    78***
Loads    96  -.2057E+04   .3433E+04  -.1471E+04  -.2106E+04  -.1835E+05   .1252E+04
Loads    97   .2490E+03  -.2030E+04  -.3324E+02  -.9632E+03   .2220E+05  -.3973E+04
Loads   107   .1842E+04  -.3195E+04   .1927E+04   .3185E+04   .1990E+05  -.7266E+03
Loads   106  -.3411E+02   .1791E+04  -.4230E+03  -.5362E+03  -.1942E+05   .2510E+04
Stress  TOP   .3999E+02   .2061E+01  -.9018E+01   .4202E+02   .2580E-01     -12.7
Stress  MID  -.6473E+01  -.7299E+01  -.1207E+02   .5193E+01  -.1896E+02     -44.0
Stress  BOT  -.5294E+02  -.1666E+02  -.1513E+02  -.1118E+02  -.5841E+02     -70.1
                            ***PLATE    79***
Loads    97  -.2377E+04   .3095E+04  -.1450E+04  -.4034E+04  -.2110E+05   .9529E+03
Loads    98   .7634E+02  -.2917E+04   .2822E+03  -.4333E+03   .2090E+05  -.3009E+04
Loads   108   .2184E+04  -.2778E+04   .1700E+04   .1595E+04   .1810E+05  -.1688E+04
Loads   107   .1166E+03   .2600E+04  -.5316E+03   .6376E+03  -.2040E+05   .3037E+04
Stress  TOP   .4014E+02   .8719E+01  -.1067E+02   .4342E+02   .5437E+01     -17.1
Stress  MID  -.6665E+01  -.9627E+00  -.1328E+02   .9770E+01  -.1740E+02     -51.1
Stress  BOT  -.5347E+02  -.1064E+02  -.1589E+02  -.5391E+01  -.5872E+02     -71.7
                            ***PLATE    80***
Loads    98  -.2022E+04   .1708E+04  -.1181E+04  -.5710E+04  -.2006E+05   .9987E+03
Loads    99  -.1439E+03  -.3494E+04   .7336E+03   .1064E+04   .1420E+05  -.1504E+04
Loads   109   .1820E+04  -.1310E+04   .1101E+04  -.7842E+03   .1186E+05  -.1706E+04
Loads   108   .3452E+03   .3096E+04  -.6535E+03   .1961E+04  -.1803E+05   .2797E+04
```

COOLING TOWER VIBRATION: INCREASED NODE MODEL

```
IMAGES-3D   s/n:800189                                03-02-87
                                                      PAGE   21
         ================ I M A G E S  3 D ================
         = Copyright (c) 1984   Celestial Software Inc. =
         ================================================

         SOLVE PLATE LOADS/STRESSES   Version 1.4  12/01/86

         COOLING TOWER VIBRATION - NODE INCREASE

Mode  1 -

GLoads Node    Fx          Fy          Fz          Mx          My          Mz
Stress Surf  Sigma  X    Sigma  Y    Tau XY     Sigma  1    Sigma  2     Angle
Stress       Shear XZ    Shear YZ
------ ----  ----------  ----------  ----------  ----------  ----------  ----------
Stress  TOP  .3244E+02   .1700E+02  -.1021E+02   .3752E+02   .1192E+02     -26.5
Stress  MID -.4827E+01   .9212E+01  -.1137E+02   .1555E+02  -.1117E+02     -60.8
Stress  BOT -.4210E+02   .1422E+01  -.1253E+02   .4773E+01  -.4545E+02     -75.0
                          ***PLATE   81***
Loads    99 -.9980E+03  -.9126E+03  -.1031E+04  -.7065E+04  -.1393E+05   .1085E+04
Loads   100 -.1073E+03  -.3527E+04   .1390E+04   .4142E+04  -.2826E+03   .3979E+03
Loads   110  .7124E+03   .1330E+04   .4503E+03  -.3672E+04   .1145E+04  -.4615E+03
Loads   109  .3930E+03   .3110E+04  -.8094E+03   .3259E+04  -.1086E+05   .1611E+04
Stress  TOP  .1316E+02   .2618E+02  -.5996E+01   .2852E+02   .1082E+02     -68.7
Stress  MID -.1745E+01   .2298E+02  -.5510E+01   .2415E+02  -.2918E+01     -78.0
Stress  BOT -.1666E+02   .1978E+02  -.5025E+01   .2046E+02  -.1734E+02     -82.3
                          ***PLATE   82***
Loads   101  .1295E+04  -.2517E+04   .1465E+04  -.7846E+04  -.3812E+05   .4224E+05
Loads   102 -.2717E+04   .1217E+04  -.2820E+03  -.2071E+04  -.4180E+04  -.1086E+05
Loads   112  .1234E+04   .1310E+04   .5409E+03   .2763E+04   .8304E+04   .7713E+03
Loads   111  .1882E+04  -.1004E+02  -.1724E+04   .5828E-02   .2493E-01   .3594E+04
Stress  TOP  .2698E+02  -.2331E+02  -.2192E+02   .3519E+02  -.3152E+02     -20.5
Stress  MID -.5999E+00   .8074E+01   .5056E+01   .1040E+02  -.2924E+01      65.3
Stress  BOT -.2818E+02   .3946E+02   .3203E+02   .5222E+02  -.4094E+02      68.3
                          ***PLATE   83***
Loads   102 -.2246E+03  -.1429E+04   .1896E+04   .1017E+04  -.1548E+04   .5636E+04
Loads   103  .6122E+03   .1546E+04  -.3149E+03   .1901E+04   .8193E+04  -.2378E+04
Loads   113  .4350E+03   .1221E+04  -.1039E+04   .2866E+04   .1011E+05  -.1043E+04
Loads   112 -.8226E+03  -.1338E+04  -.5426E+03  -.2763E+04  -.8304E+04  -.7712E+03
Stress  TOP  .2174E+02  -.7268E+01   .9255E+00   .2177E+02  -.7298E+01       1.8
Stress  MID  .2561E+01  -.4905E+00   .7805E+01   .8988E+01  -.6918E+01      39.5
Stress  BOT -.1662E+02   .6288E+01   .1468E+02   .1346E+02  -.2379E+02      64.0
                          ***PLATE   84***
Loads   103  .3222E+03  -.6883E+03   .1606E+04  -.1275E+04  -.8579E+04   .3392E+04
Loads   104  .3287E+03   .1270E+04  -.6748E+03   .1805E+04   .1072E+05  -.3768E+04
Loads   114 -.6577E+03   .6652E+03  -.1960E+04   .3037E+04   .1156E+05  -.1726E+04
Loads   113  .6726E+01  -.1246E+04   .1029E+04  -.2866E+04  -.1011E+05   .1043E+04
```

STRUCTURAL MODELS

```
IMAGES-3D   s/n:800189                              03-02-87
                                                    PAGE   22
        =============== I M A G E S   3 D ===============
        = Copyright (c) 1984   Celestial Software Inc. =
        ================================================

        SOLVE PLATE LOADS/STRESSES     Version 1.4  12/01/86

        COOLING TOWER VIBRATION - NODE INCREASE

Mode  1 -
```

GLoads	Node	Fx	Fy	Fz	Mx	My	Mz
Stress	Surf	Sigma X	Sigma Y	Tau XY	Sigma 1	Sigma 2	Angle
Stress		Shear XZ	Shear YZ				
------	----	----------	----------	----------	----------	----------	----------
Stress	TOP	.3418E+02	-.3460E+01	.3488E+01	.3450E+02	-.3780E+01	5.2
Stress	MID	.6561E+01	-.2940E+01	.5366E+01	.8976E+01	-.5356E+01	24.2
Stress	BOT	-.2106E+02	-.2421E+01	.7243E+01	.6277E-01	-.2354E+02	71.1
				PLATE 85			
Loads	104	.4675E+03	.6754E+02	.1236E+04	-.2357E+04	-.1038E+05	.2086E+04
Loads	105	.6809E+01	.5766E+03	-.1085E+04	.1703E+04	.1399E+05	-.4910E+04
Loads	115	-.1596E+04	.3764E+02	-.2091E+04	.3002E+04	.1294E+05	-.2507E+04
Loads	114	.1121E+04	-.6818E+03	.1940E+04	-.3037E+04	-.1156E+05	.1727E+04
Stress	TOP	.4207E+02	-.9191E+00	.3254E+01	.4232E+02	-.1164E+01	4.3
Stress	MID	.9205E+01	-.3375E+01	.1897E+01	.9484E+01	-.3655E+01	8.4
Stress	BOT	-.2366E+02	-.5832E+01	.5394E+00	-.5815E+01	-.2368E+02	88.3
				PLATE 86			
Loads	105	.5068E+02	.7372E+03	.6248E+03	-.3741E+04	-.1319E+05	.1877E+04
Loads	106	-.5070E+03	-.1171E+03	-.1116E+04	.1380E+04	.1619E+05	-.5514E+04
Loads	116	-.2072E+04	-.5818E+03	-.1568E+04	.2742E+04	.1411E+05	-.3267E+04
Loads	115	.2072E+04	-.3822E+02	.2059E+04	-.3002E+04	-.1294E+05	.2507E+04
Stress	TOP	.4768E+02	.9331E+00	.1130E+01	.4770E+02	.9058E+00	1.4
Stress	MID	.9794E+01	-.3327E+01	-.1665E+01	.1000E+02	-.3535E+01	-7.1
Stress	BOT	-.2809E+02	-.7587E+01	-.4460E+01	-.6658E+01	-.2902E+02	-78.2
				PLATE 87			
Loads	106	.2639E+03	.1260E+04	-.8962E+02	-.4657E+04	-.1539E+05	.2397E+04
Loads	107	-.9156E+03	-.7715E+03	-.7637E+03	.1004E+04	.1684E+05	-.5491E+04
Loads	117	-.1903E+04	-.1091E+04	-.6755E+03	.2166E+04	.1435E+05	-.3758E+04
Loads	116	.2554E+04	.6030E+03	.1529E+04	-.2742E+04	-.1411E+05	.3267E+04
Stress	TOP	.4899E+02	.2164E+01	-.1976E+01	.4907E+02	.2081E+01	-2.4
Stress	MID	.8282E+01	-.2692E+01	-.4907E+01	.1016E+02	-.4566E+01	-20.9
Stress	BOT	-.3243E+02	-.7547E+01	-.7838E+01	-.5283E+01	-.3469E+02	-73.9
				PLATE 88			
Loads	107	-.1992E+03	.1483E+04	-.7074E+03	-.4827E+04	-.1634E+05	.3181E+04
Loads	108	-.1065E+04	-.1302E+04	-.1247E+03	.6254E+03	.1481E+05	-.4631E+04
Loads	118	-.1120E+04	-.1320E+04	.1973E+03	.1299E+04	.1260E+05	-.3581E+04
Loads	117	.2384E+04	.1139E+04	.6348E+03	-.2166E+04	-.1435E+05	.3758E+04

COOLING TOWER VIBRATION: INCREASED NODE MODEL

```
IMAGES-3D  s/n:800189                                      03-02-87
                                                           PAGE   23
               =============== I M A G E S  3 D ===============
               = Copyright (c) 1984   Celestial Software Inc. =
               ================================================

               SOLVE PLATE LOADS/STRESSES    Version 1.4  12/01/86

          COOLING TOWER VIBRATION - NODE INCREASE

Mode   1 -

GLoads Node     Fx          Fy          Fz          Mx          My          Mz
Stress Surf   Sigma  X    Sigma  Y     Tau XY     Sigma  1    Sigma  2     Angle
Stress        Shear XZ    Shear YZ
------ ----  ----------  ----------  ----------  ----------  ----------  ----------
Stress TOP   .4414E+02   .3355E+01  -.4887E+01   .4472E+02   .2777E+01      -6.7
Stress MID   .5188E+01  -.1133E+01  -.7180E+01   .9872E+01  -.5818E+01     -33.1
Stress BOT  -.3377E+02  -.5621E+01  -.9474E+01  -.2729E+01  -.3666E+02     -73.0
                             ***PLATE    89***
Loads  108  -.6250E+03   .1156E+04  -.9844E+03  -.4181E+04  -.1489E+05   .3522E+04
Loads  109  -.8501E+03  -.1541E+04   .5811E+03   .5476E+03   .9310E+04  -.2757E+04
Loads  119  -.1211E+03  -.1014E+04   .6346E+03   .4124E+03   .7802E+04  -.2322E+04
Loads  118   .1596E+04   .1399E+04  -.2313E+03  -.1299E+04  -.1260E+05   .3581E+04
Stress TOP   .3159E+02   .5123E+01  -.6005E+01   .3289E+02   .3824E+01     -12.2
Stress MID   .1703E+01   .1759E+01  -.7333E+01   .9064E+01  -.5602E+01     -45.1
Stress BOT  -.2818E+02  -.1604E+01  -.8660E+01   .9690E+00  -.3075E+02     -73.5
                             ***PLATE    90***
Loads  109  -.5308E+03  -.2456E+02  -.9088E+03  -.3022E+04  -.1031E+05   .2852E+04
Loads  110  -.2974E+03  -.1179E+04   .1147E+04   .1315E+04  -.1145E+04   .4615E+03
Loads  120   .2351E+03   .7603E+02   .4157E+03  -.3041E+04   .2757E-01  -.6803E-02
Loads  119   .5931E+03   .1128E+04  -.6539E+03  -.4124E+03  -.7802E+04   .2322E+04
Stress TOP   .1122E+02   .7376E+01  -.3464E+01   .1326E+02   .5337E+01     -30.5
Stress MID  -.1500E+00   .5956E+01  -.3857E+01   .7822E+01  -.2016E+01     -64.2
Stress BOT  -.1152E+02   .4536E+01  -.4250E+01   .5592E+01  -.1258E+02     -76.1



IMAGES-3D  s/n:800189                                      03-02-87
                                                           PAGE   24
               =============== I M A G E S  3 D ===============
               = Copyright (c) 1984   Celestial Software Inc. =
               ================================================

               SOLVE PLATE LOADS/STRESSES    Version 1.4  12/01/86

          COOLING TOWER VIBRATION - NODE INCREASE

   Mode   1 -


                    MAXIMUM STRESS SUMMARY FOR PLATES
                    WITHIN SPECIFIED RANGE    1-   90

         Maximum (absolute) Stress =  .1503E+03 at Plate    9

               Plate   Sigma  X    Sigma  Y     Tau XY
               -----  ----------  ----------  ----------
                 9    .1715E+02   .1503E+03  -.1352E+02
```

```
IMAGES-3D   s/n:800189                              03-02-87
                                                   PAGE    1
        =============== I M A G E S  3 D ===============
        = Copyright (c) 1984   Celestial Software Inc. =
        ================================================

        SOLVE REACTIONS                Version 1.4  12/01/86

        COOLING TOWER VIBRATION - NODE INCREASE

  Mode  1 -

                          REACTIONS

  Node      Fx          Fy          Fz          Mx          My          Mz
  ----   ----------  ----------  ----------  ----------  ----------  ----------
    1   -.5213E+04   .1414E+05  -.1379E+04   .4500E+05   .2386E+05   .2968E+05
    2   -.6975E+04   .1910E+05   .6311E+03   .8194E+05   .3227E+05   .4429E+05
    3   -.6227E+04   .1726E+05   .1599E+04   .7746E+05   .2754E+05   .3437E+05
    4   -.4898E+04   .1374E+05   .1872E+04   .5938E+05   .1925E+05   .2734E+05
    5   -.3284E+04   .8538E+04   .1349E+04   .3338E+05   .8147E+04   .3097E+05
    6   -.1701E+04   .1553E+04  -.1109E+03   .4005E+04  -.5903E+04   .5022E+05
    7   -.5664E+03  -.7333E+04  -.2604E+04  -.2376E+05  -.2352E+05   .8831E+05
    8    .2807E+03  -.1909E+05  -.6286E+04  -.4757E+05  -.4684E+05   .1473E+06
    9   -.3042E+04  -.2834E+05  -.8562E+04  -.2380E+05  -.3430E+05   .1814E+06
   10   -.3163E+04  -.2472E+05  -.4152E+04   .4534E+05  -.2629E+05   .1130E+06
   11    .0000E+00   .0000E+00  -.2142E+04   .0000E+00   .0000E+00   .1369E+05
   20    .0000E+00   .0000E+00  -.4985E+04   .6853E+05   .0000E+00   .0000E+00
   21    .0000E+00   .0000E+00  -.4319E+04   .0000E+00   .0000E+00   .4551E+04
   30    .0000E+00   .0000E+00   .5929E+03   .8018E+04   .0000E+00   .0000E+00
   31    .0000E+00   .0000E+00  -.2410E+04   .0000E+00   .0000E+00   .2949E+04
   40    .0000E+00   .0000E+00   .8414E+03   .5276E+03   .0000E+00   .0000E+00
   41    .0000E+00   .0000E+00  -.2934E+02   .0000E+00   .0000E+00   .3239E+04
   50    .0000E+00   .0000E+00   .4286E+03   .6958E+03   .0000E+00   .0000E+00
   51    .0000E+00   .0000E+00   .1166E+04   .0000E+00   .0000E+00   .3723E+04
   60    .0000E+00   .0000E+00   .4749E+03   .4682E+03   .0000E+00   .0000E+00
   61    .0000E+00   .0000E+00   .1770E+04   .0000E+00   .0000E+00   .7887E+04
   70    .0000E+00   .0000E+00   .1346E+04   .1052E+04   .0000E+00   .0000E+00
   71    .0000E+00   .0000E+00   .2534E+04   .0000E+00   .0000E+00   .4553E+05
   80    .0000E+00   .0000E+00   .2679E+04   .2701E+04   .0000E+00   .0000E+00
   81    .0000E+00   .0000E+00   .3807E+04   .0000E+00   .0000E+00   .4569E+05
   90    .0000E+00   .0000E+00   .2358E+04   .6534E+03   .0000E+00   .0000E+00
   91    .0000E+00   .0000E+00   .7380E+04   .0000E+00   .0000E+00   .8118E+05
  100    .0000E+00   .0000E+00   .2532E+04  -.5322E+03   .0000E+00   .0000E+00
  101    .0000E+00   .0000E+00   .5821E+04   .0000E+00   .0000E+00   .7954E+05
  110    .0000E+00   .0000E+00   .1597E+04  -.2357E+04   .0000E+00   .0000E+00
  111    .0000E+00   .0000E+00  -.1724E+04   .0000E+00   .0000E+00   .3594E+04
  120    .0000E+00   .0000E+00   .4157E+03  -.3041E+02   .0000E+00   .0000E+00
```

COOLING TOWER VIBRATION: INCREASED NODE MODEL

```
IMAGES-3D   s/n:800189                                        03-02-87
                                                             PAGE    1
            =============== I M A G E S   3 D ===============
            = Copyright (c) 1984   Celestial Software Inc. =
            =================================================

            SOLVE BEAM LOADS/STRESSES     Version 1.4  12/01/86

        COOLING TOWER VIBRATION - NODE INCREASE

ABS

                        BEAM LOADS AND/OR STRESSES
```

GLoads LLoads /Stress	Node Node	Fx Axial	Fy Y-Shear	Fz Z-Shear	Mx Torsion	My Y-Bending	Mz Z-Bending
				BEAM NO. 1			
GLoads	1	.3588E+04	.1030E+05	.3526E+01	.1331E+03	.6594E+03	.1437E+05
GLoads	11	.3588E+04	.1030E+05	.3526E+01	.2594E+03	.6253E+03	.1437E+05
LLoads	1	.1088E+05	.7750E+03	.3526E+01	.6713E+03	.4359E+02	.1437E+05
LLoads	11	.1088E+05	.7750E+03	.3526E+01	.6713E+03	.8719E+02	.1437E+05
Stress	1	.1857E+02	.1323E+01	.6021E-02	.2759E-01	.5906E+00	.1792E-02
Stress	11	.1857E+02	.1323E+01	.6021E-02	.2759E-01	.5906E+00	.3584E-02
				BEAM NO. 2			
GLoads	1	.1625E+04	.3841E+04	.1383E+04	.4513E+05	.2320E+05	.1531E+05
GLoads	12	.1625E+04	.3841E+04	.1383E+04	.9611E+05	.4239E+04	.3424E+04
LLoads	1	.4185E+04	.9719E+03	.9240E+03	.2770E+04	.3968E+05	.3503E+05
LLoads	12	.4185E+04	.9719E+03	.9240E+03	.2770E+04	.3301E+04	.1017E+05
Stress	1	.7145E+01	.1659E+01	.1578E+01	.1138E+00	.1440E+01	.1631E+01
Stress	12	.7145E+01	.1659E+01	.1578E+01	.1138E+00	.4181E+00	.1357E+00
				BEAM NO. 3			
GLoads	2	.5233E+04	.1571E+05	.1542E+04	.3841E+05	.9413E+04	.3374E+05
GLoads	12	.5233E+04	.1571E+05	.1542E+04	.9624E+04	.3514E+04	.3875E+04
LLoads	2	.1658E+05	.1134E+04	.6094E+03	.7425E+03	.3332E+05	.3990E+05
LLoads	12	.1658E+05	.1134E+04	.6094E+03	.7425E+03	.1072E+05	.2145E+04
Stress	2	.2830E+02	.1936E+01	.1041E+01	.3052E-01	.1640E+01	.1369E+01
Stress	12	.2830E+02	.1936E+01	.1041E+01	.3052E-01	.8817E-01	.4405E+00
				BEAM NO. 4			
GLoads	2	.1742E+04	.3393E+04	.9105E+03	.4353E+05	.2286E+05	.1055E+05
GLoads	13	.1742E+04	.3393E+04	.9105E+03	.7443E+04	.4951E+04	.4450E+04
LLoads	2	.3715E+04	.8358E+03	.9366E+03	.3514E+04	.4004E+05	.3021E+05
LLoads	13	.3715E+04	.8358E+03	.9366E+03	.3514E+04	.3519E+04	.8659E+04
Stress	2	.6343E+01	.1427E+01	.1599E+01	.1444E+00	.1242E+01	.1646E+01
Stress	13	.6343E+01	.1427E+01	.1599E+01	.1444E+00	.3559E+00	.1446E+00
				BEAM NO. 5			
GLoads	3	.4489E+04	.1426E+05	.2237E+04	.4167E+05	.8185E+04	.2860E+05
GLoads	13	.4489E+04	.1426E+05	.2237E+04	.8793E+04	.2696E+04	.2386E+04
LLoads	3	.1507E+05	.1088E+04	.5670E+03	.2331E+03	.3050E+05	.4113E+05
LLoads	13	.1507E+05	.1088E+04	.5670E+03	.2331E+03	.9468E+04	.7653E+03
Stress	3	.2573E+02	.1858E+01	.9681E+00	.9583E-02	.1691E+01	.1253E+01
Stress	13	.2573E+02	.1858E+01	.9681E+00	.9583E-02	.3146E-01	.3892E+00
				BEAM NO. 6			
GLoads	3	.1738E+04	.3002E+04	.6380E+03	.3579E+05	.1935E+05	.5767E+04

```
IMAGES-3D   s/n:800189                              03-02-87
                                                    PAGE   2
             =============== I M A G E S  3 D ===============
             = Copyright (c) 1984   Celestial Software Inc. =
             ================================================

             SOLVE BEAM LOADS/STRESSES     Version 1.4  12/01/86

             COOLING TOWER VIBRATION - NODE INCREASE

ABS
```

GLoads LLoads /Stress	Node Node	Fx Axial	Fy Y-Shear	Fz Z-Shear	Mx Torsion	My Y-Bending	Mz Z-Bending
GLoads	14	.1738E+04	.3002E+04	.6380E+03	.5535E+04	.4654E+04	.5398E+04
LLoads	3	.3374E+04	.6341E+03	.8086E+03	.3611E+04	.3467E+05	.2176E+05
LLoads	14	.3374E+04	.6341E+03	.8086E+03	.3611E+04	.2937E+04	.7731E+04
Stress	3	.5761E+01	.1083E+01	.1381E+01	.1484E+00	.8945E+00	.1425E+01
Stress	14	.5761E+01	.1083E+01	.1381E+01	.1484E+00	.3178E+00	.1207E+00
				BEAM NO. 7			
GLoads	4	.3306E+04	.1136E+05	.2289E+04	.3565E+05	.5322E+04	.2193E+05
GLoads	14	.3306E+04	.1136E+05	.2289E+04	.8675E+04	.2139E+04	.1244E+04
LLoads	4	.1201E+05	.9048E+03	.3288E+03	.5854E+02	.2061E+05	.3682E+05
LLoads	14	.1201E+05	.9048E+03	.3288E+03	.5854E+02	.8411E+04	.3261E+04
Stress	4	.2050E+02	.1545E+01	.5614E+00	.2406E-02	.1513E+01	.8469E+00
Stress	14	.2050E+02	.1545E+01	.5614E+00	.2406E-02	.1340E+00	.3457E+00
				BEAM NO. 8			
GLoads	4	.1591E+04	.2386E+04	.4169E+03	.2374E+05	.1393E+05	.5409E+04
GLoads	15	.1591E+04	.2386E+04	.4169E+03	.3019E+04	.3863E+04	.5711E+04
LLoads	4	.2814E+04	.3438E+03	.5993E+03	.3561E+04	.2611E+05	.9595E+04
LLoads	15	.2814E+04	.3438E+03	.5993E+03	.3561E+04	.1758E+04	.6394E+04
Stress	4	.4805E+01	.5869E+00	.1023E+01	.1464E+00	.3944E+00	.1073E+01
Stress	15	.4805E+01	.5869E+00	.1023E+01	.1464E+00	.2628E+00	.7227E-01
				BEAM NO. 9			
GLoads	5	.1968E+04	.7032E+04	.1639E+04	.2402E+05	.1317E+04	.1892E+05
GLoads	15	.1968E+04	.7032E+04	.1639E+04	.9106E+04	.1413E+04	.5909E+03
LLoads	5	.7456E+04	.6371E+03	.9834E+01	.3579E+03	.6371E+04	.2993E+05
LLoads	15	.7456E+04	.6371E+03	.9834E+01	.3579E+03	.6736E+04	.6306E+04
Stress	5	.1273E+02	.1088E+01	.1679E-01	.1471E-01	.1230E+01	.2619E+00
Stress	15	.1273E+02	.1088E+01	.1679E-01	.1471E-01	.2592E+00	.2769E+00
				BEAM NO. 10			
GLoads	5	.1316E+04	.1506E+04	.2899E+03	.9356E+04	.6829E+04	.1206E+05
GLoads	16	.1316E+04	.1506E+04	.2899E+03	.9638E+02	.2590E+04	.5141E+04
LLoads	5	.1995E+04	.4620E+02	.3172E+03	.3490E+04	.1491E+05	.6725E+04
LLoads	16	.1995E+04	.4620E+02	.3172E+03	.3490E+04	.1509E+03	.4576E+04
Stress	5	.3407E+01	.7889E-01	.5416E+00	.1434E+00	.2764E+00	.6126E+00
Stress	16	.3407E+01	.7889E-01	.5416E+00	.1434E+00	.1881E+00	.6204E-02
				BEAM NO. 11			
GLoads	6	.7577E+03	.1215E+04	.2043E+03	.8915E+04	.3862E+04	.2275E+05
GLoads	16	.7577E+03	.1215E+04	.2043E+03	.1061E+05	.4851E+03	.3182E+04
LLoads	6	.1341E+04	.3041E+03	.4491E+03	.6755E+03	.1230E+05	.2145E+05
LLoads	16	.1341E+04	.3041E+03	.4491E+03	.6755E+03	.4357E+04	.1017E+05
Stress	6	.2289E+01	.5193E+00	.7667E+00	.2777E-01	.8817E+00	.5054E+00

COOLING TOWER VIBRATION: INCREASED NODE MODEL

```
IMAGES-3D   s/n:800189                                    03-02-87
                                                          PAGE    3
           =============== I M A G E S  3 D ===============
           = Copyright (c) 1984   Celestial Software Inc. =
           ================================================

             SOLVE BEAM LOADS/STRESSES     Version 1.4  12/01/86

          COOLING TOWER VIBRATION - NODE INCREASE

ABS
```

GLoads LLoads /Stress	Node Node	Fx Axial	Fy Y-Shear	Fz Z-Shear	Mx Torsion	My Y-Bending	Mz Z-Bending
Stress	16	.2289E+01	.5193E+00	.7667E+00	.2777E-01	.4181E+00	.1791E+00
			***BEAM	NO. 12***			
GLoads	6	.9430E+03	.3382E+03	.3152E+03	.4911E+04	.2041E+04	.2747E+05
GLoads	17	.9430E+03	.3382E+03	.3152E+03	.3597E+04	.7874E+03	.3364E+04
LLoads	6	.8944E+03	.5487E+03	.4223E+02	.3407E+04	.9099E+03	.2776E+05
LLoads	17	.8944E+03	.5487E+03	.4223E+02	.3407E+04	.2874E+04	.2239E+04
Stress	6	.1527E+01	.9368E+00	.7210E-01	.1400E+00	.1141E+01	.3740E-01
Stress	17	.1527E+01	.9368E+00	.7210E-01	.1400E+00	.9201E-01	.1181E+00
			***BEAM	NO. 13***			
GLoads	7	.2453E+02	.6269E+04	.2049E+04	.7233E+04	.1035E+05	.3563E+05
GLoads	17	.2453E+02	.6269E+04	.2049E+04	.1356E+05	.6431E+03	.6170E+04
LLoads	7	.6518E+04	.8824E+02	.1003E+04	.9953E+03	.3598E+05	.1155E+05
LLoads	17	.6518E+04	.8824E+02	.1003E+04	.9953E+03	.1217E+04	.1483E+05
Stress	7	.1113E+02	.1507E+00	.1713E+01	.4091E-01	.4749E+00	.1479E+01
Stress	17	.1113E+02	.1507E+00	.1713E+01	.4091E-01	.6094E+00	.5004E-01
			***BEAM	NO. 14***			
GLoads	7	.5910E+03	.1063E+04	.5553E+03	.1652E+05	.1316E+05	.5267E+05
GLoads	18	.5910E+03	.1063E+04	.5553E+03	.6681E+04	.1383E+04	.1178E+03
LLoads	7	.4060E+03	.1176E+04	.4899E+03	.3185E+04	.1678E+05	.5412E+05
LLoads	18	.4060E+03	.1176E+04	.4899E+03	.3185E+04	.6006E+04	.5868E+03
Stress	7	.6932E+00	.2008E+01	.8365E+00	.1309E+00	.2224E+01	.6897E+00
Stress	18	.6932E+00	.2008E+01	.8365E+00	.1309E+00	.2412E-01	.2469E+00
			***BEAM	NO. 15***			
GLoads	8	.2919E+02	.1563E+05	.5099E+04	.2251E+05	.1856E+05	.5940E+05
GLoads	18	.2919E+02	.1563E+05	.5099E+04	.1790E+05	.1939E+04	.8711E+04
LLoads	8	.1634E+05	.5563E+03	.1717E+04	.1333E+04	.6616E+05	.8346E+03
LLoads	18	.1634E+05	.5563E+03	.1717E+04	.1333E+04	.2499E+04	.1980E+05
Stress	8	.2789E+02	.9499E+00	.2931E+01	.5479E-01	.3430E-01	.2719E+01
Stress	18	.2789E+02	.9499E+00	.2931E+01	.5479E-01	.8137E+00	.1027E+00
			***BEAM	NO. 16***			
GLoads	8	.2515E+03	.3464E+04	.1187E+04	.2506E+05	.2829E+05	.8787E+05
GLoads	19	.2515E+03	.3464E+04	.1187E+04	.1036E+05	.6370E+04	.5738E+04
LLoads	8	.2879E+04	.1955E+04	.1167E+04	.1947E+04	.4196E+05	.8593E+05
LLoads	19	.2879E+04	.1955E+04	.1167E+04	.1947E+04	.1233E+05	.5011E+04
Stress	8	.4915E+01	.3338E+01	.1993E+01	.8001E-01	.3532E+01	.1725E+01
Stress	19	.4915E+01	.3338E+01	.1993E+01	.8001E-01	.2059E+00	.5066E+00
			***BEAM	NO. 17***			
GLoads	9	.8948E+03	.2794E+05	.8984E+04	.3443E+05	.2698E+05	.9125E+05
GLoads	19	.8948E+03	.2794E+05	.8984E+04	.2082E+05	.3347E+04	.1225E+05

```
IMAGES-3D  s/n:800189                               03-02-87
                                                    PAGE    4
            =============== I M A G E S  3 D ===============
            = Copyright (c) 1984   Celestial Software Inc. =
            ================================================

            SOLVE BEAM LOADS/STRESSES     Version 1.4  12/01/86

        COOLING TOWER VIBRATION - NODE INCREASE

ABS

GLoads Node   Fx          Fy          Fz          Mx          My          Mz
LLoads Node   Axial       Y-Shear     Z-Shear     Torsion     Y-Bending   Z-Bending
/Stress
------ ----  ----------  ----------  ----------  ----------  ----------  ----------
LLoads   9   .2924E+05   .1097E+04   .2441E+04   .1027E+04   .9957E+05   .1806E+05
LLoads  19   .2924E+05   .1097E+04   .2441E+04   .1027E+04   .9026E+04   .2263E+05
Stress   9   .4993E+02   .1873E+01   .4168E+01   .4221E-01   .7423E+00   .4092E+01
Stress  19   .4993E+02   .1873E+01   .4168E+01   .4221E-01   .9301E+00   .3710E+00
                        ***BEAM   NO.   18***
GLoads   9   .2148E+04   .4011E+03   .4220E+03   .1063E+05   .7318E+04   .9016E+05
GLoads  20   .2148E+04   .4011E+03   .4220E+03   .7346E+04   .4104E+04   .1723E+04
LLoads   9   .1087E+04   .1939E+04   .1083E+03   .1658E+04   .1346E+05   .9006E+05
LLoads  20   .1087E+04   .1939E+04   .1083E+03   .1658E+04   .8427E+04   .9956E+02
Stress   9   .1855E+01   .3310E+01   .1848E+00   .6813E-01   .3702E+01   .5533E+00
Stress  20   .1855E+01   .3310E+01   .1848E+00   .6813E-01   .4092E-02   .3464E+00
                        ***BEAM   NO.   19***
GLoads  10   .3163E+04   .2472E+05   .4152E+04   .4534E+05   .2629E+05   .1130E+06
GLoads  20   .3163E+04   .2472E+05   .4152E+04   .4534E+05   .4335E+04   .2206E+03
LLoads  10   .2495E+05   .2445E+04   .3163E+04   .4127E+04   .1160E+06   .4534E+05
LLoads  20   .2495E+05   .2445E+04   .3163E+04   .4127E+04   .1344E+04   .4534E+05
Stress  10   .4260E+02   .4175E+01   .5401E+01   .1696E+00   .1864E+01   .4767E+01
Stress  20   .4260E+02   .4175E+01   .5401E+01   .1696E+00   .1864E+01   .5526E-01


IMAGES-3D  s/n:800189                               03-02-87
                                                    PAGE    5
            =============== I M A G E S  3 D ===============
            = Copyright (c) 1984   Celestial Software Inc. =
            ================================================

            SOLVE BEAM LOADS/STRESSES     Version 1.4  12/01/86

        COOLING TOWER VIBRATION - NODE INCREASE

ABS


                MAXIMUM STRESS SUMMARY FOR BEAMS/TRUSSES
                   WITHIN SPECIFIED RANGE    1-    19

            Maximum (absolute) Stress = .4993E+02 at BEAM    17

    Beam    Axial       Y-Shear     Z-Shear     Torsion     Y-Bending   Z-Bending
    ----  ----------  ----------  ----------  ----------  ----------  ----------
      17   .4993E+02   .1873E+01   .4168E+01   .4221E-01   .7423E+00   .4092E+01
```

COOLING TOWER VIBRATION: INCREASED NODE MODEL

```
?
 IMAGES-3D  s/n:800189                                    03-03-87
                                                         PAGE   1
              =============== I M A G E S   3 D ===============
              = Copyright (c) 1984   Celestial Software Inc. =
              ================================================

                 SOLVE PLATE LOADS/STRESSES   Version 1.4  12/01/86

         COOLING TOWER VIBRATION - NODE INCREASE

    Mode  1 -

                        PLATE LOADS AND/OR STRESSES

    Stress Surf  Sigma  X    Sigma  Y    Tau XY     Sigma  1    Sigma  2      Angle
    Stress       Shear XZ    Shear YZ
    ------ ----  ----------  ----------  ----------  ----------  ----------  ----------
                             ***PLATE    1***
    Stress  TOP -.1211E+02 -.7446E+02 -.3617E+01 -.1190E+02 -.7467E+02      -3.3
    Stress  MID -.9947E+01 -.6641E+02 -.1521E+01 -.9906E+01 -.6645E+02      -1.5
    Stress  BOT -.7788E+01 -.5835E+02  .5746E+00 -.7782E+01 -.5836E+02       .7
                             ***PLATE    2***
    Stress  TOP -.1210E+02 -.7203E+02 -.4312E+01 -.1179E+02 -.7234E+02      -4.1
    Stress  MID -.1089E+02 -.6211E+02 -.2900E+01 -.1073E+02 -.6227E+02      -3.2
    Stress  BOT -.9679E+01 -.5218E+02 -.1489E+01 -.9627E+01 -.5223E+02      -2.0
                             ***PLATE    3***
    Stress  TOP -.1177E+02 -.6539E+02 -.5716E+01 -.1117E+02 -.6599E+02      -6.0
    Stress  MID -.9769E+01 -.5362E+02 -.4314E+01 -.9348E+01 -.5404E+02      -5.6
    Stress  BOT -.7764E+01 -.4185E+02 -.2913E+01 -.7517E+01 -.4209E+02      -4.9
                             ***PLATE    4***
    Stress  TOP -.1008E+02 -.5330E+02 -.7440E+01 -.8839E+01 -.5454E+02      -9.5
    Stress  MID -.7577E+01 -.3987E+02 -.5818E+01 -.6560E+01 -.4089E+02      -9.9
    Stress  BOT -.5069E+01 -.2645E+02 -.4197E+01 -.4274E+01 -.2725E+02     -10.7
                             ***PLATE    5***
    Stress  TOP -.7315E+01 -.3554E+02 -.9288E+01 -.4533E+01 -.3832E+02     -16.7
    Stress  MID -.4455E+01 -.2048E+02 -.7353E+01 -.1592E+01 -.2334E+02     -21.3
    Stress  BOT -.1595E+01 -.5425E+01 -.5419E+01  .2237E+01 -.9257E+01     -35.3
                             ***PLATE    6***
    Stress  TOP -.3429E+01 -.1151E+02 -.1111E+02  .4351E+01 -.1929E+02     -35.0
    Stress  MID -.5302E+00  .5016E+01 -.8967E+01  .1163E+02 -.7143E+01     -53.6
    Stress  BOT  .2369E+01  .2154E+02 -.6824E+01  .2372E+02  .1882E+00     -72.3
                             ***PLATE    7***
```

```
IMAGES-3D  s/n:800189                          03-03-87
                                               PAGE    2
     =============== I M A G E S  3 D ===============
     = Copyright (c) 1984   Celestial Software Inc. =
     ================================================

        SOLVE PLATE LOADS/STRESSES   Version 1.4  12/01/86

     COOLING TOWER VIBRATION - NODE INCREASE

Mode   1 -
```

Stress Stress	Surf	Sigma X Shear XZ	Sigma Y Shear YZ	Tau XY	Sigma 1	Sigma 2	Angle
Stress	TOP	.1577E+01	.2011E+02	-.1258E+02	.2647E+02	-.4777E+01	-63.2
Stress	MID	.3984E+01	.3744E+02	-.1062E+02	.4053E+02	.8974E+00	-73.8
Stress	BOT	.6391E+01	.5477E+02	-.8666E+01	.5628E+02	.4886E+01	-80.1
			***PLATE	8***			
Stress	TOP	.7030E+01	.6101E+02	-.1382E+02	.6434E+02	.3696E+01	-76.4
Stress	MID	.8374E+01	.7753E+02	-.1276E+02	.7981E+02	.6094E+01	-79.9
Stress	BOT	.9717E+01	.9405E+02	-.1170E+02	.9564E+02	.8125E+01	-82.2
			***PLATE	9***			
Stress	TOP	.8245E+01	.1240E+03	-.1345E+02	.1255E+03	.6703E+01	-83.5
Stress	MID	.1270E+02	.1372E+03	-.1348E+02	.1386E+03	.1125E+02	-83.9
Stress	BOT	.1715E+02	.1503E+03	-.1352E+02	.1517E+03	.1579E+02	-84.3
			***PLATE	10***			
Stress	TOP	-.7575E+01	-.8014E+02	-.4458E+01	-.7302E+01	-.8041E+02	-3.5
Stress	MID	-.8102E+01	-.6686E+02	-.1393E+01	-.8069E+01	-.6690E+02	-1.4
Stress	BOT	-.8629E+01	-.5359E+02	.1671E+01	-.8567E+01	-.5365E+02	2.1
			***PLATE	11***			
Stress	TOP	-.8369E+01	-.7586E+02	-.7457E+01	-.7555E+01	-.7667E+02	-6.2
Stress	MID	-.7241E+01	-.6279E+02	-.4595E+01	-.6863E+01	-.6317E+02	-4.7
Stress	BOT	-.6112E+01	-.4972E+02	-.1733E+01	-.6044E+01	-.4979E+02	-2.3
			***PLATE	12***			
Stress	TOP	-.7398E+01	-.6603E+02	-.1095E+02	-.5421E+01	-.6801E+02	-10.2
Stress	MID	-.5845E+01	-.5363E+02	-.7816E+01	-.4599E+01	-.5488E+02	-9.1
Stress	BOT	-.4291E+01	-.4123E+02	-.4685E+01	-.3706E+01	-.4182E+02	-7.1
			***PLATE	13***			
Stress	TOP	-.5575E+01	-.5054E+02	-.1416E+02	-.1485E+01	-.5463E+02	-16.1
Stress	MID	-.3773E+01	-.3907E+02	-.1092E+02	-.6669E+00	-.4218E+02	-15.9
Stress	BOT	-.1970E+01	-.2761E+02	-.7678E+01	.1536E+00	-.2973E+02	-15.5
			***PLATE	14***			

COOLING TOWER VIBRATION: INCREASED NODE MODEL

```
IMAGES-3D   s/n:800189                                    03-03-87
                                                         PAGE    3
            =============== I M A G E S   3 D ===============
            = Copyright (c) 1984   Celestial Software Inc. =
            ================================================

            SOLVE PLATE LOADS/STRESSES   Version 1.4  12/01/86

       COOLING TOWER VIBRATION - NODE INCREASE

Mode  1 -

Stress Surf  Sigma  X    Sigma  Y    Tau XY     Sigma  1    Sigma  2     Angle
Stress       Shear XZ    Shear YZ
------ ----  ----------  ----------  ----------  ----------  ----------  ----------
Stress  TOP -.2439E+01 -.2803E+02 -.1689E+02  .5951E+01 -.3642E+02    -26.4
Stress  MID -.1226E+01 -.1861E+02 -.1377E+02  .6368E+01 -.2620E+02    -28.9
Stress  BOT -.1229E-01 -.9191E+01  .1066E+02  .7003E+01 -.1621E+02    -33.4
                        ***PLATE    15***
Stress  TOP  .2037E+01  .2425E+01 -.1861E+02  .2084E+02 -.1637E+02    -45.3
Stress  MID  .1379E+01  .8257E+01 -.1612E+02  .2130E+02 -.1166E+02    -51.0
Stress  BOT  .7200E+00  .1409E+02 -.1363E+02  .2259E+02 -.7779E+01    -58.1
                        ***PLATE    16***
Stress  TOP  .7261E+01  .4098E+02 -.1866E+02  .4927E+02 -.1030E+01    -66.0
Stress  MID  .3284E+01  .4174E+02 -.1748E+02  .4849E+02 -.3474E+01    -68.9
Stress  BOT -.6933E+00  .4249E+02 -.1629E+02  .4795E+02 -.6151E+01    -71.5
                        ***PLATE    17***
Stress  TOP  .1060E+02  .8791E+02 -.1569E+02  .9097E+02  .7534E+01    -79.0
Stress  MID  .3248E+01  .8351E+02 -.1632E+02  .8670E+02  .5795E-01    -78.9
Stress  BOT -.4101E+01  .7912E+02 -.1694E+02  .8244E+02 -.7418E+01    -78.9
                        ***PLATE    18***
Stress  TOP  .5671E+01  .1310E+03 -.3022E+01  .1311E+03  .5598E+01    -88.6
Stress  MID -.1441E+01  .1249E+03 -.5493E+01  .1251E+03 -.1680E+01    -87.5
Stress  BOT -.8554E+01  .1187E+03 -.7963E+01  .1192E+03 -.9050E+01    -86.4
                        ***PLATE    19***
Stress  TOP -.2239E+01 -.7192E+02 -.6154E+01 -.1700E+01 -.7246E+02     -5.0
Stress  MID -.2291E+01 -.6502E+02 -.2397E+01 -.2200E+01 -.6511E+02     -2.2
Stress  BOT -.2343E+01 -.5812E+02  .1360E+02 -.2310E+01 -.5815E+02      1.4
                        ***PLATE    20***
Stress  TOP -.1457E+01 -.6775E+02 -.1070E+02  .2277E+00 -.6944E+02     -8.9
Stress  MID -.1649E+01 -.6052E+02 -.7068E+01 -.8127E+00 -.6136E+02     -6.8
Stress  BOT -.1841E+01 -.5329E+02 -.3433E+01  .1613E+01 -.5352E+02     -3.8
                        ***PLATE    21***
```

```
IMAGES-3D   s/n:800189                               03-03-87
                                                     PAGE    4
            =============== I M A G E S  3 D ===============
            = Copyright (c) 1984   Celestial Software Inc. =
            ================================================

            SOLVE PLATE LOADS/STRESSES    Version 1.4  12/01/86

        COOLING TOWER VIBRATION - NODE INCREASE

Mode  1 -

Stress Surf  Sigma  X    Sigma  Y    Tau XY      Sigma  1    Sigma  2      Angle
Stress       Shear XZ    Shear YZ
------ ----  ----------  ----------  ----------  ----------  ----------  ----------
Stress TOP  -.6748E+00  -.5826E+02  -.1588E+02   .3413E+01  -.6235E+02    -14.4
Stress MID  -.5487E+00  -.5107E+02  -.1161E+02   .1993E+01  -.5361E+02    -12.3
Stress BOT  -.4225E+00  -.4388E+02  -.7350E+01   .7870E+00  -.4509E+02     -9.3
                          ***PLATE    22***
Stress TOP   .9759E+00  -.4303E+02  -.1993E+02   .8662E+01  -.5072E+02    -21.1
Stress MID   .7646E+00  -.3625E+02  -.1574E+02   .6554E+01  -.4203E+02    -20.2
Stress BOT   .5532E+00  -.2946E+02  -.1155E+02   .4483E+01  -.3339E+02    -18.8
                          ***PLATE    23***
Stress TOP   .3622E+01  -.2143E+02  -.2239E+02   .1675E+02  -.3456E+02    -30.4
Stress MID   .1921E+01  -.1570E+02  -.1897E+02   .1402E+02  -.2780E+02    -32.5
Stress BOT   .2202E+00  -.9960E+01  -.1554E+02   .1148E+02  -.2122E+02    -35.9
                          ***PLATE    24***
Stress TOP   .6961E+01   .6757E+01  -.2258E+02   .2944E+02  -.1572E+02    -44.9
Stress MID   .2480E+01   .1074E+02  -.2064E+02   .2766E+02  -.1443E+02    -50.7
Stress BOT  -.2001E+01   .1473E+02  -.1869E+02   .2684E+02  -.1411E+02    -57.1
                          ***PLATE    25***
Stress TOP   .9836E+01   .4103E+02  -.1986E+02   .5069E+02   .1806E+00    -64.1
Stress MID   .1993E+01   .4298E+02  -.1972E+02   .5092E+02  -.5954E+01    -68.0
Stress BOT  -.5850E+01   .4492E+02  -.1958E+02   .5160E+02  -.1252E+02    -71.2
                          ***PLATE    26***
Stress TOP   .9681E+01   .7778E+02  -.1317E+02   .8024E+02   .7222E+01    -79.4
Stress MID   .2356E+00   .7830E+02  -.1426E+02   .8082E+02  -.2288E+01    -80.0
Stress BOT  -.9210E+01   .7881E+02  -.1535E+02   .8141E+02  -.1181E+02    -80.4
                          ***PLATE    27***
Stress TOP   .4670E+01   .1145E+03  -.3984E+01   .1146E+03   .4526E+01    -87.9
Stress MID   .1844E+00   .1149E+03  -.5539E+01   .1152E+03  -.8236E-01    -87.2
Stress BOT  -.4301E+01   .1153E+03  -.7094E+01   .1157E+03  -.4721E+01    -86.6
                          ***PLATE    28***
```

COOLING TOWER VIBRATION: INCREASED NODE MODEL

```
IMAGES-3D   s/n:800189                           03-03-87
                                                 PAGE    5
         =============== I M A G E S   3 D ===============
         = Copyright (c) 1984   Celestial Software Inc. =
         ================================================

         SOLVE PLATE LOADS/STRESSES   Version 1.4  12/01/86

      COOLING TOWER VIBRATION - NODE INCREASE

Mode  1 -

Stress Surf  Sigma  X    Sigma  Y    Tau XY     Sigma  1   Sigma  2     Angle
Stress       Shear XZ    Shear YZ
------ ----  ----------  ----------  ----------  ----------  ----------  ----------
Stress TOP  .2534E+01 -.6830E+02 -.8267E+01  .3486E+01 -.6926E+02    -6.6
Stress MID  .1777E+01 -.6004E+02 -.3178E+01  .1940E+01 -.6020E+02    -2.9
Stress BOT  .1020E+01 -.5177E+02  .1911E+01  .1089E+01 -.5184E+02     2.1
                      ***PLATE   29***
Stress TOP  .1173E+01 -.6427E+02 -.1383E+02  .3976E+01 -.6707E+02   -11.5
Stress MID  .2155E+01 -.5565E+02 -.9367E+01  .3635E+01 -.5713E+02    -9.0
Stress BOT  .3138E+01 -.4704E+02 -.4904E+01  .3613E+01 -.4751E+02    -5.5
                      ***PLATE   30***
Stress TOP  .2607E+01 -.5433E+02 -.2038E+02  .9151E+01 -.6088E+02   -17.8
Stress MID  .2796E+01 -.4656E+02 -.1497E+02  .6982E+01 -.5074E+02   -15.6
Stress BOT  .2985E+01 -.3879E+02 -.9560E+01  .5069E+01 -.4087E+02   -12.3
                      ***PLATE   31***
Stress TOP  .5186E+01 -.3924E+02 -.2471E+02  .1620E+02 -.5025E+02   -24.0
Stress MID  .3220E+01 -.3255E+02 -.1936E+02  .1169E+02 -.4102E+02   -23.6
Stress BOT  .1253E+01 -.2586E+02 -.1400E+02  .7187E+01 -.3180E+02   -23.0
                      ***PLATE   32***
Stress TOP  .8737E+01 -.1859E+02 -.2640E+02  .2480E+02 -.3465E+02   -31.3
Stress MID  .3143E+01 -.1331E+02 -.2207E+02  .1847E+02 -.2864E+02   -34.8
Stress BOT -.2451E+01 -.8023E+01 -.1775E+02  .1273E+02 -.2320E+02   -40.5
                      ***PLATE   33***
Stress TOP  .1232E+02  .7395E+01 -.2491E+02  .3489E+02 -.1517E+02   -42.2
Stress MID  .2423E+01  .1104E+02 -.2250E+02  .2964E+02 -.1617E+02   -50.4
Stress BOT -.7471E+01  .1469E+02 -.2009E+02  .2656E+02 -.1933E+02   -59.4
                      ***PLATE   34***
Stress TOP  .1418E+02  .3758E+02 -.2005E+02  .4909E+02  .2664E+01   -60.1
Stress MID  .1186E+01  .3958E+02 -.1990E+02  .4804E+02 -.7270E+01   -67.0
Stress BOT -.1181E+02  .4158E+02 -.1976E+02  .4810E+02 -.1832E+02   -71.7
                      ***PLATE   35***
```

```
IMAGES-3D    s/n:800189                                    03-03-87
                                                          PAGE    6
            =============== I M A G E S   3 D ===============
            = Copyright (c) 1984   Celestial Software Inc. =
            ================================================

            SOLVE PLATE LOADS/STRESSES    Version 1.4  12/01/86

      COOLING TOWER VIBRATION - NODE INCREASE

Mode   1 -

Stress Surf  Sigma  X    Sigma  Y    Tau XY      Sigma  1   Sigma  2    Angle
Stress       Shear XZ    Shear YZ
------ ----  ----------  ----------  ----------  ---------- ---------- ----------
Stress TOP   .1243E+02   .7076E+02 -.1271E+02   .7341E+02  .9782E+01    -78.2
Stress MID   .2814E+00   .7129E+02 -.1434E+02   .7408E+02 -.2505E+01    -79.0
Stress BOT -.1187E+02   .7182E+02 -.1597E+02   .7476E+02 -.1481E+02    -79.6
                        ***PLATE    36***
Stress TOP   .5213E+01   .1052E+03 -.2679E+01   .1052E+03  .5141E+01    -88.5
Stress MID -.2441E+00   .1045E+03 -.5392E+01   .1048E+03 -.5209E+00    -87.1
Stress BOT -.5701E+01   .1039E+03 -.8106E+01   .1045E+03 -.6298E+01    -85.8
                        ***PLATE    37***
Stress TOP -.3149E+01 -.6166E+02 -.1186E+02 -.8358E+00 -.6397E+02    -11.0
Stress MID   .3518E+01 -.5092E+02 -.4126E+01   .3829E+01 -.5123E+02     -4.3
Stress BOT   .1018E+02 -.4019E+02   .3610E+01   .1044E+02 -.4045E+02      4.1
                        ***PLATE    38***
Stress TOP   .1239E+01 -.5612E+02 -.1694E+02   .5867E+01 -.6075E+02    -15.3
Stress MID   .3990E+01 -.4770E+02 -.1193E+02   .6610E+01 -.5032E+02    -12.4
Stress BOT   .6742E+01 -.3928E+02 -.6918E+01   .7760E+01 -.4030E+02     -8.4
                        ***PLATE    39***
Stress TOP   .3108E+01 -.4819E+02 -.2424E+02   .1275E+02 -.5783E+02    -21.7
Stress MID   .4152E+01 -.4071E+02 -.1792E+02   .1043E+02 -.4698E+02    -19.3
Stress BOT   .5196E+01 -.3322E+02 -.1160E+02   .8427E+01 -.3645E+02    -15.6
                        ***PLATE    40***
Stress TOP   .7783E+01 -.3481E+02 -.2857E+02   .2212E+02 -.4915E+02    -26.6
Stress MID   .3965E+01 -.2890E+02 -.2208E+02   .1505E+02 -.3999E+02    -26.7
Stress BOT   .1475E+00 -.2299E+02 -.1558E+02   .7988E+01 -.3083E+02    -26.7
                        ***PLATE    41***
Stress TOP   .1407E+02 -.1613E+02 -.2949E+02   .3211E+02 -.3416E+02    -31.4
Stress MID   .3344E+01 -.1234E+02 -.2401E+02   .2076E+02 -.2976E+02    -36.0
Stress BOT -.7386E+01 -.8551E+01 -.1852E+02   .1056E+02 -.2650E+02    -44.1
                        ***PLATE    42***
```

COOLING TOWER VIBRATION: INCREASED NODE MODEL

```
IMAGES-3D  s/n:800189                                    03-03-87
                                                         PAGE    7
             =============== I M A G E S  3 D ===============
             = Copyright (c) 1984   Celestial Software Inc. =
             ===============================================

             SOLVE PLATE LOADS/STRESSES    Version 1.4  12/01/86

       COOLING TOWER VIBRATION - NODE INCREASE

Mode  1 -
```

| Stress | Surf | Sigma X | Sigma Y | Tau XY | Sigma 1 | Sigma 2 | Angle |
Stress		Shear XZ	Shear YZ				
Stress	TOP	.1992E+02	.7172E+01	-.2668E+02	.4098E+02	-.1388E+02	-38.3
Stress	MID	.2324E+01	.8788E+01	-.2337E+02	.2915E+02	-.1803E+02	-48.9
Stress	BOT	-.1527E+02	.1040E+02	-.2006E+02	.2138E+02	-.2625E+02	-61.3
			***PLATE	43***			
Stress	TOP	.2263E+02	.3416E+02	-.2041E+02	.4960E+02	.7182E+01	-52.9
Stress	MID	.1182E+01	.3404E+02	-.2004E+02	.4353E+02	-.8306E+01	-64.7
Stress	BOT	-.2026E+02	.3392E+02	-.1968E+02	.4031E+02	-.2665E+02	-72.0
			***PLATE	44***			
Stress	TOP	.1914E+02	.6355E+02	-.1124E+02	.6624E+02	.1646E+02	-76.6
Stress	MID	.1668E+00	.6251E+02	-.1385E+02	.6545E+02	-.2773E+01	-78.0
Stress	BOT	-.1881E+02	.6146E+02	-.1646E+02	.6471E+02	-.2206E+02	-78.8
			***PLATE	45***			
Stress	TOP	.7153E+01	.9338E+02	-.1681E+02	.9338E+02	.7153E+01	-89.9
Stress	MID	-.1684E+00	.9267E+02	-.5034E+01	.9294E+02	-.4406E+00	-86.9
Stress	BOT	-.7490E+01	.9197E+02	-.9900E+01	.9294E+02	-.8466E+01	-84.4
			***PLATE	46***			
Stress	TOP	-.7709E+01	-.3713E+02	-.2193E+02	.3990E+01	-.4883E+02	-28.1
Stress	MID	.4862E+01	-.3618E+02	-.5522E+01	.5592E+01	-.3691E+02	-7.5
Stress	BOT	.1743E+02	-.3523E+02	.1089E+02	.1959E+02	-.3739E+02	11.2
			***PLATE	47***			
Stress	TOP	.1234E+00	-.4620E+02	-.2035E+02	.7790E+01	-.5387E+02	-20.6
Stress	MID	.4611E+01	-.3710E+02	-.1354E+02	.8622E+01	-.4111E+02	-16.5
Stress	BOT	.9099E+01	-.2799E+02	-.6736E+01	.1028E+02	-.2918E+02	-10.0
			***PLATE	48***			
Stress	TOP	.1532E+01	-.4289E+02	-.2725E+02	.1447E+02	-.5583E+02	-25.4
Stress	MID	.4095E+01	-.3323E+02	-.2003E+02	.1281E+02	-.4194E+02	-23.5
Stress	BOT	.6658E+01	-.2356E+02	-.1281E+02	.1136E+02	-.2826E+02	-20.1
			***PLATE	49***			

IMAGES-3D   s/n:800189   03-03-87
                                                  PAGE   8
        =============== I M A G E S 3 D ===============
        = Copyright (c) 1984   Celestial Software Inc. =
        ==

        SOLVE PLATE LOADS/STRESSES   Version 1.4 12/01/86

        COOLING TOWER VIBRATION - NODE INCREASE

Mode 1 -

Stress	Surf	Sigma X	Sigma Y	Tau XY	Sigma 1	Sigma 2	Angle
Stress		Shear XZ	Shear YZ				
Stress	TOP	.9233E+01	-.3199E+02	-.3075E+02	.2564E+02	-.4840E+02	-28.1
Stress	MID	.3312E+01	-.2536E+02	-.2415E+02	.1706E+02	-.3911E+02	-29.7
Stress	BOT	-.2610E+01	-.1874E+02	-.1754E+02	.8633E+01	-.2998E+02	-32.7
			***PLATE	50***			
Stress	TOP	.1916E+02	-.1514E+02	-.3067E+02	.3715E+02	-.3313E+02	-30.4
Stress	MID	.2217E+01	-.1293E+02	-.2551E+02	.2126E+02	-.3197E+02	-36.7
Stress	BOT	-.1472E+02	-.1071E+02	-.2035E+02	.7736E+01	-.3317E+02	-47.8
			***PLATE	51***			
Stress	TOP	.2858E+02	.6762E+01	-.2686E+02	.4666E+02	-.1132E+02	-33.9
Stress	MID	.9544E+00	.4230E+01	-.2404E+02	.2669E+02	-.2151E+02	-46.9
Stress	BOT	-.2667E+02	.1698E+01	-.2123E+02	.1305E+02	-.3802E+02	-61.9
			***PLATE	52***			
Stress	TOP	.3359E+02	.3206E+01	-.1964E+02	.5248E+02	.1316E+02	-43.9
Stress	MID	-.2246E+00	.2587E+02	-.1981E+02	.3654E+02	-.1090E+02	-61.7
Stress	BOT	-.3404E+02	.1968E+02	-.1998E+02	.2630E+02	-.4066E+02	-71.7
			***PLATE	53***			
Stress	TOP	.2989E+02	.5843E+02	-.9854E+01	.6150E+02	.2682E+02	-72.7
Stress	MID	-.9082E+00	.5113E+02	-.1310E+02	.5424E+02	-.4019E+01	-76.6
Stress	BOT	-.3171E+02	.4382E+02	-.1634E+02	.4721E+02	-.3509E+02	-78.3
			***PLATE	54***			
Stress	TOP	.1233E+02	.8351E+02	.1750E+01	.8356E+02	.1229E+02	88.6
Stress	MID	-.6222E+00	.7878E+02	-.4014E+01	.7898E+02	-.8246E+00	-87.1
Stress	BOT	-.1358E+02	.7404E+02	-.9778E+01	.7511E+02	-.1465E+02	-83.7
			***PLATE	55***			
Stress	TOP	-.8969E+00	-.3306E+02	-.2907E+02	.1624E+02	-.5020E+02	-30.5
Stress	MID	.6446E+01	-.1735E+02	-.5149E+01	.7513E+01	-.1842E+02	-11.7
Stress	BOT	.1379E+02	-.1637E+01	.1877E+02	.2637E+02	-.1422E+02	33.8
			***PLATE	56***			

COOLING TOWER VIBRATION: INCREASED NODE MODEL

```
IMAGES-3D  s/n:800189                                    03-03-87
                                                         PAGE    9
             =============== I M A G E S   3 D ===============
             = Copyright (c) 1984   Celestial Software Inc. =
             ================================================

             SOLVE PLATE LOADS/STRESSES   Version 1.4  12/01/86

        COOLING TOWER VIBRATION - NODE INCREASE

Mode  1 -

Stress Surf  Sigma  X   Sigma  Y    Tau XY     Sigma  1   Sigma  2     Angle
Stress       Shear XZ   Shear YZ
------ ----  ---------- ---------- ---------- ---------- ---------- ----------
Stress TOP   .1686E+01 -.3356E+02 -.2055E+02  .1113E+02 -.4300E+02   -24.7
Stress MID   .6179E+01 -.2211E+02 -.1357E+02  .1164E+02 -.2757E+02   -21.9
Stress BOT   .1067E+02 -.1066E+02 -.6597E+01  .1255E+02 -.1253E+02   -15.9
                        ***PLATE    57***
Stress TOP   .5513E+01 -.3248E+02 -.2661E+02  .1921E+02 -.4618E+02   -27.2
Stress MID   .5315E+01 -.2387E+02 -.2034E+02  .1576E+02 -.3431E+02   -27.2
Stress BOT   .5118E+01 -.1525E+02 -.1407E+02  .1230E+02 -.2244E+02   -27.1
                        ***PLATE    58***
Stress TOP   .1470E+02 -.2631E+02 -.2835E+02  .2918E+02 -.4079E+02   -27.1
Stress MID   .3685E+01 -.2159E+02 -.2393E+02  .1811E+02 -.3602E+02   -31.1
Stress BOT  -.7327E+01 -.1687E+02 -.1952E+02  .7996E+01 -.3219E+02   -38.1
                        ***PLATE    59***
Stress TOP   .2646E+02 -.1436E+02 -.2697E+02  .3986E+02 -.2777E+02   -26.4
Stress MID   .1650E+01 -.1449E+02 -.2476E+02  .1963E+02 -.3246E+02   -36.0
Stress BOT  -.2316E+02 -.1461E+02 -.2256E+02  .4075E+01 -.4185E+02   -50.4
                        ***PLATE    60***
Stress TOP   .3715E+02  .2737E+01 -.2286E+02  .4855E+02 -.8667E+01   -26.5
Stress MID  -.2591E+00 -.2215E+01 -.2296E+02  .2174E+02 -.2422E+02   -43.8
Stress BOT  -.3766E+02 -.7168E+01 -.2306E+02  .5226E+01 -.5006E+02   -61.7
                        ***PLATE    61***
Stress TOP   .4237E+02  .2347E+02 -.1669E+02  .5210E+02  .1374E+02   -30.2
Stress MID  -.1562E+01  .1514E+02 -.1868E+02  .2725E+02 -.1367E+02   -57.0
Stress BOT  -.4550E+02  .6818E+01 -.2067E+02  .1400E+02 -.5268E+02   -70.8
                        ***PLATE    62***
Stress TOP   .3701E+02  .4571E+02 -.9136E+01  .5148E+02  .3124E+02   -57.7
Stress MID  -.1906E+01  .3706E+02 -.1216E+02  .4055E+02 -.5390E+01   -74.0
Stress BOT  -.4082E+02  .2841E+02 -.1519E+02  .3159E+02 -.4401E+02   -78.2
                        ***PLATE    63***
```

```
IMAGES-3D   s/n:800189                                    03-03-87
                                                          PAGE    10
     =============== I M A G E S   3 D ===============
     = Copyright (c) 1984   Celestial Software Inc. =
     ================================================

     SOLVE PLATE LOADS/STRESSES   Version 1.4  12/01/86

   COOLING TOWER VIBRATION - NODE INCREASE

Mode   1 -

Stress Surf  Sigma  X    Sigma  Y    Tau XY      Sigma  1    Sigma  2   Angle
Stress       Shear XZ    Shear YZ
------ ----  ----------  ----------  ----------  ----------  ----------  ----------
Stress TOP   .1586E+02   .6603E+02  -.2296E+00   .6603E+02   .1586E+02   -89.7
Stress MID  -.1050E+01   .6156E+02  -.3831E+01   .6179E+02  -.1283E+01   -86.5
Stress BOT  -.1796E+02   .5708E+02  -.7433E+01   .5781E+02  -.1869E+02   -84.4
                        ***PLATE    64***
Stress TOP   .1210E+02   .1674E+02  -.3585E+02   .5035E+02  -.2151E+02   -46.9
Stress MID   .1282E+02   .3505E+01  -.5542E+01   .1540E+02   .9224E+00   -25.0
Stress BOT   .1354E+02  -.9734E+01   .2477E+02   .2927E+02  -.2546E+02    32.4
                        ***PLATE    65***
Stress TOP   .1152E+02  -.6872E+01  -.1960E+02   .2398E+02  -.1933E+02   -32.4
Stress MID   .1185E+02  -.7961E+01  -.1174E+02   .1730E+02  -.1342E+02   -24.9
Stress BOT   .1217E+02  -.9050E+01  -.3881E+01   .1286E+02  -.9737E+01   -10.0
                        ***PLATE    66***
Stress TOP   .1514E+02  -.1561E+02  -.2147E+02   .2617E+02  -.2665E+02   -27.2
Stress MID   .8236E+01  -.1557E+02  -.1560E+02   .1595E+02  -.2329E+02   -26.3
Stress BOT   .1336E+01  -.1553E+02  -.9725E+01   .5776E+01  -.1996E+02   -24.5
                        ***PLATE    67***
Stress TOP   .2299E+02  -.1636E+02  -.2081E+02   .3195E+02  -.2532E+02   -23.3
Stress MID   .3659E+01  -.1786E+02  -.1798E+02   .1268E+02  -.2805E+02   -29.6
Stress BOT  -.1567E+02  -.1935E+02  -.1515E+02  -.2250E+01  -.3278E+02   -41.5
                        ***PLATE    68***
Stress TOP   .3288E+02  -.1073E+02  -.1924E+02   .4015E+02  -.1800E+02   -20.7
Stress MID  -.4485E+00  -.1522E+02  -.1914E+02   .1268E+02  -.2835E+02   -34.4
Stress BOT  -.3377E+02  -.1972E+02  -.1904E+02  -.6453E+01  -.4704E+02   -55.1
                        ***PLATE    69***
Stress TOP   .4132E+02  -.3401E+00  -.1681E+02   .4726E+02  -.6279E+01   -19.5
Stress MID  -.3191E+01  -.7792E+01  -.1864E+02   .1329E+02  -.2427E+02   -41.5
Stress BOT  -.4770E+02  -.1524E+02  -.2046E+02  -.5356E+01  -.5759E+02   -64.2
                        ***PLATE    70***
```

COOLING TOWER VIBRATION: INCREASED NODE MODEL

```
IMAGES-3D  s/n:800189                                    03-03-87
                                                         PAGE   11
            =============== I M A G E S  3 D ===============
            = Copyright (c) 1984  Celestial Software Inc. =
            ================================================

            SOLVE PLATE LOADS/STRESSES   Version 1.4  12/01/86

       COOLING TOWER VIBRATION - NODE INCREASE

Mode  1 -
```

Stress	Surf	Sigma X	Sigma Y	Tau XY	Sigma 1	Sigma 2	Angle
Stress		Shear XZ	Shear YZ				
Stress	TOP	.4416E+02	.1363E+02	-.1337E+02	.4918E+02	.8604E+01	-20.6
Stress	MID	-.4199E+01	.4596E+01	-.1598E+02	.1677E+02	-.1637E+02	-52.7
Stress	BOT	-.5256E+02	-.4437E+01	-.1859E+02	.1908E+01	-.5890E+02	-71.2
			***PLATE	71***			
Stress	TOP	.3688E+02	.3001E+02	-.8546E+01	.4265E+02	.2423E+02	-34.1
Stress	MID	-.3501E+01	.2183E+02	-.1093E+02	.2590E+02	-.7567E+01	-69.6
Stress	BOT	-.4388E+02	.1366E+02	-.1332E+02	.1660E+02	-.4681E+02	-77.6
			***PLATE	72***			
Stress	TOP	.1551E+02	.4746E+02	-.1895E+02	.4757E+02	.1540E+02	-86.6
Stress	MID	-.1360E+01	.4342E+02	-.3731E+01	.4373E+02	-.1668E+01	-85.3
Stress	BOT	-.1823E+02	.3938E+02	-.5567E+01	.3991E+02	-.1876E+02	-84.5
			***PLATE	73***			
Stress	TOP	.2378E+02	.1880E+02	-.3867E+02	.6004E+02	-.1746E+02	-43.2
Stress	MID	.1873E+02	.1553E+02	-.2010E+01	.1970E+02	.1456E+02	-25.7
Stress	BOT	.1367E+02	.1225E+02	.3465E+02	.4762E+02	-.2169E+02	44.4
			***PLATE	74***			
Stress	TOP	.2135E+02	-.1412E+01	-.1040E+02	.2538E+02	-.5449E+01	-21.2
Stress	MID	.1257E+02	-.2222E+01	.5509E-01	.1257E+02	-.2222E+01	.2
Stress	BOT	.3783E+01	-.3031E+01	.1051E+02	.1143E+02	-.1067E+02	36.0
			***PLATE	75***			
Stress	TOP	.2374E+02	-.7488E+01	-.7056E+01	.2526E+02	-.9008E+01	-12.2
Stress	MID	.5738E+01	-.9073E+01	-.1430E+01	.5875E+01	-.9210E+01	-5.5
Stress	BOT	-.1226E+02	-.1066E+02	.4195E+01	-.7190E+01	-.1573E+02	50.4
			***PLATE	76***			
Stress	TOP	.2901E+02	-.6819E+01	-.5704E+01	.2989E+02	-.7705E+01	-8.8
Stress	MID	-.1167E-01	-.1121E+02	-.4756E+01	.1736E+01	-.1296E+02	-20.2
Stress	BOT	-.2903E+02	-.1560E+02	-.3809E+01	-.1460E+02	-.3003E+02	-75.2
			***PLATE	77***			

STRUCTURAL MODELS

        SOLVE PLATE LOADS/STRESSES   Version 1.4 12/01/86

      COOLING TOWER VIBRATION - NODE INCREASE

Mode 1 -

| Stress | Surf | Sigma X | Sigma Y | Tau XY | Sigma 1 | Sigma 2 | Angle |
Stress		Shear XZ	Shear YZ				
Stress	TOP	.3527E+02	-.3175E+01	-.6857E+01	.3646E+02	-.4361E+01	-9.8
Stress	MID	-.4176E+01	-.1052E+02	-.8811E+01	.2015E+01	-.1671E+02	-35.1
Stress	BOT	-.4363E+02	-.1787E+02	-.1076E+02	-.1396E+02	-.4753E+02	-70.1
			***PLATE	78***			
Stress	TOP	.3999E+02	.2061E+01	-.9018E+01	.4202E+02	.2580E-01	-12.7
Stress	MID	-.6473E+01	-.7299E+01	-.1207E+02	.5193E+01	-.1896E+02	-44.0
Stress	BOT	-.5294E+02	-.1666E+02	-.1513E+02	-.1118E+02	-.5841E+02	-70.1
			***PLATE	79***			
Stress	TOP	.4014E+02	.8719E+01	-.1067E+02	.4342E+02	.5437E+01	-17.1
Stress	MID	-.6665E+01	-.9627E+00	-.1328E+02	.9770E+01	-.1740E+02	-51.1
Stress	BOT	-.5347E+02	-.1064E+02	-.1589E+02	-.5391E+01	-.5872E+02	-71.7
			***PLATE	80***			
Stress	TOP	.3244E+02	.1700E+02	-.1021E+02	.3752E+02	.1192E+02	-26.5
Stress	MID	-.4827E+01	.9212E+01	-.1137E+02	.1555E+02	-.1117E+02	-60.8
Stress	BOT	-.4210E+02	.1422E+01	-.1253E+02	.4773E+01	-.4545E+02	-75.0
			***PLATE	81***			
Stress	TOP	.1316E+02	.2618E+02	-.5996E+01	.2852E+02	.1082E+02	-68.7
Stress	MID	-.1745E+01	.2298E+02	-.5510E+01	.2415E+02	-.2918E+01	-78.0
Stress	BOT	-.1666E+02	.1978E+02	-.5025E+01	.2046E+02	-.1734E+02	-82.3
			***PLATE	82***			
Stress	TOP	.2698E+02	-.2331E+02	-.2192E+02	.3519E+02	-.3152E+02	-20.5
Stress	MID	-.5999E+00	.8074E+01	.5056E+01	.1040E+02	-.2924E+01	65.3
Stress	BOT	-.2818E+02	.3946E+02	.3203E+02	.5222E+02	-.4094E+02	68.3
			***PLATE	83***			
Stress	TOP	.2174E+02	-.7268E+01	.9255E+00	.2177E+02	-.7298E+01	1.8
Stress	MID	.2561E+01	-.4905E+00	.7805E+01	.8988E+01	-.6918E+01	39.5
Stress	BOT	-.1662E+02	.6288E+01	.1468E+02	.1346E+02	-.2379E+02	64.0
			***PLATE	84***			

COOLING TOWER VIBRATION: INCREASED NODE MODEL

```
IMAGES-3D  s/n:800189                                          03-03-87
                                                              PAGE    13
          =============== I M A G E S  3 D ===============
          = Copyright (c) 1984   Celestial Software Inc. =
          ================================================

          SOLVE PLATE LOADS/STRESSES   Version 1.4  12/01/86

     COOLING TOWER VIBRATION - NODE INCREASE
```

Mode   1 -

Stress	Surf	Sigma X	Sigma Y	Tau XY	Sigma 1	Sigma 2	Angle
Stress		Shear XZ	Shear YZ				
Stress	TOP	.3418E+02	-.3460E+01	.3488E+01	.3450E+02	-.3780E+01	5.2
Stress	MID	.6561E+01	-.2940E+01	.5366E+01	.8976E+01	-.5356E+01	24.2
Stress	BOT	-.2106E+02	-.2421E+01	.7243E+01	.6277E-01	-.2354E+02	71.1
			PLATE 85				
Stress	TOP	.4207E+02	-.9191E+00	.3254E+01	.4232E+02	-.1164E+01	4.3
Stress	MID	.9205E+01	-.3375E+01	.1897E+01	.9484E+01	-.3655E+01	8.4
Stress	BOT	-.2366E+02	-.5832E+01	.5394E+00	-.5815E+01	-.2368E+02	88.3
			PLATE 86				
Stress	TOP	.4768E+02	.9331E+00	.1130E+01	.4770E+02	.9058E+00	1.4
Stress	MID	.9794E+01	-.3327E+01	-.1665E+01	.1000E+02	-.3535E+01	-7.1
Stress	BOT	-.2809E+02	-.7587E+01	-.4460E+01	-.6658E+01	-.2902E+02	-78.2
			PLATE 87				
Stress	TOP	.4899E+02	.2164E+01	-.1976E+01	.4907E+02	.2081E+01	-2.4
Stress	MID	.8282E+01	-.2692E+01	-.4907E+01	.1016E+02	-.4566E+01	-20.9
Stress	BOT	-.3243E+02	-.7547E+01	-.7838E+01	-.5283E+01	-.3469E+02	-73.9
			PLATE 88				
Stress	TOP	.4414E+02	.3355E+01	-.4887E+01	.4472E+02	.2777E+01	-6.7
Stress	MID	.5188E+01	-.1133E+01	-.7180E+01	.9872E+01	-.5818E+01	-33.1
Stress	BOT	-.3377E+02	-.5621E+01	-.9474E+01	-.2729E+01	-.3666E+02	-73.0
			PLATE 89				
Stress	TOP	.3159E+02	.5123E+01	-.6005E+01	.3289E+02	.3824E+01	-12.2
Stress	MID	.1703E+01	.1759E+01	-.7333E+01	.9064E+01	-.5602E+01	-45.1
Stress	BOT	-.2818E+02	-.1604E+01	-.8660E+01	.9690E+00	-.3075E+02	-73.5
			PLATE 90				
Stress	TOP	.1122E+02	.7376E+01	-.3464E+01	.1326E+02	.5337E+01	-30.5
Stress	MID	-.1500E+00	.5956E+01	-.3857E+01	.7822E+01	-.2016E+01	-64.2
Stress	BOT	-.1152E+02	.4536E+01	-.4250E+01	.5592E+01	-.1258E+02	-76.1

STRUCTURAL MODELS   199

```
IMAGES-3D   s/n:800189                              03-03-87
                                                   PAGE   14
            =============== I M A G E S  3 D ===============
            = Copyright (c) 1984  Celestial Software Inc. =
            ===============================================

            SOLVE PLATE LOADS/STRESSES    Version 1.4  12/01/86

        COOLING TOWER VIBRATION - NODE INCREASE

  Mode  1 -
```

```
                    MAXIMUM STRESS SUMMARY FOR PLATES
                    WITHIN SPECIFIED RANGE    1-  90


            Maximum (absolute) Stress = .1503E+03 at Plate   9

                Plate   Sigma  X   Sigma  Y    Tau XY
                -----  ---------- ---------- ----------
                   9   .1715E+02  .1503E+03 -.1352E+02
```

```
IMAGES-3D   s/n:800189                              03-03-87
                                                   PAGE    1
            =============== I M A G E S  3 D ===============
            = Copyright (c) 1984  Celestial Software Inc. =
            ===============================================

            SOLVE PLATE LOADS/STRESSES    Version 1.4  12/01/86

        COOLING TOWER VIBRATION - NODE INCREASE

  ABS
```

```
                    PLATE LOADS AND/OR STRESSES

Stress Surf  Sigma  X   Sigma  Y    Tau XY    Sigma  1   Sigma  2    Angle
Stress       Shear XZ   Shear YZ
------ ----  ---------- ---------- ---------- ---------- ---------- ----------
                       ***PLATE    1***
Stress TOP   .1211E+02  .7446E+02  .3617E+01  .7467E+02  .1190E+02    86.7
Stress MID   .9947E+01  .6641E+02  .1521E+01  .6645E+02  .9906E+01    88.5
Stress BOT   .7788E+01  .5835E+02 -.5746E+00  .5836E+02  .7782E+01   -89.3
                       ***PLATE    2***
Stress TOP   .1210E+02  .7203E+02  .4312E+01  .7234E+02  .1179E+02    85.9
Stress MID   .1089E+02  .6211E+02  .2900E+01  .6227E+02  .1073E+02    86.8
Stress BOT   .9679E+01  .5218E+02  .1489E+01  .5223E+02  .9627E+01    88.0
                       ***PLATE    3***
Stress TOP   .1177E+02  .6539E+02  .5716E+01  .6599E+02  .1117E+02    84.0
Stress MID   .9769E+01  .5362E+02  .4314E+01  .5404E+02  .9348E+01    84.4
Stress BOT   .7764E+01  .4185E+02  .2913E+01  .4209E+02  .7517E+01    85.1
                       ***PLATE    4***
Stress TOP   .1008E+02  .5330E+02  .7440E+01  .5454E+02  .8839E+01    80.5
Stress MID   .7577E+01  .3987E+02  .5818E+01  .4089E+02  .6560E+01    80.1
Stress BOT   .5069E+01  .2645E+02  .4197E+01  .2725E+02  .4274E+01    79.3
                       ***PLATE    5***
Stress TOP   .7315E+01  .3554E+02  .9288E+01  .3832E+02  .4533E+01    73.3
Stress MID   .4455E+01  .2048E+02  .7353E+01  .2334E+02  .1592E+01    68.7
Stress BOT   .1595E+01  .5425E+01  .5419E+01  .9257E+01 -.2237E+01    54.7
                       ***PLATE    6***
Stress TOP   .3429E+01  .2154E+02  .1111E+02  .2682E+02 -.1847E+01    64.6
Stress MID   .5302E+00  .5016E+01  .8967E+01  .1202E+02 -.6470E+01    52.0
Stress BOT  -.2369E+01 -.1151E+02  .6824E+01  .1274E+01 -.1515E+02    28.1
                       ***PLATE    7***
```

COOLING TOWER VIBRATION: INCREASED NODE MODEL

```
IMAGES-3D   s/n:800189                                    03-03-87
                                                         PAGE    2
          =============== I M A G E S   3 D ===============
          = Copyright (c) 1984   Celestial Software Inc. =
          ================================================

          SOLVE PLATE LOADS/STRESSES   Version 1.4  12/01/86

        COOLING TOWER VIBRATION - NODE INCREASE

ABS
```

Stress	Surf	Sigma X	Sigma Y	Tau XY	Sigma 1	Sigma 2	Angle
Stress		Shear XZ	Shear YZ				
Stress	TOP	.6391E+01	.5477E+02	.1258E+02	.5785E+02	.3318E+01	76.3
Stress	MID	.3984E+01	.3744E+02	.1062E+02	.4053E+02	.8974E+00	73.8
Stress	BOT	.1577E+01	.2011E+02	.8666E+01	.2353E+02	-.1844E+01	68.5
			***PLATE	8***			
Stress	TOP	.9717E+01	.9405E+02	.1382E+02	.9626E+02	.7509E+01	80.9
Stress	MID	.8374E+01	.7753E+02	.1276E+02	.7981E+02	.6094E+01	79.9
Stress	BOT	.7030E+01	.6101E+02	.1170E+02	.6343E+02	.4605E+01	78.3
			***PLATE	9***			
Stress	TOP	.1715E+02	.1503E+03	.1352E+02	.1517E+03	.1579E+02	84.3
Stress	MID	.1270E+02	.1372E+03	.1348E+02	.1386E+03	.1125E+02	83.9
Stress	BOT	.8245E+01	.1240E+03	.1345E+02	.1255E+03	.6703E+01	83.5
			***PLATE	10***			
Stress	TOP	.8629E+01	.8014E+02	.4458E+01	.8042E+02	.8353E+01	86.4
Stress	MID	.8102E+01	.6686E+02	.1393E+02	.6690E+02	.8069E+01	88.6
Stress	BOT	.7575E+01	.5359E+02	-.1671E+01	.5365E+02	.7514E+01	-87.9
			***PLATE	11***			
Stress	TOP	.8369E+01	.7586E+02	.7457E+01	.7667E+02	.7555E+01	83.8
Stress	MID	.7241E+01	.6279E+02	.4595E+01	.6317E+02	.6863E+01	85.3
Stress	BOT	.6112E+01	.4972E+02	.1733E+01	.4979E+02	.6044E+01	87.7
			***PLATE	12***			
Stress	TOP	.7398E+01	.6603E+02	.1095E+02	.6801E+02	.5421E+01	79.8
Stress	MID	.5845E+01	.5363E+02	.7816E+01	.5488E+02	.4599E+01	80.9
Stress	BOT	.4291E+01	.4123E+02	.4685E+01	.4182E+02	.3706E+01	82.9
			***PLATE	13***			
Stress	TOP	.5575E+01	.5054E+02	.1416E+02	.5463E+02	.1485E+01	73.9
Stress	MID	.3773E+01	.3907E+02	.1092E+02	.4218E+02	.6669E+00	74.1
Stress	BOT	.1970E+01	.2761E+02	.7678E+01	.2973E+02	-.1536E+00	74.5
			***PLATE	14***			

IMAGES-3D   s/n:800189   03-03-87
                                                          PAGE   3
        =============== I M A G E S 3 D ===============
        = Copyright (c) 1984 Celestial Software Inc. =
        ===

        SOLVE PLATE LOADS/STRESSES   Version 1.4 12/01/86

    COOLING TOWER VIBRATION - NODE INCREASE

ABS

| Stress | Surf | Sigma X | Sigma Y | Tau XY | Sigma 1 | Sigma 2 | Angle |
Stress		Shear XZ	Shear YZ				
Stress	TOP	.2439E+01	.2803E+02	.1689E+02	.3642E+02	-.5951E+01	63.6
Stress	MID	.1226E+01	.1861E+02	.1377E+02	.2620E+02	-.6368E+01	61.1
Stress	BOT	.1229E-01	.9191E+01	.1066E+02	.1621E+02	-.7003E+01	56.6
			PLATE 15				
Stress	TOP	.2037E+01	.1409E+02	.1861E+02	.2762E+02	-.1149E+02	54.0
Stress	MID	.1379E+01	.8257E+01	.1612E+02	.2130E+02	-.1166E+02	51.0
Stress	BOT	.7200E+00	.2425E+01	.1363E+02	.1523E+02	-.1209E+02	46.8
			PLATE 16				
Stress	TOP	.7261E+01	.4249E+02	.1866E+02	.5054E+02	-.7882E+00	66.7
Stress	MID	.3284E+01	.4174E+02	.1748E+02	.4849E+02	-.3474E+01	68.9
Stress	BOT	-.6933E+00	.4098E+02	.1629E+02	.4660E+02	-.6307E+01	71.0
			PLATE 17				
Stress	TOP	.1060E+02	.8791E+02	.1694E+02	.9146E+02	.7046E+01	78.2
Stress	MID	.3248E+01	.8351E+02	.1632E+02	.8670E+02	.5795E-01	78.9
Stress	BOT	-.4101E+01	.7912E+02	.1569E+02	.8198E+02	-.6960E+01	79.7
			PLATE 18				
Stress	TOP	.8554E+01	.1310E+03	.7963E+01	.1315E+03	.8038E+01	86.3
Stress	MID	.1441E+01	.1249E+03	.5493E+01	.1251E+03	.1197E+01	87.5
Stress	BOT	-.5671E+01	.1187E+03	.3022E+01	.1188E+03	-.5745E+01	88.6
			PLATE 19				
Stress	TOP	.2343E+01	.7192E+02	.6154E+01	.7246E+02	.1803E+01	85.0
Stress	MID	.2291E+01	.6502E+02	.2397E+01	.6511E+02	.2200E+01	87.8
Stress	BOT	.2239E+01	.5812E+02	-.1360E+01	.5815E+02	.2206E+01	-88.6
			PLATE 20				
Stress	TOP	.1841E+01	.6775E+02	.1070E+02	.6945E+02	.1468E+00	81.0
Stress	MID	.1649E+01	.6052E+02	.7068E+01	.6136E+02	.8127E+00	83.2
Stress	BOT	.1457E+01	.5329E+02	.3433E+01	.5352E+02	.1231E+02	86.2
			PLATE 21				

COOLING TOWER VIBRATION: INCREASED NODE MODEL

IMAGES-3D   s/n:800189   03-03-87
                                                             PAGE   4
                =============== I M A G E S 3 D ===============
                = Copyright (c) 1984   Celestial Software Inc. =
                ==

                SOLVE PLATE LOADS/STRESSES   Version 1.4 12/01/86

                COOLING TOWER VIBRATION - NODE INCREASE

ABS

Stress	Surf	Sigma X	Sigma Y	Tau XY	Sigma 1	Sigma 2	Angle
Stress		Shear XZ	Shear YZ				
Stress	TOP	.6748E+00	.5826E+02	.1588E+02	.6235E+02	-.3413E+01	75.6
Stress	MID	.5487E+00	.5107E+02	.1161E+02	.5361E+02	-.1993E+01	77.7
Stress	BOT	.4225E+00	.4388E+02	.7350E+01	.4509E+02	-.7870E+00	80.7
				PLATE 22			
Stress	TOP	.9759E+00	.4303E+02	.1993E+02	.5098E+02	-.6970E+01	68.3
Stress	MID	.7646E+00	.3625E+02	.1574E+02	.4222E+02	-.5212E+01	69.2
Stress	BOT	.5532E+00	.2946E+02	.1155E+02	.3351E+02	-.3494E+01	70.7
				PLATE 23			
Stress	TOP	.3622E+01	.2143E+02	.2239E+02	.3662E+02	-.1157E+02	55.8
Stress	MID	.1921E+01	.1570E+02	.1897E+02	.2899E+02	-.1137E+02	55.0
Stress	BOT	.2202E+00	.9960E+01	.1554E+02	.2137E+02	-.1119E+02	53.7
				PLATE 24			
Stress	TOP	.6961E+01	.1473E+02	.2258E+02	.3376E+02	-.1207E+02	49.9
Stress	MID	.2480E+01	.1074E+02	.2064E+02	.2766E+02	-.1443E+02	50.7
Stress	BOT	-.2001E+01	.6757E+01	.1869E+02	.2158E+02	-.1682E+02	51.6
				PLATE 25			
Stress	TOP	.9836E+01	.4492E+02	.1986E+02	.5388E+02	.8802E+00	65.7
Stress	MID	.1993E+01	.4298E+02	.1972E+02	.5092E+02	-.5954E+01	68.0
Stress	BOT	-.5850E+01	.4103E+02	.1958E+02	.4813E+02	-.1295E+02	70.1
				PLATE 26			
Stress	TOP	.9681E+01	.7881E+02	.1535E+02	.8206E+02	.6427E+01	78.0
Stress	MID	.2356E+00	.7830E+02	.1426E+02	.8082E+02	-.2288E+01	80.0
Stress	BOT	-.9210E+01	.7778E+02	.1317E+02	.7973E+02	-.1116E+02	81.6
				PLATE 27			
Stress	TOP	.4670E+01	.1153E+03	.7094E+01	.1158E+03	.4217E+01	86.3
Stress	MID	.1844E+00	.1149E+03	.5539E+01	.1152E+03	-.8236E-01	87.2
Stress	BOT	-.4301E+01	.1145E+03	.3984E+01	.1146E+03	-.4435E+01	88.1
				PLATE 28			

```
IMAGES-3D  s/n:800189                              03-03-87
                                                   PAGE    5
           =============== I M A G E S  3 D ===============
           = Copyright (c) 1984  Celestial Software Inc. =
           ================================================

           SOLVE PLATE LOADS/STRESSES    Version 1.4  12/01/86

        COOLING TOWER VIBRATION - NODE INCREASE

ABS

Stress Surf  Sigma  X    Sigma  Y    Tau XY      Sigma  1    Sigma  2    Angle
Stress       Shear XZ    Shear YZ
------ ----  ----------  ----------  ----------  ----------  ----------  ----------
Stress TOP  .2534E+01   .6830E+02   .8267E+01   .6933E+02   .1511E+01    82.9
Stress MID  .1777E+01   .6004E+02   .3178E+01   .6021E+02   .1604E+01    86.9
Stress BOT  .1020E+01   .5177E+02  -.1911E+01   .5184E+02   .9483E+00   -87.8
                        ***PLATE    29***
Stress TOP  .3138E+01   .6427E+02   .1383E+02   .6725E+02   .1542E+00    77.8
Stress MID  .2155E+01   .5565E+02   .9367E+01   .5725E+02   .5625E+00    80.3
Stress BOT  .1173E+01   .4704E+02   .4904E+01   .4756E+02   .6544E+00    84.0
                        ***PLATE    30***
Stress TOP  .2985E+01   .5433E+02   .2038E+02   .6144E+02  -.4122E+01    70.8
Stress MID  .2796E+01   .4656E+02   .1497E+02   .5119E+02  -.1835E+01    72.8
Stress BOT  .2607E+01   .3879E+02   .9560E+01   .4116E+02   .2362E+00    76.1
                        ***PLATE    31***
Stress TOP  .5186E+01   .3924E+02   .2471E+02   .5222E+02  -.7793E+01    62.3
Stress MID  .3220E+01   .3255E+02   .1936E+02   .4217E+02  -.6399E+01    63.6
Stress BOT  .1253E+01   .2586E+02   .1400E+02   .3220E+02  -.5084E+01    65.7
                        ***PLATE    32***
Stress TOP  .8737E+01   .1859E+02   .2640E+02   .4052E+02  -.1319E+02    50.3
Stress MID  .3143E+01   .1331E+02   .2207E+02   .3088E+02  -.1442E+02    51.5
Stress BOT -.2451E+01   .8023E+01   .1775E+02   .2129E+02  -.1572E+02    53.2
                        ***PLATE    33***
Stress TOP  .1232E+02   .1469E+02   .2491E+02   .3844E+02  -.1143E+02    46.4
Stress MID  .2423E+01   .1104E+02   .2250E+02   .2964E+02  -.1617E+02    50.4
Stress BOT -.7471E+01   .7395E+01   .2009E+02   .2138E+02  -.2146E+02    55.2
                        ***PLATE    34***
Stress TOP  .1418E+02   .4158E+02   .2005E+02   .5217E+02   .3596E+01    62.2
Stress MID  .1186E+01   .3958E+02   .1990E+02   .4804E+02  -.7270E+01    67.0
Stress BOT -.1181E+02   .3758E+02   .1976E+02   .4451E+02  -.1874E+02    70.7
                        ***PLATE    35***
```

COOLING TOWER VIBRATION: INCREASED NODE MODEL

```
                ================ I M A G E S   3 D ================
                = Copyright (c) 1984   Celestial Software Inc. =
                ==================================================
```

            SOLVE PLATE LOADS/STRESSES   Version 1.4 12/01/86

        COOLING TOWER VIBRATION - NODE INCREASE

ABS

Stress	Surf	Sigma X / Shear XZ	Sigma Y / Shear YZ	Tau XY	Sigma 1	Sigma 2	Angle
Stress	TOP	.1243E+02	.7182E+02	.1597E+02	.7584E+02	.8405E+01	75.9
Stress	MID	.2814E+00	.7129E+02	.1434E+02	.7408E+02	-.2505E+01	79.0
Stress	BOT	-.1187E+02	.7076E+02	.1271E+02	.7267E+02	-.1378E+02	81.5
			PLATE 36				
Stress	TOP	.5701E+01	.1052E+03	.8106E+01	.1058E+03	.5045E+01	85.4
Stress	MID	.2441E+00	.1045E+03	.5392E+01	.1048E+03	-.3393E-01	87.0
Stress	BOT	-.5213E+01	.1039E+03	.2679E+01	.1040E+03	-.5279E+01	88.6
			PLATE 37				
Stress	TOP	.1018E+02	.6166E+02	.1186E+02	.6426E+02	.7583E+01	77.6
Stress	MID	.3518E+01	.5092E+02	.4126E+01	.5128E+02	.3162E+01	85.1
Stress	BOT	-.3149E+01	.4019E+02	-.3610E+01	.4049E+02	-.3448E+01	-85.3
			PLATE 38				
Stress	TOP	.6742E+01	.5612E+02	.1694E+02	.6137E+02	.1491E+01	72.8
Stress	MID	.3990E+01	.4770E+02	.1193E+02	.5075E+02	.9473E+00	75.7
Stress	BOT	.1239E+01	.3928E+02	.6918E+01	.4050E+02	.1959E-01	80.0
			PLATE 39				
Stress	TOP	.5196E+01	.4819E+02	.2424E+02	.5909E+02	-.5703E+01	65.8
Stress	MID	.4152E+01	.4071E+02	.1792E+02	.4802E+02	-.3167E+01	67.8
Stress	BOT	.3108E+01	.3322E+02	.1160E+02	.3717E+02	-.8433E+00	71.2
			PLATE 40				
Stress	TOP	.7783E+01	.3481E+02	.2857E+02	.5290E+02	-.1030E+02	57.7
Stress	MID	.3965E+01	.2890E+02	.2208E+02	.4179E+02	-.8920E+01	59.7
Stress	BOT	.1475E+00	.2299E+02	.1558E+02	.3089E+02	-.7753E+01	63.1
			PLATE 41				
Stress	TOP	.1407E+02	.1613E+02	.2949E+02	.4461E+02	-.1441E+02	46.0
Stress	MID	.3344E+01	.1234E+02	.2401E+02	.3227E+02	-.1658E+02	50.3
Stress	BOT	-.7386E+01	.8551E+01	.1852E+02	.2075E+02	-.1958E+02	56.6
			PLATE 42				

STRUCTURAL MODELS

```
IMAGES-3D   s/n:800189                              03-03-87
                                                    PAGE    7
        ================ I M A G E S   3 D ================
        = Copyright (c) 1984   Celestial Software Inc. =
        =================================================

        SOLVE PLATE LOADS/STRESSES    Version 1.4  12/01/86
```

COOLING TOWER VIBRATION - NODE INCREASE

ABS

Stress	Surf	Sigma X	Sigma Y	Tau XY	Sigma 1	Sigma 2	Angle
Stress		Shear XZ	Shear YZ				
Stress	TOP	.1992E+02	.1040E+02	.2668E+02	.4226E+02	-.1193E+02	39.9
Stress	MID	.2324E+01	.8788E+01	.2337E+02	.2915E+02	-.1803E+02	48.9
Stress	BOT	-.1527E+02	.7172E+01	.2006E+02	.1894E+02	-.2704E+02	59.6
		***PLATE	43***				
Stress	TOP	.2263E+02	.3416E+02	.2041E+02	.4960E+02	.7182E+01	52.9
Stress	MID	.1182E+01	.3404E+02	.2004E+02	.4353E+02	-.8306E+01	64.7
Stress	BOT	-.2026E+02	.3392E+02	.1968E+02	.4031E+02	-.2665E+02	72.0
		***PLATE	44***				
Stress	TOP	.1914E+02	.6355E+02	.1646E+02	.6899E+02	.1371E+02	71.7
Stress	MID	.1668E+00	.6251E+02	.1385E+02	.6545E+02	-.2773E+01	78.0
Stress	BOT	-.1881E+02	.6146E+02	.1124E+02	.6301E+02	-.2036E+02	82.2
		***PLATE	45***				
Stress	TOP	.7490E+01	.9338E+02	.9900E+01	.9450E+02	.6363E+01	83.5
Stress	MID	.1684E+00	.9267E+02	.5034E+01	.9294E+02	-.1047E+00	86.9
Stress	BOT	-.7153E+01	.9197E+02	.1681E+00	.9197E+02	-.7153E+01	89.9
		***PLATE	46***				
Stress	TOP	.1743E+02	.3713E+02	.2193E+02	.5132E+02	.3238E+01	57.1
Stress	MID	.4862E+01	.3618E+02	.5522E+01	.3712E+02	.3916E+01	80.3
Stress	BOT	-.7709E+01	.3523E+02	-.1089E+02	.3783E+02	-.1031E+02	-76.6
		***PLATE	47***				
Stress	TOP	.9099E+01	.4620E+02	.2035E+02	.5519E+02	.1168E+00	66.2
Stress	MID	.4611E+01	.3710E+02	.1354E+02	.4200E+02	-.2929E+00	70.1
Stress	BOT	.1234E+00	.2799E+02	.6736E+01	.2953E+02	-.1419E+01	77.1
		***PLATE	48***				
Stress	TOP	.6658E+01	.4289E+02	.2725E+02	.5749E+02	-.7946E+01	61.8
Stress	MID	.4095E+01	.3323E+02	.2003E+02	.4343E+02	-.6103E+01	63.0
Stress	BOT	.1532E+01	.2356E+02	.1281E+02	.2944E+02	-.4345E+01	65.4
		***PLATE	49***				

COOLING TOWER VIBRATION: INCREASED NODE MODEL

```
IMAGES-3D   s/n:800189                                03-03-87
                                                      PAGE    8
          =============== I M A G E S   3 D ===============
          = Copyright (c) 1984   Celestial Software Inc. =
          ================================================

          SOLVE PLATE LOADS/STRESSES    Version 1.4  12/01/86

       COOLING TOWER VIBRATION - NODE INCREASE

ABS

Stress Surf  Sigma  X    Sigma  Y    Tau XY     Sigma  1    Sigma  2     Angle
Stress       Shear XZ    Shear YZ
------ ----  ----------  ----------  ----------  ----------  ----------  ----------
Stress TOP  .9233E+01   .3199E+02   .3075E+02   .5340E+02  -.1218E+02      55.2
Stress MID  .3312E+01   .2536E+02   .2415E+02   .4088E+02  -.1221E+02      57.3
Stress BOT -.2610E+01   .1874E+02   .1754E+02   .2860E+02  -.1247E+02      60.7
                         ***PLATE   50***
Stress TOP  .1916E+02   .1514E+02   .3067E+02   .4788E+02  -.1359E+02      43.1
Stress MID  .2217E+01   .1293E+02   .2551E+02   .3364E+02  -.1850E+02      50.9
Stress BOT -.1472E+02   .1071E+02   .2035E+02   .2199E+02  -.2600E+02      61.0
                         ***PLATE   51***
Stress TOP  .2858E+02   .6762E+01   .2686E+02   .4666E+02  -.1132E+02      33.9
Stress MID  .9544E+00   .4230E+01   .2404E+02   .2669E+02  -.2151E+02      46.9
Stress BOT -.2667E+02   .1698E+01   .2123E+02   .1305E+02  -.3802E+02      61.9
                         ***PLATE   52***
Stress TOP  .3404E+02   .3206E+02   .1998E+02   .5305E+02   .1304E+02      43.6
Stress MID  .2246E+00   .2587E+02   .1981E+02   .3665E+02  -.1055E+02      61.5
Stress BOT -.3359E+02   .1968E+02   .1964E+02   .2614E+02  -.4005E+02      71.8
                         ***PLATE   53***
Stress TOP  .3171E+02   .5843E+02   .1634E+02   .6618E+02   .2396E+02      64.6
Stress MID  .9082E+00   .5113E+02   .1310E+02   .5434E+02  -.2302E+01      76.2
Stress BOT -.2989E+02   .4382E+02   .9854E+01   .4512E+02  -.3119E+02      82.5
                         ***PLATE   54***
Stress TOP  .1358E+02   .8351E+02   .9778E+01   .8485E+02   .1224E+02      82.2
Stress MID  .6222E+00   .7878E+02   .4014E+01   .7898E+02   .4166E+00      87.1
Stress BOT -.1233E+02   .7404E+02  -.1750E+02   .7407E+02  -.1237E+02     -88.8
                         ***PLATE   55***
Stress TOP  .1379E+02   .3306E+02   .2907E+02   .5405E+02  -.7201E+01      54.2
Stress MID  .6446E+01   .1735E+02   .5149E+01   .1940E+02   .4399E+01      68.3
Stress BOT -.8969E+00   .1637E+01  -.1877E+02   .1919E+02  -.1845E+02     -46.9
                         ***PLATE   56***
```

STRUCTURAL MODELS

IMAGES-3D   s/n:800189

03-03-87
PAGE   9
=============== I M A G E S   3 D ===============
= Copyright (c) 1984   Celestial Software Inc. =
==

SOLVE PLATE LOADS/STRESSES   Version 1.4   12/01/86

COOLING TOWER VIBRATION - NODE INCREASE

ABS

Stress	Surf	Sigma X Shear XZ	Sigma Y Shear YZ	Tau XY	Sigma 1	Sigma 2	Angle
Stress	TOP	.1067E+02	.3356E+02	.2055E+02	.4563E+02	-.1406E+01	59.6
Stress	MID	.6179E+01	.2211E+02	.1357E+02	.2988E+02	-.1594E+01	60.2
Stress	BOT	.1686E+01	.1066E+02	.6597E+01	.1415E+02	-.1806E+01	62.1
		PLATE 57					
Stress	TOP	.5513E+01	.3248E+02	.2661E+02	.4883E+02	-.1084E+02	58.4
Stress	MID	.5315E+01	.2387E+02	.2034E+02	.3695E+02	-.7765E+01	57.3
Stress	BOT	.5118E+01	.1525E+02	.1407E+02	.2514E+02	-.4771E+01	54.9
		PLATE 58					
Stress	TOP	.1470E+02	.2631E+02	.2835E+02	.4944E+02	-.8432E+01	50.8
Stress	MID	.3685E+01	.2159E+02	.2393E+02	.3819E+02	-.1292E+02	55.3
Stress	BOT	-.7327E+01	.1687E+02	.1952E+02	.2773E+02	-.1819E+02	60.9
		PLATE 59					
Stress	TOP	.2646E+02	.1461E+02	.2697E+02	.4814E+02	-.7072E+01	38.8
Stress	MID	.1650E+01	.1449E+02	.2476E+02	.3365E+02	-.1751E+02	52.3
Stress	BOT	-.2316E+02	.1436E+02	.2256E+02	.2494E+02	-.3374E+02	64.9
		PLATE 60					
Stress	TOP	.3766E+02	.7168E+01	.2306E+02	.5006E+02	-.5226E+01	28.3
Stress	MID	.2591E+00	.2215E+01	.2296E+02	.2422E+02	-.2174E+02	46.2
Stress	BOT	-.3715E+02	-.2737E+01	.2286E+02	.8667E+01	-.4855E+02	63.5
		PLATE 61					
Stress	TOP	.4550E+02	.2347E+02	.2067E+02	.5790E+02	.1106E+02	31.0
Stress	MID	.1562E+01	.1514E+02	.1868E+02	.2823E+02	-.1152E+02	55.0
Stress	BOT	-.4237E+02	.6818E+01	.1669E+02	.1194E+02	-.4750E+02	72.9
		PLATE 62					
Stress	TOP	.4082E+02	.4571E+02	.1519E+02	.5865E+02	.2788E+02	49.6
Stress	MID	.1906E+01	.3706E+02	.1216E+02	.4086E+02	-.1892E+01	72.7
Stress	BOT	-.3701E+02	.2841E+02	.9136E+01	.2966E+02	-.3826E+02	82.2
		PLATE 63					

COOLING TOWER VIBRATION: INCREASED NODE MODEL

```
IMAGES-3D   s/n:800189                                  03-03-87
                                                        PAGE   10
               =============== I M A G E S   3 D ===============
               = Copyright (c) 1984   Celestial Software Inc. =
               ================================================

               SOLVE PLATE LOADS/STRESSES   Version 1.4  12/01/86

          COOLING TOWER VIBRATION - NODE INCREASE

ABS

Stress Surf  Sigma  X   Sigma  Y   Tau XY    Sigma  1   Sigma  2    Angle
Stress       Shear XZ   Shear YZ
------ ----  ---------- ---------- ---------- ---------- ---------- ----------
Stress TOP  .1796E+02  .6603E+02  .7433E+01  .6716E+02  .1684E+02    81.4
Stress MID  .1050E+01  .6156E+02  .3831E+01  .6180E+02  .8082E+00    86.4
Stress BOT -.1586E+02  .5708E+02  .2296E+00  .5708E+02 -.1586E+02    89.8
                        ***PLATE    64***
Stress TOP  .1354E+02  .1674E+02  .3585E+02  .5103E+02 -.2075E+02    46.3
Stress MID  .1282E+02  .3505E+01  .5542E+01  .1540E+02  .9224E+00    25.0
Stress BOT  .1210E+02 -.9734E+01 -.2477E+02  .2825E+02 -.2589E+02   -33.1
                        ***PLATE    65***
Stress TOP  .1217E+02  .9050E+01  .1960E+02  .3028E+02 -.9056E+01    42.7
Stress MID  .1185E+02  .7961E+01  .1174E+02  .2181E+02 -.1999E+01    40.3
Stress BOT  .1152E+02  .6872E+01  .3881E+01  .1372E+02  .4672E+01    29.5
                        ***PLATE    66***
Stress TOP  .1514E+02  .1561E+02  .2147E+02  .3685E+02 -.6100E+01    45.3
Stress MID  .8236E+01  .1557E+02  .1560E+02  .2793E+02 -.4122E+01    51.6
Stress BOT  .1336E+01  .1553E+02  .9725E+01  .2047E+02 -.3607E+01    63.1
                        ***PLATE    67***
Stress TOP  .2299E+02  .1935E+02  .2081E+02  .4206E+02  .2875E+00    42.5
Stress MID  .3659E+01  .1786E+02  .1798E+02  .3009E+02 -.8572E+01    55.8
Stress BOT -.1567E+02  .1636E+02  .1515E+02  .2239E+02 -.2171E+02    68.3
                        ***PLATE    68***
Stress TOP  .3377E+02  .1972E+02  .1924E+02  .4723E+02  .6262E+01    35.0
Stress MID  .4485E+00  .1522E+02  .1914E+02  .2835E+02 -.1268E+02    55.6
Stress BOT -.3288E+02  .1073E+02  .1904E+02  .1787E+02 -.4002E+02    69.4
                        ***PLATE    69***
Stress TOP  .4770E+02  .1524E+02  .2046E+02  .5759E+02  .5356E+01    25.8
Stress MID  .3191E+01  .7792E+01  .1864E+02  .2427E+02 -.1329E+02    48.5
Stress BOT -.4132E+02  .3401E+00  .1681E+02  .6279E+01 -.4726E+02    70.5
                        ***PLATE    70***
```

IMAGES-3D   s/n:800189   03-03-87
                              PAGE   11
```
================ I M A G E S  3 D ================
= Copyright (c) 1984  Celestial Software Inc. =
===================================================
```

SOLVE PLATE LOADS/STRESSES   Version 1.4 12/01/86

COOLING TOWER VIBRATION - NODE INCREASE

ABS

Stress	Surf	Sigma X	Sigma Y	Tau XY	Sigma 1	Sigma 2	Angle
Stress		Shear XZ	Shear YZ				
Stress	TOP	.5256E+02	.1363E+02	.1859E+02	.6001E+02	.6179E+01	21.8
Stress	MID	.4199E+01	.4596E+01	.1598E+02	.2038E+02	-.1158E+02	45.4
Stress	BOT	-.4416E+02	-.4437E+01	.1337E+02	-.3578E+00	-.4824E+02	73.0
			***PLATE	71***			
Stress	TOP	.4388E+02	.3001E+02	.1332E+02	.5196E+02	.2192E+02	31.2
Stress	MID	.3501E+01	.2183E+02	.1093E+02	.2693E+02	-.1600E+01	65.0
Stress	BOT	-.3688E+02	.1366E+02	.8546E+01	.1507E+02	-.3828E+02	80.7
			***PLATE	72***			
Stress	TOP	.1823E+02	.4746E+02	.5567E+01	.4848E+02	.1720E+02	79.6
Stress	MID	.1360E+01	.4342E+02	.3731E+01	.4375E+02	.1031E+01	85.0
Stress	BOT	-.1551E+02	.3938E+02	.1895E+01	.3944E+02	-.1558E+02	88.0
			***PLATE	73***			
Stress	TOP	.2378E+02	.1880E+02	.3867E+02	.6004E+02	-.1746E+02	43.2
Stress	MID	.1873E+02	.1553E+02	.2010E+01	.1970E+02	.1456E+02	25.7
Stress	BOT	.1367E+02	.1225E+02	-.3465E+02	.4762E+02	-.2169E+02	-44.4
			***PLATE	74***			
Stress	TOP	.2135E+02	.3031E+01	.1051E+02	.2613E+02	-.1752E+01	24.5
Stress	MID	.1257E+02	.2222E+01	.5509E-01	.1257E+02	.2221E+01	.3
Stress	BOT	.3783E+01	.1412E+01	-.1040E+02	.1307E+02	-.7870E+01	-41.7
			***PLATE	75***			
Stress	TOP	.2374E+02	.1066E+02	.7056E+01	.2682E+02	.7578E+01	23.6
Stress	MID	.5738E+01	.9073E+01	.1430E+01	.9603E+01	.5209E+01	69.7
Stress	BOT	-.1226E+02	.7488E+01	-.4195E+01	.8342E+01	-.1312E+02	-78.5
			***PLATE	76***			
Stress	TOP	.2903E+02	.1560E+02	.5704E+01	.3112E+02	.1350E+02	20.2
Stress	MID	.1167E-01	.1121E+02	.4756E+01	.1296E+02	-.1736E+01	69.8
Stress	BOT	-.2901E+02	.6819E+01	.3809E+01	.7219E+01	-.2941E+02	84.0
			***PLATE	77***			

COOLING TOWER VIBRATION: INCREASED NODE MODEL

```
IMAGES-3D   s/n:800189                                    03-03-87
                                                          PAGE   12
        ================ I M A G E S   3 D ================
        = Copyright (c) 1984   Celestial Software Inc. =
        ================================================

           SOLVE PLATE LOADS/STRESSES   Version 1.4  12/01/86

     COOLING TOWER VIBRATION - NODE INCREASE

ABS
```

Stress Stress	Surf	Sigma X Shear XZ	Sigma Y Shear YZ	Tau XY	Sigma 1	Sigma 2	Angle
Stress	TOP	.4363E+02	.1787E+02	.1076E+02	.4753E+02	.1396E+02	19.9
Stress	MID	.4176E+01	.1052E+02	.8811E+01	.1671E+02	-.2015E+01	54.9
Stress	BOT	-.3527E+02	.3175E+01	.6857E+01	.4361E+01	-.3646E+02	80.2
			***PLATE	78***			
Stress	TOP	.5294E+02	.1666E+02	.1513E+02	.5841E+02	.1118E+02	19.9
Stress	MID	.6473E+01	.7299E+01	.1207E+02	.1896E+02	-.5193E+01	46.0
Stress	BOT	-.3999E+02	-.2061E+02	.9018E+01	-.2580E-01	-.4202E+02	77.3
			***PLATE	79***			
Stress	TOP	.5347E+02	.1064E+02	.1589E+02	.5872E+02	.5391E+01	18.3
Stress	MID	.6665E+01	.9627E+00	.1328E+02	.1740E+02	-.9770E+01	38.9
Stress	BOT	-.4014E+02	-.8719E+01	.1067E+02	-.5437E+01	-.4342E+02	72.9
			***PLATE	80***			
Stress	TOP	.4210E+02	.1700E+02	.1253E+02	.4728E+02	.1182E+02	22.5
Stress	MID	.4827E+01	.9212E+01	.1137E+02	.1860E+02	-.4560E+01	50.5
Stress	BOT	-.3244E+02	.1422E+01	.1021E+02	.4262E+01	-.3528E+02	74.5
			***PLATE	81***			
Stress	TOP	.1666E+02	.2618E+02	.5996E+01	.2908E+02	.1376E+02	64.2
Stress	MID	.1745E+01	.2298E+02	.5510E+01	.2433E+02	.4006E+00	76.3
Stress	BOT	-.1316E+02	.1978E+02	.5025E+01	.2053E+02	-.1391E+02	81.5
			***PLATE	82***			
Stress	TOP	.2818E+02	.3946E+02	.3203E+02	.6634E+02	.1293E+01	50.0
Stress	MID	.5999E+00	.8074E+01	.5056E+01	.1062E+02	-.1951E+01	63.2
Stress	BOT	-.2698E+02	-.2331E+02	-.2192E+02	-.3148E+01	-.4714E+02	-47.4
			***PLATE	83***			
Stress	TOP	.2174E+02	.7268E+01	.1468E+02	.3088E+02	-.1866E+01	31.9
Stress	MID	.2561E+01	.4905E+00	.7805E+01	.9399E+01	-.6348E+01	41.2
Stress	BOT	-.1662E+02	-.6288E+01	.9255E+00	-.6205E+01	-.1670E+02	84.9
			***PLATE	84***			

```
IMAGES-3D  s/n:800189                              03-03-87
                                                   PAGE   13
           =============== I M A G E S   3 D ===============
           = Copyright (c) 1984   Celestial Software Inc. =
           ================================================

           SOLVE PLATE LOADS/STRESSES   Version 1.4  12/01/86

       COOLING TOWER VIBRATION - NODE INCREASE

ABS
```

Stress Stress	Surf	Sigma X Shear XZ	Sigma Y Shear YZ	Tau XY	Sigma 1	Sigma 2	Angle
Stress	TOP	.3418E+02	.3460E+01	.7243E+01	.3580E+02	.1838E+01	12.6
Stress	MID	.6561E+01	.2940E+01	.5366E+01	.1041E+02	-.9121E+00	35.7
Stress	BOT	-.2106E+02	.2421E+01	.3488E+01	.2928E+01	-.2157E+02	81.7
			***PLATE	85***			
Stress	TOP	.4207E+02	.5832E+01	.3254E+01	.4236E+02	.5542E+01	5.1
Stress	MID	.9205E+01	.3375E+01	.1897E+01	.9768E+01	.2812E+01	16.5
Stress	BOT	-.2366E+02	.9191E+01	.5394E+00	.9310E+00	-.2368E+02	88.7
			***PLATE	86***			
Stress	TOP	.4768E+02	.7587E+01	.4460E+01	.4817E+02	.7096E+01	6.3
Stress	MID	.9794E+01	.3327E+01	.1665E+01	.1020E+02	.2923E+01	13.6
Stress	BOT	-.2809E+02	-.9331E+01	-.1130E+01	-.8862E+00	-.2814E+02	-87.6
			***PLATE	87***			
Stress	TOP	.4899E+02	.7547E+01	.7838E+01	.5042E+02	.6114E+01	10.4
Stress	MID	.8282E+01	.2692E+01	.4907E+01	.1113E+02	-.1604E+00	30.2
Stress	BOT	-.3243E+02	-.2164E+01	.1976E+01	-.2035E+01	-.3256E+02	86.3
			***PLATE	88***			
Stress	TOP	.4414E+02	.5621E+01	.9474E+01	.4635E+02	.3417E+01	13.1
Stress	MID	.5188E+01	.1133E+01	.7180E+01	.1062E+02	-.4301E+01	37.1
Stress	BOT	-.3377E+02	-.3355E+01	.4887E+01	-.2589E+01	-.3453E+02	81.1
			***PLATE	89***			
Stress	TOP	.3159E+02	.5123E+01	.8660E+01	.3417E+02	.2540E+01	16.6
Stress	MID	.1703E+01	.1759E+01	.7333E+01	.9064E+01	-.5602E+01	45.1
Stress	BOT	-.2818E+02	-.1604E+01	.6005E+01	-.3100E+00	-.2948E+02	77.8
			***PLATE	90***			
Stress	TOP	.1152E+02	.7376E+01	.4250E+01	.1418E+02	.4721E+01	32.0
Stress	MID	.1500E+00	.5956E+01	.3857E+01	.7881E+01	-.1774E+01	63.5
Stress	BOT	-.1122E+02	.4536E+01	.3464E+01	.5264E+01	-.1195E+02	78.1

```
IMAGES-3D  s/n:800189                              03-03-87
                                                   PAGE   14
           =============== I M A G E S   3 D ===============
           = Copyright (c) 1984   Celestial Software Inc. =
           ================================================

           SOLVE PLATE LOADS/STRESSES   Version 1.4  12/01/86

       COOLING TOWER VIBRATION - NODE INCREASE

ABS


                   MAXIMUM STRESS SUMMARY FOR PLATES
                   WITHIN SPECIFIED RANGE    1-  90

       Maximum (absolute) Stress = .1503E+03 at Plate    9
```

Plate	Sigma X	Sigma Y	Tau XY
9	.1715E+02	.1503E+03	.1352E+02

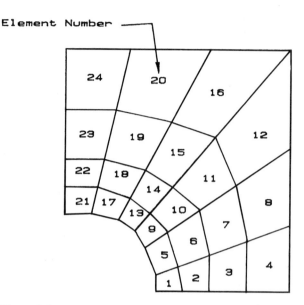

Element Number

Figure 5.3 Square plate with hole (stress analysis)

5.4 STRESS ANALYSIS OF PLATE WITH HOLE

The problem of determining the stress distribution in a plate with a circular hole is one that is treated in most FEA texts in one form or another using the two-dimensional isoparametric quadrilateral element. An application of this typical analysis is the determination of how reducing weight in a system by removing material affects the stress distribution. The problem is illustrated in Figure 5.3. Only one-quarter of the plate is shown and used in the analysis. Symmetry plays an important role in the problem solution, especially from a computational time standpoint.

5.4.1 Input Data

The input data for this problem is similar to the previous problem in that only key data is given. Nodes and elements other than those given in the text are generated using the various node and element generation commands of the particular model. Illustrated in this model is the use of node spacing bias. Node psacing bias is the ratio of the length of the element divisions near the hole to the length of the element divisions along the outer boundary. Again, attention is paid to the element aspect ratio and the use of the four-node quadrilateral elements. A refined model could be done with biasing the nodes and elements toward the high stress region (i.e. around the hole) and by using an eight-node quadrilateral element. Eight-node quadrilateral elements assume a quadratic stress gradient rather than the linear stress gradient of the four-node quadrilateral element. Some finite element analysis programs such as PCTRAN Plus utilize a four-node element with the extra quadratic shape functions of the eight-node element. Tables 5.8, 5.9, 5.10, and 5.11 contain the data necessary to run this particular example.

Table 5.8 Key Nodes for Plate with Hole

Node	x	y	z
1	2	0	0
5	6	0	0
20	6	6	0
35	0	6	0

Table 5.9 Material Properties

Modulus of elasticity	30.0 106 lb/in.^{2}
Density	0.282 lb/in.^{3}
Poisson's ratio	0.3

Table 5.10 Restraints

Node	Ux	Uy
1		0
2		0
3		0
4		0
5		0
31	0	
32	0	
33	0	
34	0	
35	0	

Table 5.11 Applied Forces (lb)

Node	Fx	Fy
5	100	0
20	100	0
10	200	0
15	200	0

5.4.2 Solution

The solution to this problem is illustrated with the program PCTRAN Plus,
following.

PLATE-WITH-HOLE SOLUTION

```
pctran

C:\PCTRAN>echo off


P C T R A N   P l u s
(C) Copyright 1985, 1986   Brooks Scientific, Inc.
Version 3.2


ENTER JOB NAME: fea5



Portions of the PCTRAN menu system are copyrighted
by Alpha Software Corporation.
C:\PCTRAN>PREP1

C:\PCTRAN>echo off

PCTRAN Plus 3.3 PREPROCESSOR
(C) Copyright 1987 Brooks Scientific, Inc.

DEVICE DRIVER IS /PCTIBME.DEV/
MODE =   4
    ... Please wait ...

New job: fea5

* ETYPE,1,6
Element type  1 is STIFF   6    2-D SOLID (MEMBRANE)
IEOPT =  0      4 Nodes; 2 DOF per node.
* EX,1,30E6
Elasticity for material number   1 is   3.0000E+07
* NUXY,1,.3
Poissons ratio for material number   1 is   3.0000E-01
* R,1,1
Real set   1. value(s) =  1.0000E+00 0.0000E+00 0.0000E+00 0.0000E+00
0.0000E+00 0.0000E+00 0.0000E+00 0.0000E+00 0.0000E+00 0.0000E+00
* N,1,2.,0
Node:     1 =   2.0000E+00  0.0000E+00  0.0000E+00
* CSYS,1
Coordinate system  1  center =    .0000    .0000    .0000
* NGEN,6,5,1,,,0.,15,
Generate   6 sets of nodes with nodal increment     5
from node pattern     1 to     1 in steps of    1
DX =     .0000 DY =    15.0000 DZ =     .0000
* CSYS,0
Coordinate system  0  center =    .0000    .0000    .0000
* N,5,6,0
Node:     5 =   6.0000E+00  0.0000E+00  0.0000E+00
* N,20,6,6
Node:    20 =   6.0000E+00  6.0000E+00  0.0000E+00
* N,35,0,6
Node:    35 =   0.0000E+00  6.0000E+00  0.0000E+00
* FILL,5,20,2,10,5
```

```
Fill      2 nodes between nodes      5
to    20     Start node numbering with     10
Incrementing node numbers by      5
Do this generation a total of      1 times
incrementing nodal numbering set by      1
Space =       1.0000
* FILL,20,35,2,25,5
Fill      2 nodes between nodes     20
to    35     Start node numbering with     25
Incrementing node numbers by      5
Do this generation a total of      1 times
incrementing nodal numbering set by      1
Space =       1.0000
* FILL,1,5,,,,7,5,.5
Fill      3 nodes between nodes      1
to     5     Start node numbering with      2
Incrementing node numbers by      1
Do this generation a total of      7 times
incrementing nodal numbering set by      5
Space =        .5000
* E,1,1,2,7,6
Element       1 =    1   1   1      1      2      7      6      0      0      0      0
* EGEN,3,1,1
Generate    3 sets of elements with nodal increment      1
from the element pattern      1 to      1 in steps of      1
* EGEN,5,5,1,4
Generate    5 sets of elements with nodal increment      5
from the element pattern      1 to      4 in steps of      1
* D,1,UY,,5
Define displacement from node      1 with label UY   and value   0.0000E+00
to node      5 in steps of      1
* D,31,UX,,35
Define displacement from node     31 with label UX   and value   0.0000E+00
to node     35 in steps of      1
* F,5,FX,100,20,15
Load case    1
Define force from node      5 with label FX   and value   1.0000E+02
to node     20 in steps of     15
* F,10,FX,200,15,5
Load case    1
Define force from node     10 with label FX   and value   2.0000E+02
to node     15 in steps of      5
* EXIT

    Statistics for job fea5
********************************

Analysis type:  0 STATIC STRESS
Active coordinate system:    0  0.0000E+00  0.0000E+00  0.0000E+00
Number of load cases:  1
Max node number:     35
Number of system nodes:    0
Max element number:     24
Number of applied displacements:     10
Half bandwidth:     0
Input file echo: ON
Reorder load cases in ascending order
Load Case, # of Loads
   1        4
Reorder displacements in ascending order
```

PLATE-WITH-HOLE SOLUTION

```
Copying "PCTX" files back to job fea5
Exit preprocessor .....
C:\PCTRAN>PREP1

C:\PCTRAN>echo off

PCTRAN Plus 3.3 PREPROCESSOR
(C) Copyright 1987 Brooks Scientific, Inc.

DEVICE DRIVER IS /PCTIBME.DEV/
MODE =   4
    ... Please wait ...

Old job: fea5


   Statistics for job fea5
*********************************

Analysis type:  0 STATIC STRESS
Active coordinate system:   0  0.0000E+00  0.0000E+00  0.0000E+00
Number of load cases:  1
Max node number:    35
Number of system nodes:    0
Max element number:    24
Number of applied displacements:    10
Half bandwidth:    0
Input file echo: ON
* SOLVER,6
Use solver number   6
* EXIT

   Statistics for job fea5
*********************************

Analysis type:  0 STATIC STRESS
Active coordinate system:   0  0.0000E+00  0.0000E+00  0.0000E+00
Number of load cases:  1
Max node number:    35
Number of system nodes:    0
Max element number:    24
Number of applied displacements:    10
Half bandwidth:    0
Input file echo: ON
Reorder load cases in ascending order
Load Case, # of Loads
    1      4
Reorder displacements in ascending order
Copying "PCTX" files back to job fea5
Exit preprocessor .....
C:\PCTRAN>STATICS

C:\PCTRAN>echo off

PCTRAN Plus 3.3 NODE RENUMBERING
(C) Copyright 1987 Brooks Scientific, Inc.

MAXN:        35
MAXE:        24
NSN:         35
```

```
NODE RENUMBERING [Y/N/(E to exit)]? Y
Enter number of starting nodes (30 max) NSTART = 1
Enter the  1 starting nodes [N1,N2, ... etc.]: 1

Nodes for element     1 renumbered
Nodes for element     2 renumbered
Nodes for element     5 renumbered
Nodes for element     6 renumbered
Nodes for element     3 renumbered
Nodes for element     7 renumbered
Nodes for element     9 renumbered
Nodes for element    10 renumbered
Nodes for element    11 renumbered
Nodes for element     4 renumbered
Nodes for element     8 renumbered
Nodes for element    12 renumbered
Nodes for element    13 renumbered
Nodes for element    14 renumbered
Nodes for element    15 renumbered
Nodes for element    16 renumbered
Nodes for element    17 renumbered
Nodes for element    18 renumbered
Nodes for element    19 renumbered
Nodes for element    20 renumbered
Nodes for element    21 renumbered
Nodes for element    22 renumbered
Nodes for element    23 renumbered
Nodes for element    24 renumbered

NUMBER OF NODES          =      35
NUMBER OF SYSTEM NODES =        35
HALF BANDWIDTH           =      22
DEGREES OF FREEDOM       =      70

NODE RENUMBERING [Y/N/(E to exit)]? E

PCTRAN Plus 3.3 STIFFNESS ASSEMBLY
(C) Copyright 1987 Brooks Scientific, Inc.

Assemble stiffness for use with solver   6
Max node number:      35
Max element number:      24
Half bandwidth:      22
Total DOF:      70
Assembling element      1
Assembling element      2
Assembling element      3
Assembling element      4
Assembling element      5
Assembling element      6
Assembling element      7
Assembling element      8
Assembling element      9
Assembling element     10
Assembling element     11
Assembling element     12
Assembling element     13
Assembling element     14
Assembling element     15
```

PLATE-WITH-HOLE SOLUTION

```
Assembling element    16
Assembling element    17
Assembling element    18
Assembling element    19
Assembling element    20
Assembling element    21
Assembling element    22
Assembling element    23
Assembling element    24
EXIT STIFF . . .

PCTRAN Plus 3.3 LOAD CONVERSION
(C) Copyright 1987 Brooks Scientific, Inc.

NUMBER OF LOAD CASES:        1
# OF LOADS IN CASE  1:      4
NUMBER OF DISPLACEMENTS:     10

Applied displacements converted:      10

Load Case:   1
Forces converted:     4

Finished load conversion . . .
                    QUICKSOLV (tm) Version 1.0
              Copyright (C) TEKTON ENGINEERING INC. 1987
                       All rights reserved.

                    44 Concord Ave., Suite 106
                       Cambridge, MA 02138
                         (617) 576-3729

                Licensed to be used with PCTRAN (tm) Plus.
                Distributed by Brooks Scientific, Inc.

Number of Equations = 70
Matrix Size         = 941 elements
Half-Bandwidth      = 22
Mean Half-Bandwidth = 13
Memory allocated    = 9K

Assemble
Impose displacements
        10 specified displacements
Triangularize
Backsubstitute
        Load vector 1

Timing
        Assemble :              0.27 sec.
        Triangularize :         0.49 sec.
        Backsubstitute :        1.87 sec.
        Total :                 2.64 sec.

              QUICKSOLV (tm) Version 1.0 (Reactions Module)
              Copyright (C) TEKTON ENGINEERING INC. 1987
                       All rights reserved.

                    44 Concord Ave., Suite 106
```

Cambridge, MA 02138
(617) 576-3729

Licensed to be used with PCTRAN (tm) Plus.
Distributed by Brooks Scientific, Inc.

```
Number of Equations     = 70
Matrix Size             = 941 elements
Half-Bandwidth          = 22
Average Half-Bandwidth  = 14
Memory allocated        = 9K
```

Compute Reactions
        Load Case 1

PCTRAN Plus 3.3 RESULT CONVERSION
(C) Copyright 1987 Brooks Scientific, Inc.

```
Deflections converted:    70
Reactions converted:      10
```

PCTRAN Plus 3.3 STRESS CALCULATION
(C) Copyright 1987 Brooks Scientific, Inc.

```
READ POST CODES FROM FILE "PCTRAN.CSD"
MAX NODE NUMBER:       35
MAX ELEMENT NUMBER:     24
HALF BANDWIDTH:      22
TOTAL DOF:       70
NUMBER OF SYSTEM NODES:    35
STRESS CALCULATION: ELEMENT     1
STRESS CALCULATION: ELEMENT     6
STRESS CALCULATION: ELEMENT    11
STRESS CALCULATION: ELEMENT    16
STRESS CALCULATION: ELEMENT    21
EXIT STRESS ....
C:\PCTRAN>POST1
```

C:\PCTRAN>echo off

PCTRAN Plus 3.3 POSTPROCESSOR
(C) Copyright 1987 Brooks Scientific, Inc.

```
DEVICE DRIVER IS /PCTIBME.DEV/
MODE =    4
        ... Please wait ...
```

JOBNAME: fea5

CALCULATE MIN,MAX POST DATA

   Statistics for fea5
****

```
Analysis type:  0 STATIC STRESS
Active coordinate system:    0  0.0000E+00  0.0000E+00  0.0000E+00
Number of load cases:  1
Current load case:  1
Max node number:      35
Max element number:     24
```

PLATE-WITH-HOLE SOLUTION

```
Number of displacemenets:      10
Half bandwidth:      22
Reading POST codes from file "PCTRAN.CSD"
* PRDISP
Print displacements for nodes NODE1 to NODE2 in steps of NINC 1,35,1
                    DISPLACEMENTS: LOAD CASE   1
```

NODE	UX	UY	UZ	ROTX	ROTY	ROTZ
1	2.5971E-05	0.0000E+00	0.0000E+00	0.0000E+00	0.0000E+00	0.0000E-
2	2.7104E-05	0.0000E+00	0.0000E+00	0.0000E+00	0.0000E+00	0.0000E
3	2.8198E-05	0.0000E+00	0.0000E+00	0.0000E+00	0.0000E+00	0.0000E-
4	3.0508E-05	0.0000E+00	0.0000E+00	0.0000E+00	0.0000E+00	0.0000E-
5	3.4340E-05	0.0000E+00	0.0000E+00	0.0000E+00	0.0000E+00	0.0000E-
6	2.5373E-05	-2.7701E-06	0.0000E+00	0.0000E+00	0.0000E+00	0.0000E-
7	2.5271E-05	-1.1985E-06	0.0000E+00	0.0000E+00	0.0000E+00	0.0000E-
8	2.5934E-05	-7.0473E-07	0.0000E+00	0.0000E+00	0.0000E+00	0.0000E-
9	2.7936E-05	-3.0725E-07	0.0000E+00	0.0000E+00	0.0000E+00	0.0000E-
10	3.0413E-05	2.2936E-06	0.0000E+00	0.0000E+00	0.0000E+00	0.0000E-
11	2.2409E-05	-5.3455E-06	0.0000E+00	0.0000E+00	0.0000E+00	0.0000E-
12	2.0715E-05	-2.9511E-06	0.0000E+00	0.0000E+00	0.0000E+00	0.0000E-
13	2.1003E-05	-2.5467E-06	0.0000E+00	0.0000E+00	0.0000E+00	0.0000E-
14	2.2278E-05	-1.5870E-06	0.0000E+00	0.0000E+00	0.0000E+00	0.0000E-
15	2.3697E-05	1.4507E-06	0.0000E+00	0.0000E+00	0.0000E+00	0.0000E-
16	1.7823E-05	-7.5404E-06	0.0000E+00	0.0000E+00	0.0000E+00	0.0000E-
17	1.5303E-05	-5.5276E-06	0.0000E+00	0.0000E+00	0.0000E+00	0.0000E-
18	1.6059E-05	-4.9993E-06	0.0000E+00	0.0000E+00	0.0000E+00	0.0000E-
19	1.6907E-05	-3.4635E-06	0.0000E+00	0.0000E+00	0.0000E+00	0.0000E-
20	1.7593E-05	-6.9544E-07	0.0000E+00	0.0000E+00	0.0000E+00	0.0000E-
21	1.2692E-05	-9.5905E-06	0.0000E+00	0.0000E+00	0.0000E+00	0.0000E-
22	1.0218E-05	-8.8087E-06	0.0000E+00	0.0000E+00	0.0000E+00	0.0000E-
23	1.0562E-05	-8.4416E-06	0.0000E+00	0.0000E+00	0.0000E+00	0.0000E-
24	1.1123E-05	-7.8414E-06	0.0000E+00	0.0000E+00	0.0000E+00	0.0000E-
25	1.0801E-05	-6.5621E-06	0.0000E+00	0.0000E+00	0.0000E+00	0.0000E-
26	6.6480E-06	-1.1261E-05	0.0000E+00	0.0000E+00	0.0000E+00	0.0000E-
27	4.9855E-06	-1.1878E-05	0.0000E+00	0.0000E+00	0.0000E+00	0.0000E-
28	5.0469E-06	-1.1987E-05	0.0000E+00	0.0000E+00	0.0000E+00	0.0000E-
29	5.3230E-06	-1.2174E-05	0.0000E+00	0.0000E+00	0.0000E+00	0.0000E-
30	4.4336E-06	-1.2383E-05	0.0000E+00	0.0000E+00	0.0000E+00	0.0000E-
31	0.0000E+00	-1.2546E-05	0.0000E+00	0.0000E+00	0.0000E+00	0.0000E-
32	0.0000E+00	-1.3301E-05	0.0000E+00	0.0000E+00	0.0000E+00	0.0000E-
33	0.0000E+00	-1.3703E-05	0.0000E+00	0.0000E+00	0.0000E+00	0.0000E-
34	0.0000E+00	-1.4380E-05	0.0000E+00	0.0000E+00	0.0000E+00	0.0000E-
35	0.0000E+00	-1.5106E-05	0.0000E+00	0.0000E+00	0.0000E+00	0.0000E-

```
                              ** MAX DEFLECTION **
NODE:         5          35          0           0           0           0
          3.4340E-05 -1.5106E-05  0.0000E+00  0.0000E+00  0.0000E+00  0.0000E-
* PRRFOR

Reaction forces for load case   1
```

NODE	LABEL	RFORCE
1	UY	4.1327E+01
2	UY	3.9691E+01
3	UY	1.9179E+00
4	UY	-3.5944E+01
5	UY	-4.6992E+01
31	UX	-1.1491E+02

```
  32      UX       -1.6769E+02
  33      UX       -1.4577E+02
  34      UX       -1.0323E+02
  35      UX       -1.5825E+02
```

TOTALS:

```
     FX          FY          FZ          MX          MY          MZ

 -6.8985E+02  3.8147E-06  0.0000E+00  0.0000E+00  0.0000E+00  0.0000E+00
```
* PRNSTR
Print nodal stresses for nodes NODE1 to NODE2 in steps of NINC 1,35,1
                    Nodal Stresses: Load Case 1

NODE	SX	SY	SZ	SXY	SYZ	SXZ	SIGE
1	.3097E+01	-.1596E+03	.0000E+00	-.1076E+02	.0000E+00	.0000E+00	.0000E+00
2	.1744E+02	-.3865E+02	.0000E+00	-.1060E+02	.0000E+00	.0000E+00	.0000E+00
3	.5222E+02	-.3217E+01	.0000E+00	-.1866E+02	.0000E+00	.0000E+00	.0000E+00
4	.8464E+02	.1937E+02	.0000E+00	-.1680E+02	.0000E+00	.0000E+00	.0000E+00
5	.9513E+02	.6294E+02	.0000E+00	-.5350E+01	.0000E+00	.0000E+00	.0000E+00
6	-.8435E+01	-.1318E+02	.0000E+00	.7279E+01	.0000E+00	.0000E+00	.0000E+00
7	.3985E+02	-.3553E+02	.0000E+00	-.1861E+02	.0000E+00	.0000E+00	.0000E+00
8	.7602E+02	-.1014E+02	.0000E+00	-.2491E+02	.0000E+00	.0000E+00	.0000E+00
9	.9080E+02	.1280E+02	.0000E+00	-.1340E+02	.0000E+00	.0000E+00	.0000E+00
10	.1008E+03	.4113E+02	.0000E+00	-.8018E+01	.0000E+00	.0000E+00	.0000E+00
11	.5875E+02	-.5467E+02	.0000E+00	-.2182E+02	.0000E+00	.0000E+00	.0000E+00
12	.9282E+02	-.3375E+02	.0000E+00	-.3089E+02	.0000E+00	.0000E+00	.0000E+00
13	.1135E+03	-.1798E+02	.0000E+00	-.2404E+02	.0000E+00	.0000E+00	.0000E+00
14	.1111E+03	.5499E+01	.0000E+00	-.1384E+02	.0000E+00	.0000E+00	.0000E+00
15	.1057E+03	.9282E+01	.0000E+00	-.6313E+01	.0000E+00	.0000E+00	.0000E+00
16	.1488E+03	-.2841E+01	.0000E+00	-.4695E+02	.0000E+00	.0000E+00	.0000E+00
17	.1418E+03	-.2862E+01	.0000E+00	-.2431E+02	.0000E+00	.0000E+00	.0000E+00
18	.1355E+03	-.1378E+02	.0000E+00	-.9504E+01	.0000E+00	.0000E+00	.0000E+00
19	.1145E+03	-.1447E+01	.0000E+00	-.3480E+01	.0000E+00	.0000E+00	.0000E+00
20	.1026E+03	-.2746E+01	.0000E+00	.8784E+00	.0000E+00	.0000E+00	.0000E+00
21	.2329E+03	.3116E+02	.0000E+00	-.6134E+02	.0000E+00	.0000E+00	.0000E+00
22	.1689E+03	-.1390E+02	.0000E+00	-.1151E+02	.0000E+00	.0000E+00	.0000E+00
23	.1436E+03	-.4274E+01	.0000E+00	.5183E+01	.0000E+00	.0000E+00	.0000E+00
24	.1121E+03	-.3501E+01	.0000E+00	.6968E+01	.0000E+00	.0000E+00	.0000E+00
25	.9670E+02	-.6646E+01	.0000E+00	.8162E+01	.0000E+00	.0000E+00	.0000E+00
26	.3439E+03	.4892E+02	.0000E+00	-.4596E+02	.0000E+00	.0000E+00	.0000E+00
27	.1963E+03	.1635E+02	.0000E+00	.5299E+01	.0000E+00	.0000E+00	.0000E+00
28	.1479E+03	.1613E+02	.0000E+00	.1354E+02	.0000E+00	.0000E+00	.0000E+00
29	.1117E+03	.5866E+01	.0000E+00	.1004E+02	.0000E+00	.0000E+00	.0000E+00
30	.8032E+02	-.2292E+01	.0000E+00	.1417E+02	.0000E+00	.0000E+00	.0000E+00
31	.4122E+03	.8969E+02	.0000E+00	.2693E+02	.0000E+00	.0000E+00	.0000E+00
32	.2071E+03	.3836E+02	.0000E+00	.2080E+02	.0000E+00	.0000E+00	.0000E+00
33	.1468E+03	.2813E+02	.0000E+00	.1787E+02	.0000E+00	.0000E+00	.0000E+00
34	.1108E+03	.1594E+02	.0000E+00	.1680E+02	.0000E+00	.0000E+00	.0000E+00
35	.6769E+02	.3967E+01	.0000E+00	.1571E+02	.0000E+00	.0000E+00	.0000E+00

```
                        ** MAX STRESS **
NODE:     31          1          0         21          0          0          0
        .4122E+03 -.1596E+03 .0000E+00 -.6134E+02 .0000E+00 .0000E+00 .0000E+00
*
```

PLATE-WITH-HOLE SOLUTION

```
    PLOT
 * EXIT
EXIT POSTPROCESSOR .....
C:\PCTRAN>PCTM/V

C:\PCTRAN>
```

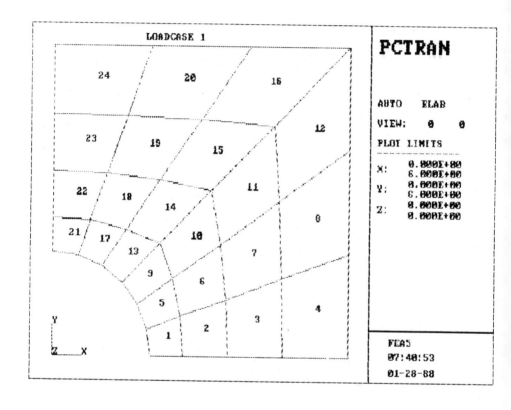

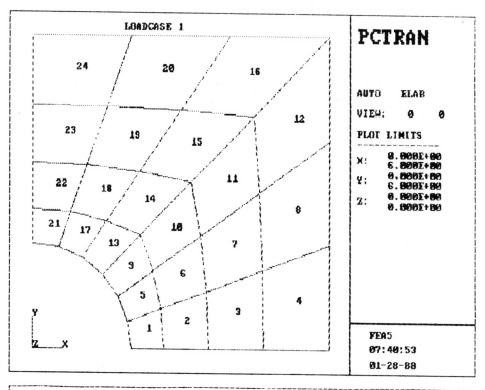

LOADCASE 1

PCTRAN

AUTO   ELAB
VIEW: 0   0

PLOT LIMITS

X: 0.000E+00
      6.000E+00
Y: 0.000E+00
      6.000E+00
Z: 0.000E+00
      0.000E+00

FEA5
07:40:53
01-28-88

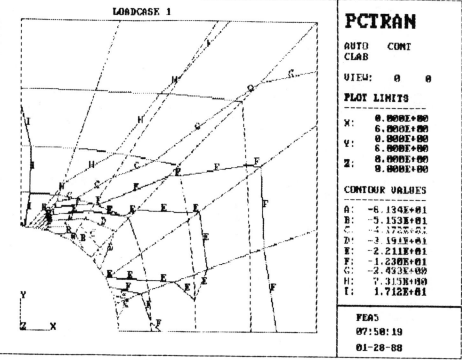

LOADCASE 1

PCTRAN

AUTO   CONT
CLAB
VIEW: 0   0

PLOT LIMITS

X: 0.000E+00
      6.000E+00
Y: 0.000E+00
      6.000E+00
Z: 0.000E+00
      0.000E+00

CONTOUR VALUES

A: -6.134E+01
B: -5.153E+01
C: -4.172E+01
D: -3.191E+01
E: -2.211E+01
F: -1.230E+01
G: -2.493E+00
H: 7.315E+00
I: 1.712E+01

FEA5
07:50:19
01-28-88

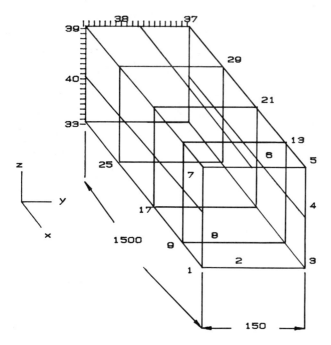

Figure 5.4 Torsion of a square box beam

5.5 TORSION OF A SQUARE BOX BEAM

This problem examines the response of a square box beam when subjected to torsion. Although simple in concept, a problem of this type can be extended to any box beam type element such as an aircraft wing. This problem uses the three-dimensional plate element. Figure 5.4 illustrates the problem as posed.

5.5.1 Input Data

All input data for the model is given in Tables 5.12, 5.13, and 5.14. Only the key nodes are given since the remainder of the problem can be generated from the key nodes. The problem can be run with either the triangular plate element or the quad-plate element.

5.5.2 Solution

The solution to the box beam problem is illustrated with the use of the Algor SUPERSAP FEA program. The "classical" solution for this problem is as follows:

Table 5.12 Key Nodes for Box Beam

Node	x	y	z
1	1500	-75	-75
2	1500	0	-75
3	1500	75	-75
4	1500	75	0
5	1500	75	75
6	1500	0	75
7	1500	-75	75
8	1500	-75	0

Table 5.13 Material Properties

Modulus of elasticity	$7.5 \ \text{lb/in.}^2$
Poisson's ratio	0.3

Table 5.14 Applied Forces for Box Beam

Node	Dir	Value
1	y	.25
1	z	-.25
2	y	.25
2	y	.50
2	z	.25
3	z	.25
4	z	.50
5	y	-.25
6	y	-.50
7	y	-.25
7	z	-.25
8	z	-.50

$$= T/2at$$
$$= T/taG$$
$$O = 1$$

Here,

a   =   150
t   =   3
1   =   1500

and

$$T = 2 (1 \times 150) = 300$$

Substituting the values in the first three equations gives

$$T = 0.002222$$
$$O = 0.0154074$$

The Algor results are given from the calculated data.

$$T = 0.00222$$
$$0 = \tan-1 ((1.1554 + 1.1554)/150) = 0.015405$$

SOLUTION TO BOX BEAM PROBLEM

MENUS

```
                        S U P E R S A P
                            AEDIT
        (c) Copyright Algor Interactive Systems, Inc.
                        1984,1985,1986


**** FILE NAME >> VE03001
**** EXISTING FILE:  VE03001
**** READ THIS ONE  (Y/N) >> YES
**** READING FILE . . .

**** ACTION (-1 GIVES OPTIONS) >> -1

    ACTIONS:

        1. HEADER CARD
        2. MASTER CONTROL CARD
        3. NODE CARDS
        4. ELEMENT CARDS
        5. LOAD CASE CARDS
        6. ELEMENT LOAD MULTIPLIERS
        7. WEIGHT - CENTER OF GRAVITY
        8. NATURAL FREQUENCY
       11. CENTRIFUGAL LOADS
       12. PIECUT BOUNDARY ELEMENT GENERATION
       13. FOUR NODE BOUNDARY ELEMENT GENERATION (FOR PLATES)
       14. PRESSURE LOAD GENERATION (MEMBRANES)
       19. UNUSED NODE AND ELEMENT TRIMMING
       20. STORE DATA
       21. CREATE DRAWING FILE AND QUIT
       22. STORE DATA, CREATE DRAWING FILE AND QUIT
       23. QUIT WITHOUT STORING ANY DATA
       29. STORE DATA AND QUIT

**** ACTION (-1 GIVES OPTIONS) >> 3

**** SECTION III - NODAL DATA ****

    CONSIDER TEMPERATURES  (Y/N) >> NO

**** NODAL DATA (-1 GIVES OPTIONS; <CR> TO CONTINUE) >> -1

    NODAL DATA OPTIONS:

        1. INPUT INDIVIDUAL NODES
        2. GENERATE NODES
        3. CHANGE NODAL DATA
        4. VALUE SELECTED CHANGES
        5. NODAL DATA DISPLAY
        6. COORDINATE DATA INPUT FROM AUXILARY FILE
        7. COPY NODE DATA
```

SOLUTION TO BOX BEAM PROBLEM

```
**** NODAL DATA (-1 GIVES OPTIONS; <CR> TO CONTINUE) >> 5
**** SUBACTION 5: NODAL DATA DISPLAY

        DISPLAY ONE, SOME, OR ALL (<CR> TO CONTINUE) >> ALL
```

NODE NO.	DEGREES OF FREEDOM	------X------	------Y------	------Z------	TEMPERATURE	SYS
1	0 0 0 1 1 1	1500.00000000	-75.00000000	-75.00000000	.0000	0
2	0 0 0 1 1 1	1500.00000000	.00000000	-75.00000000	.0000	0
3	0 0 0 1 1 1	1500.00000000	75.00000000	-75.00000000	.0000	0
4	0 0 0 1 1 1	1500.00000000	75.00000000	.00000000	.0000	0
5	0 0 0 1 1 1	1500.00000000	75.00000000	75.00000000	.0000	0
6	0 0 0 1 1 1	1500.00000000	.00000000	75.00000000	.0000	0
7	0 0 0 1 1 1	1500.00000000	-75.00000000	75.00000000	.0000	0
8	0 0 0 1 1 1	1500.00000000	-75.00000000	.00000000	.0000	0
9	0 0 0 1 1 1	1125.00000000	-75.00000000	-75.00000000	.0000	0
10	0 0 0 1 1 1	1125.00000000	.00000000	-75.00000000	.0000	0
11	0 0 0 1 1 1	1125.00000000	75.00000000	-75.00000000	.0000	0
12	0 0 0 1 1 1	1125.00000000	75.00000000	.00000000	.0000	0
13	0 0 0 1 1 1	1125.00000000	75.00000000	75.00000000	.0000	0
14	0 0 0 1 1 1	1125.00000000	.00000000	75.00000000	.0000	0
15	0 0 0 1 1 1	1125.00000000	-75.00000000	75.00000000	.0000	0
16	0 0 0 1 1 1	1125.00000000	-75.00000000	.00000000	.0000	0
17	0 0 0 1 1 1	750.00000000	-75.00000000	-75.00000000	.0000	0
18	0 0 0 1 1 1	750.00000000	.00000000	-75.00000000	.0000	0
19	0 0 0 1 1 1	750.00000000	75.00000000	-75.00000000	.0000	0
20	0 0 0 1 1 1	750.00000000	75.00000000	.00000000	.0000	0
21	0 0 0 1 1 1	750.00000000	75.00000000	75.00000000	.0000	0
22	0 0 0 1 1 1	750.00000000	.00000000	75.00000000	.0000	0
23	0 0 0 1 1 1	750.00000000	-75.00000000	75.00000000	.0000	0
24	0 0 0 1 1 1	750.00000000	-75.00000000	.00000000	.0000	0
25	0 0 0 1 1 1	375.00000000	-75.00000000	-75.00000000	.0000	0
26	0 0 0 1 1 1	375.00000000	.00000000	-75.00000000	.0000	0
27	0 0 0 1 1 1	375.00000000	75.00000000	-75.00000000	.0000	0
28	0 0 0 1 1 1	375.00000000	75.00000000	.00000000	.0000	0
29	0 0 0 1 1 1	375.00000000	75.00000000	75.00000000	.0000	0
30	0 0 0 1 1 1	375.00000000	.00000000	75.00000000	.0000	0
31	0 0 0 1 1 1	375.00000000	-75.00000000	75.00000000	.0000	0
32	0 0 0 1 1 1	375.00000000	-75.00000000	.00000000	.0000	0
33	1 1 1 1 1 1	.00000000	-75.00000000	-75.00000000	.0000	0
34	1 1 1 1 1 1	.00000000	.00000000	-75.00000000	.0000	0
35	1 1 1 1 1 1	.00000000	75.00000000	-75.00000000	.0000	0
36	1 1 1 1 1 1	.00000000	75.00000000	.00000000	.0000	0
37	1 1 1 1 1 1	.00000000	75.00000000	75.00000000	.0000	0
38	1 1 1 1 1 1	.00000000	.00000000	75.00000000	.0000	0
39	1 1 1 1 1 1	.00000000	-75.00000000	75.00000000	.0000	0
40	1 1 1 1 1 1	.00000000	-75.00000000	.00000000	.0000	0

```
        DISPLAY ONE, SOME, OR ALL (<CR> TO CONTINUE) >>

**** NODAL DATA (-1 GIVES OPTIONS; <CR> TO CONTINUE) >>
**** ACTION (-1 GIVES OPTIONS) >> 4

**** SECTION IV - ELEMENT DATA ****
```

```
**** ELEMENT SUBACTION (-1 GIVES OPTIONS; <CR> TO CONTINUE) >> -1

    AVAILABLE SUBACTIONS FOR ELEMENT DATA:

        1. SWITCH TWO GROUPS
        2. MOVE GROUP TO POSITION
        3. DELETE GROUP
        4. UNDELETE DELETED GROUP
        5. ADD NEW ELEMENT GROUP
        6. EDIT EXISTING GROUP
        7. LIST ACTIVE GROUPS
        8. LIST DELETED GROUPS
        9. EXPUNGE DELETED GROUPS


**** ELEMENT SUBACTION (-1 GIVES OPTIONS; <CR> TO CONTINUE) >> 7

    CURRENT SAP-FILE WRITE ORDER:

    1.   32 Stress elements.


**** ELEMENT SUBACTION (-1 GIVES OPTIONS; <CR> TO CONTINUE) >>

**** ACTION (-1 GIVES OPTIONS) >> 5


**** SECTION V - NODAL LOAD DATA ****


**** NODAL LOAD DATA (-1 GIVES OPTIONS; <CR> TO CONTINUE) >> -1

    NODAL LOAD DATA OPTIONS:

        1. INPUT NODAL LOADS/MASSES ONE NODE AT A TIME
        2. NODAL LOAD/MASS GENERATION
        3. DELETE NODAL LOAD/MASS DATA
        4. CHANGE NODAL LOAD/MASS DATA
        5. DISPLAY NODAL LOAD/MASS DATA


**** NODAL LOAD DATA (-1 GIVES OPTIONS; <CR> TO CONTINUE) >> 5

**** SUBACTION 5: DISPLAY NODAL LOAD/MASS DATA

    DISPLAY ONE, SOME, OR ALL (<CR> TO CONTINUE) >> ALL
```

NODE	LOAD	----------FORCES----------			----------MOMENTS----------			
1	1	.00E+00	2.50E-01	-2.50E-01	.00E+00	.00E+00	.00E+00	FORCE
2	1	.00E+00	5.00E-01	.00E+00	.00E+00	.00E+00	.00E+00	FORCE
3	1	.00E+00	2.50E-01	2.50E-01	.00E+00	.00E+00	.00E+00	FORCE
4	1	.00E+00	.00E+00	5.00E-01	.00E+00	.00E+00	.00E+00	FORCE
5	1	.00E+00	-2.50E-01	2.50E-01	.00E+00	.00E+00	.00E+00	FORCE
6	1	.00E+00	-5.00E-01	.00E+00	.00E+00	.00E+00	.00E+00	FORCE
7	1	.00E+00	-2.50E-01	-2.50E-01	.00E+00	.00E+00	.00E+00	FORCE
8	1	.00E+00	.00E+00	-5.00E-01	.00E+00	.00E+00	.00E+00	FORCE

SOLUTION TO BOX BEAM PROBLEM

```
DISPLAY ONE, SOME, OR ALL (<CR> TO CONTINUE) >>

**** NODAL LOAD DATA (-1 GIVES OPTIONS; <CR> TO CONTINUE) >>

**** ACTION (-1 GIVES OPTIONS) >> 22

**** EDIT OPTION 22 - STORE DATA, CREATE DRAWING FILE AND QUIT ****

**** FILE NAME (<CR> GIVES CURRENT NAME) >>
**** OK TO SAVE DATA IN FILE:   VE03001 >> NO
**** DATA HAS BEEN STORED IN FILE:   VE03001
**** DRAWING FILE CREATED:   VE03001.DI
**** DELETING 'ELE001.A01'.
**** DELETING 'ELE001.M01'.
**** DELETING 'ELE001.N01'.
**** DELETING 'LOD001'.
     BEGIN EDITING A DIFFERENT FILE  (Y/N) >> NO


**** TOTAL ELAPSED TIME >> 11.328 MINUTES


Done executing program. Hit any key.
                    A L G O R   S U P E R S A P
          Linear Stress Analysis - Version 7.613
        (c) Copyright Algor Interactive Systems, Inc.
                    1984,1985,1986,1987


**** Enter SUPERSAP input file name (<CR> to quit) >> VE03001
**** Linear stress anaylsis
**** Enter run option (HELP- gives list)
                    (CLEAR-start again)
                    (QUIT- to escape)
                    (RUN-  to execute)  >> RUN

     Options executed are:



     processing ...

**** OPENING TEMPORARY FILES
     NDYN = 0
**** INITIALIZING MAIN ARRAY
IDAMAX=  3        Percent capacity: 0.6330
**** BEGIN NODAL DATA INPUT
**** END   NODAL DATA INPUT
**** BEGIN TYPE-3 DATA INPUT
     32 ELEMENTS
     Percent capacity: 2.0444
**** begin bandwidth optimization
     Percent capacity: 0.5275
```

```
      Percent capacity: 0.6216
      Percent capacity: 0.7157
     _computing . . .        Percent capacity: 0.4248
      bandwidth after  resequencing    = 45
**** end    bandwidth optimization
**** EQUATION PARAMETERS
      total number of equations     = 96
      bandwidth                     = 45
      number of equations in a block = 96
      number of blocks              = 1
      blocking memory (kilobytes)   = 280
      available memory (kilobytes)  = 280
**** BEGIN LOAD INPUT
      Percent capacity: 0.7927
**** END    LOAD INPUT
**** PRINT ASSEMBLY BLOCKING EFFICIENCY:
      Percent efficiency: 25.741
**** BEGIN GLOBAL STIFFNESS FORMATION
**** PRINT SOLUTION BLOCKING EFFICIENCY:
      Percent efficiency: 25.388
**** BEGIN BLOCK SOLUTION

      Percent capacity: 0.9352
**** BEGIN DISPLACEMENT OUTPUT
**** END    DISPLACEMENT OUTPUT
**** BEGIN TYPE-3 STRESS OUTPUT
**** END    TYPE-3 STRESS OUTPUT

      total temporary disk storage (megabytes) = 0.2823

**** BEGIN DELETING TEMPORARY FILES
**** TEMPORARY FILES DELETED
**** END OF SUCCESSFUL EXECUTION


**** TOTAL ELAPSED TIME: 4.0911 MINUTES


Stop - Program terminated.


Done executing program. Hit any key.

                  S U P E R S A P
                       POSTD
      (c) Copyright Algor Interactive Systems, Inc.
               1984,1985,1986,1987


**** ENTER SUPERSAP INPUT FILE NAME (<CR> TO QUIT) >> VE03001

**** STATIC ANALYSIS

      NO. OF NODES.............. 40
```

SOLUTION TO BOX BEAM PROBLEM

```
        NO. OF LOAD CASES......... 1

**** OUTPUT TO TERMINAL...............0
                   FILE...................1
                   BINARY GRAPHIC (.DO)...2 >> 0

**** NODE SPECIFICATIONS

**** RESTORE NODE SPECIFICATIONS FROM FILE (Y/N) >> N

        ENTER PAIR OF NODE NUMBERS TO SPECIFY
        A NODE RANGE.
        ENTER A SINGLE NODE NUMBER TO SPECIFY
        A SINGLE NODE.

        NODES (CL......CLEAR LIST)
              (<CR>....CONTINUE  ) >> 1 40

        1-40

        NODES (CL......CLEAR LIST)
              (<CR>....CONTINUE  ) >>

        1-40

**** STORE NODE SPECIFICATIONS (Y/N) >> Y

**** ENTER OUTPUT FILE NAME (<CR> TO ESCAPE) >> VEO3001

**** FILE ALREADY EXISTS
     OK TO DELETE (Y/N) >> N

**** ENTER OUTPUT FILE NAME (<CR> TO ESCAPE) >>

**** STORE NODE SPECIFICATIONS (Y/N) >> N


**** ENTER SUPERSAP INPUT FILE NAME (<CR> TO QUIT) >> VEO3001

**** STATIC ANALYSIS

        NO. OF NODES............... 40
        NO. OF LOAD CASES.......... 1

**** OUTPUT TO TERMINAL...............0
                   FILE...................1
                   BINARY GRAPHIC (.DO)...2 >> 0

**** NODE SPECIFICATIONS

**** RESTORE NODE SPECIFICATIONS FROM FILE (Y/N) >> Y

**** ENTER INPUT FILE NAME (<CR> TO ESCAPE) >> VEO3001

!!!! ERROR: BAD ELEMENT SPECIFICATION HEADER
            AN INTEGER WAS EXPECTED

**** NODE SPECIFICATIONS

**** RESTORE NODE SPECIFICATIONS FROM FILE (Y/N) >> N
```

```
ENTER PAIR OF NODE NUMBERS TO SPECIFY
A NODE RANGE.
ENTER A SINGLE NODE NUMBER TO SPECIFY
A SINGLE NODE.

NODES (CL......CLEAR LIST)
      ((CR)....CONTINUE  ) >>

1-40
```

**** STORE NODE SPECIFICATIONS (Y/N) >> Y

**** ENTER OUTPUT FILE NAME ((CR) TO ESCAPE) >> VEO3NS

**** ENTER DISPLACEMENT COMPONENT (-1 FOR LIST; (CR) TO CONTINUE) >> -1

```
GLOBAL X TRANSLATION........ 1
GLOBAL Y TRANSLATION........ 2
GLOBAL Z TRANSLATION........ 3
GLOBAL X ROTATION........... 4
GLOBAL Y ROTATION........... 5
GLOBAL Z ROTATION........... 6
ALL COMPONENTS.............. 7
CLEAR....................... 8
```

**** ENTER DISPLACEMENT COMPONENT (-1 FOR LIST; (CR) TO CONTINUE) >> 7

**** DISPLACEMENT COMPONENT SPECIFICATIONS:

```
GLOBAL X TRANSLATION
GLOBAL Y TRANSLATION
GLOBAL Z TRANSLATION
GLOBAL X ROTATION
GLOBAL Y ROTATION
GLOBAL Z ROTATION
```

**** DEFINE LOAD CASE COMBINATION

```
ENTER LOAD CASE (1 - 1; (CR) TO CONTINUE) >> 1
ENTER SCALE FACTOR FOR LOAD CASE 1 >>
LOAD CASE = 1, SCALE FACTOR = 1.0000

ENTER LOAD CASE (1 - 1; (CR) TO CONTINUE) >>
```

**** DISPLAY MAXIMA ONLY (Y/N) >> N

**** DISPLACEMENT POST-PROCESSING OF VEO3001.DO

LOAD CASE   1   SCALE FACTOR   1.000E+00

NODE	GLOBAL X TRANSLATION	GLOBAL Y TRANSLATION	GLOBAL Z TRANSLATION	GLOBAL X ROTATION (DEGREE)	GLOBAL Y ROTATION (DEGREE)	GLOBAL Z ROTATION (DEGREE)
1	-9.0546E-14	1.1554E+00	-1.1554E+00	.0000E+00	.0000E+00	.0000E+00

SOLUTION TO BOX BEAM PROBLEM

2	4.8639E-15	1.1554E+00	.0000E+00	.0000E+00	.0000E+00	.0000E+00
3	1.0045E-13	1.1554E+00	1.1554E+00	.0000E+00	.0000E+00	.0000E+00
4	7.2057E-14	.0000E+00	1.1554E+00	.0000E+00	.0000E+00	.0000E+00
5	4.2748E-14	-1.1554E+00	1.1554E+00	.0000E+00	.0000E+00	.0000E+00
6	-3.9406E-15	-1.1554E+00	.0000E+00	.0000E+00	.0000E+00	.0000E+00
7	-5.0736E-14	-1.1554E+00	-1.1554E+00	.0000E+00	.0000E+00	.0000E+00
8	-7.1195E-14	.0000E+00	-1.1554E+00	.0000E+00	.0000E+00	.0000E+00
9	-9.0978E-14	8.6655E-01	-8.6655E-01	.0000E+00	.0000E+00	.0000E+00
10	4.5005E-15	8.6655E-01	.0000E+00	.0000E+00	.0000E+00	.0000E+00
11	9.9983E-14	8.6655E-01	8.6655E-01	.0000E+00	.0000E+00	.0000E+00
12	7.0422E-14	.0000E+00	8.6655E-01	.0000E+00	.0000E+00	.0000E+00
13	4.2012E-14	-8.6655E-01	8.6655E-01	.0000E+00	.0000E+00	.0000E+00
14	-3.6454E-15	-8.6655E-01	.0000E+00	.0000E+00	.0000E+00	.0000E+00
15	-4.9376E-14	-8.6655E-01	-8.6655E-01	.0000E+00	.0000E+00	.0000E+00
16	-6.9482E-14	.0000E+00	-8.6655E-01	.0000E+00	.0000E+00	.0000E+00
17	-8.5107E-14	5.7770E-01	-5.7770E-01	.0000E+00	.0000E+00	.0000E+00
18	3.6072E-15	5.7770E-01	.0000E+00	.0000E+00	.0000E+00	.0000E+00
19	9.2236E-14	5.7770E-01	5.7770E-01	.0000E+00	.0000E+00	.0000E+00
20	6.3873E-14	.0000E+00	5.7770E-01	.0000E+00	.0000E+00	.0000E+00
21	3.5875E-14	-5.7770E-01	5.7770E-01	.0000E+00	.0000E+00	.0000E+00
22	-2.9689E-15	-5.7770E-01	.0000E+00	.0000E+00	.0000E+00	.0000E+00
23	-4.1720E-14	-5.7770E-01	-5.7770E-01	.0000E+00	.0000E+00	.0000E+00
24	-6.3247E-14	.0000E+00	-5.7770E-01	.0000E+00	.0000E+00	.0000E+00
25	-5.7846E-14	2.8885E-01	-2.8885E-01	.0000E+00	.0000E+00	.0000E+00
26	2.0243E-15	2.8885E-01	.0000E+00	.0000E+00	.0000E+00	.0000E+00
27	6.2107E-14	2.8885E-01	2.8885E-01	.0000E+00	.0000E+00	.0000E+00
28	3.9914E-14	.0000E+00	2.8885E-01	.0000E+00	.0000E+00	.0000E+00
29	2.1705E-14	-2.8885E-01	2.8885E-01	.0000E+00	.0000E+00	.0000E+00
30	-1.6701E-15	-2.8885E-01	.0000E+00	.0000E+00	.0000E+00	.0000E+00
31	-2.5248E-14	-2.8885E-01	-2.8885E-01	.0000E+00	.0000E+00	.0000E+00
32	-3.9578E-14	.0000E+00	-2.8885E-01	.0000E+00	.0000E+00	.0000E+00
33	.0000E+00	.0000E+00	.0000E+00	.0000E+00	.0000E+00	.0000E+00
34	.0000E+00	.0000E+00	.0000E+00	.0000E+00	.0000E+00	.0000E+00
35	.0000E+00	.0000E+00	.0000E+00	.0000E+00	.0000E+00	.0000E+00
36	.0000E+00	.0000E+00	.0000E+00	.0000E+00	.0000E+00	.0000E+00
37	.0000E+00	.0000E+00	.0000E+00	.0000E+00	.0000E+00	.0000E+00
38	.0000E+00	.0000E+00	.0000E+00	.0000E+00	.0000E+00	.0000E+00
39	.0000E+00	.0000E+00	.0000E+00	.0000E+00	.0000E+00	.0000E+00
40	.0000E+00	.0000E+00	.0000E+00	.0000E+00	.0000E+00	.0000E+00

**** ALGEBRAIC MAXIMA (Y/N) >> Y

NODE	GLOBAL X TRANSLATION	GLOBAL Y TRANSLATION	GLOBAL Z TRANSLATION	GLOBAL X ROTATION (DEGREE)	GLOBAL Y ROTATION (DEGREE)	GLOBAL Z ROTATION (DEGREE)
NODE	3	3	5	40	40	40
	1.0045E-13	1.1554E+00	1.1554E+00	.0000E+00	.0000E+00	.0000E+00
NODE	11	2	4	39	39	39
	9.9983E-14	1.1554E+00	1.1554E+00	.0000E+00	.0000E+00	.0000E+00
NODE	19	1	3	38	38	38
	9.2236E-14	1.1554E+00	1.1554E+00	.0000E+00	.0000E+00	.0000E+00
NODE	4	11	13	37	37	37
	7.2057E-14	8.6655E-01	8.6655E-01	.0000E+00	.0000E+00	.0000E+00
NODE	12	10	12	36	36	36
	7.0422E-14	8.6655E-01	8.6655E-01	.0000E+00	.0000E+00	.0000E+00
NODE	20	9	11	35	35	35
	6.3873E-14	8.6655E-01	8.6655E-01	.0000E+00	.0000E+00	.0000E+00
NODE	27	19	21	34	34	34
	6.2107E-14	5.7770E-01	5.7770E-01	.0000E+00	.0000E+00	.0000E+00

NODE	5	18	20	33	33	33
	4.2748E-14	5.7770E-01	5.7770E-01	.0000E+00	.0000E+00	.0000E+00
NODE	13	17	19	32	32	32
	4.2012E-14	5.7770E-01	5.7770E-01	.0000E+00	.0000E+00	.0000E+00

**** ALGEBRAIC MINIMA (Y/N) >> Y

NODE	GLOBAL X TRANSLATION	GLOBAL Y TRANSLATION	GLOBAL Z TRANSLATION	GLOBAL X ROTATION (DEGREE)	GLOBAL Y ROTATION (DEGREE)	GLOBAL Z ROTATION (DEGREE)
NODE	9	7	8	40	40	40
	-9.0978E-14	-1.1554E+00	-1.1554E+00	.0000E+00	.0000E+00	.0000E+00
NODE	1	6	7	39	39	39
	-9.0546E-14	-1.1554E+00	-1.1554E+00	.0000E+00	.0000E+00	.0000E+00
NODE	17	5	1	38	38	38
	-8.5107E-14	-1.1554E+00	-1.1554E+00	.0000E+00	.0000E+00	.0000E+00
NODE	8	15	16	37	37	37
	-7.1195E-14	-8.6655E-01	-8.6655E-01	.0000E+00	.0000E+00	.0000E+00
NODE	16	14	15	36	36	36
	-6.9482E-14	-8.6655E-01	-8.6655E-01	.0000E+00	.0000E+00	.0000E+00
NODE	24	13	9	35	35	35
	-6.3247E-14	-8.6655E-01	-8.6655E-01	.0000E+00	.0000E+00	.0000E+00
NODE	25	23	24	34	34	34
	-5.7846E-14	-5.7770E-01	-5.7770E-01	.0000E+00	.0000E+00	.0000E+00
NODE	7	22	23	33	33	33
	-5.0736E-14	-5.7770E-01	-5.7770E-01	.0000E+00	.0000E+00	.0000E+00
NODE	15	21	17	32	32	32
	-4.9376E-14	-5.7770E-01	-5.7770E-01	.0000E+00	.0000E+00	.0000E+00

**** MAGNITUDE MAXIMA (Y/N) >> Y

NODE	GLOBAL X TRANSLATION	GLOBAL Y TRANSLATION	GLOBAL Z TRANSLATION	GLOBAL X ROTATION (DEGREE)	GLOBAL Y ROTATION (DEGREE)	GLOBAL Z ROTATION (DEGREE)
NODE	3	7	8	40	40	40
	1.0045E-13	-1.1554E+00	-1.1554E+00	.0000E+00	.0000E+00	.0000E+00
NODE	11	6	7	39	39	39
	9.9983E-14	-1.1554E+00	-1.1554E+00	.0000E+00	.0000E+00	.0000E+00
NODE	19	5	5	38	38	38
	9.2236E-14	-1.1554E+00	1.1554E+00	.0000E+00	.0000E+00	.0000E+00
NODE	9	3	4	37	37	37
	-9.0978E-14	1.1554E+00	1.1554E+00	.0000E+00	.0000E+00	.0000E+00
NODE	1	2	3	36	36	36
	-9.0546E-14	1.1554E+00	1.1554E+00	.0000E+00	.0000E+00	.0000E+00
NODE	17	1	1	35	35	35
	-8.5107E-14	1.1554E+00	-1.1554E+00	.0000E+00	.0000E+00	.0000E+00
NODE	4	15	16	34	34	34
	7.2057E-14	-8.6655E-01	-8.6655E-01	.0000E+00	.0000E+00	.0000E+00
NODE	8	14	15	33	33	33
	-7.1195E-14	-8.6655E-01	-8.6655E-01	.0000E+00	.0000E+00	.0000E+00
NODE	12	13	13	32	32	32
	7.0422E-14	-8.6655E-01	8.6655E-01	.0000E+00	.0000E+00	.0000E+00

**** ANOTHER CASE (Y/N) >> N

**** ANOTHER FILE (Y/N) >> N

SOLUTION TO BOX BEAM PROBLEM

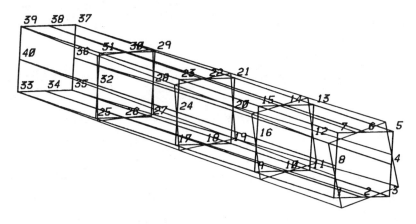

5.6 FINITE ELEMENT MODEL COMPARISONS

This section contains a comparison of the various models and the problems described above. Set-up times for each model were approximately the same for each model. Times reported in Table 5.15 include data input and problem execution. These times are based on an XT compatible with an 8088-2 CPU running at 8 MHz and with an 8087-2 co-processor. The same programs were run with the 8088-2 CPU replaced by an NEC V-20 (8 MHz). Execution times decreased by 43.4 percent. Set-up times varied for each program. The easiest to use were Algor, SUPERSAP, and PCTRAN Plus. The most difficult was SAP86.

Table 5.15 Comparison of Execution Times (Time in Minutes)

	Problem			
Model	5.1	5.2	5.3	5.4
ANSYS-PC/Linear	5.13	13.43	5.33	7.89
PCTRAN Plus	4.02	13.21	4.54	5.43
IMAGES3D	3.86	12.29	4.65	5.54
MSC/Pal 2	4.45	13.41	4.76	5.46
COSMOS/M	4.21	12.96	4.50	5.34
SAP 86	6.47	13.65	5.01	6.75
LIBRA	6.21	NA	5.76	7.54
SUPERSAP	3.62	12.56	4.45	5.21

6

Thermal Models

6.1 INTRODUCTION

The objective of this chapter is to introduce the FEA user to several heat transfer related problems in order to grasp the basics of setting up thermal problems on the microcomputer. A large body of problems exist which require solutions dealing with thermal properties of an object. We will discuss, in general, how to approach a thermal problem using FEA and discuss methods used to set up and solve problems of this nature. We shall briefly discuss the thermal properties of materials and, of course, the mechanisms of heat transport. The problems used as examples in this chapter will address the three heat-transport mechanisms. These mechanisms are conduction, convection, and radiation. These are addressed individually in the following sections. Additionally, a discussion of the FEA programs used to solve the example problems will follow at the end of the chapter.

6.2 TYPES OF HEAT TRANSFER ANALYSIS

6.2.1 Conduction Analysis

Conduction is probably the most well-known type of heat transfer. This form of heat transfer involves the transport of kinetic energy through a solid. The rate at which the kinetic energy (heat) "flows" is proportional to the cross-sectional area of a solid and two inherent properties of materials called thermal conductivity and specific heat. The thermal conductivity of a material is directly proportional to its ability to transfer heat. Therefore, a solid with a small thermal conductivity

Table 6.1 Material Conductivities

Material (room temperature)	English units $\dfrac{\text{Btu}}{\text{hr ft } ^{\circ}\text{F}}$	Metric units $\dfrac{\text{cal}}{\text{sec cm } ^{\circ}\text{C}}$
Metal		
Aluminum		
pure	125	0.52
5052	83	0.34
6061 T6	90	0.37
2024 T4	70	0.29
7075 T6	70	0.29
356 T6	87	0.36
Beryllium	95	0.39
Beryllium copper	50	0.21
Copper		
pure	230	0.95
drawn wire	166	0.68
bronze	130	0.54
red brass	64	0.26
yellow brass	54	0.22
5% phosphor bronze	30	0.12
Gold	170	0.70
Iron		
wrought	34	0.14
cast	32	0.13
Kovar	9	0.037
Lead	19	0.078
Magnesium		
pure	92	0.38
cast	41	0.17
Molybdenum	75	0.31
Nickel		
pure	46	0.19
inconel	10	0.041
monel	15	0.062
Silver	242	1.00
Steel		
SAE 1010	34	0.14
1020	32	0.13
1045	26	0.11
4130	24	0.10
Tin	36	0.15
Titanium	9	0.037
Zinc	59	0.24

would not be conducive to the flow of heat and would therefore insulate. Likewise, a material with a high value for thermal conductivity would be a good conductor of heat. The specific heat of a substance is the quantity of heat required to raise the temperature of a substance by one unit of temperature for one unit of mass. Table 6.1 displays the thermal conductivities and specific heats for several materials.

To visualize the process of heat conduction, consider a rod of some arbitrary length which is composed of many smaller elements of length dx. The ends of this rod are maintained at two different temperatures. At a later time the rod reaches an equilibrium temperature. If we measure the temperature at each dx element, we see that the change in temperature results in the following relationship:

$$dQ/dt = -(KA)dT/dx$$

where K is the thermal conductivity and A is the cross-sectional area of the rod. It is clear that the rate of heat conduction through a solid is proportional to the temperature difference between two points within a solid as well as the thermal conductivity and cross-sectional area as mentioned previously. The first example in Section 6.3.1 is a problem which solves for the temperature distribution in a rod.

The FEA solution to the previous equation is represented by the steady state equation, in matrix form, as described in Chapter 3 and is given again by:

$$(Q) = [K](T)$$

For the one-dimensional case the classical solution to heat conduction in this rod is the solution to the equation:

$$KA(T_1 - T_2)/x$$

6.2.2 Convection Analysis

Convection is the process whereby heat is transferred via a fluid, by that fluid absorbing heat at one point and moving to another point where it is diluted in a larger and usually cooler mass of fluid. Generally, this occurs in the practical world when a fluid is flowing across the surface of a solid. The heat in the solid is transferred to the fluid and subsequently transported away. This method of transport may be one of two types: natural or forced convection. Natural convection, obviously occurs when the driving force for heat transfer is the temperature difference between the hotter body and the cooler ambient fluid. Forced convection occurs when the fluid is moved by mechanical means such as one might find in the fan of a home air-conditioning system. The fundamental analytical equation for convective heat flow is given by:

$$dQ/dt = hAdT$$

where h is the convection coefficient.

There are many factors one must consider when developing an FEA model which involves convection. Some of these items are:

1. The geometry of the solid surface
2. The fluid velocity
3. The properties of the fluid

6.2.3 Radiation

All bodies give off thermal radiation. The extent of that radiation is a function of a body's temperature with respect to its surroundings and the properties of the surface of that body. The fundamental equation for heat transfer by radiation is:

$$dQ/dt = A \ (TB4\text{-}T4)$$

where A is the area of the body, ? is the absorptivity of the body, ? is the Stefan-Boltzmann constant, Ts is the temperature of the body, and T is the surrounding temperature. The above equation is more widely known as the Stefan-Boltzmann law.

6.3 SAMPLE THERMAL PROBLEMS

The example problems given in this section will be solved using some of the FEA packages which were available at the time of writing this book. The output from the FEA thermal software package will be given.

We shall now examine several problems using FEA as our solution technique. The first problem will address heat conduction in a two-dimensional bar. Problem two will examine heat conduction and heat convection and problem three will address heat transfer via radiation by examining heat transfer in a solar collector.

6.3.1 Heat Conduction in a Two-Dimensional Bar

Our first example will examine the steady state heat conduction in a two-dimensional bar. This problem is solved using the FEA programs with temperatures of 100 F and 350 F maintained at opposite ends. The thermal conductivity of the material iron is approximately 50 Btu/hr-ft/ F and the length of the bar is 4.5 inches.

Input Data

Table 6.2 gives the data used to model this problem. The table displays the node number and the x, y, z coordinates of the nodes. Once the input geometry data is entered, we shall enter the material properties of the bar; in this case, it is nothing more than the thermal conductivity. The first analysis presents the necessary input to perform this analysis using the ANSYS FEA program. The second

Table 6.2 Input Geometry for Two-Dimensional
Conducting Bar

Node	x	y	z
1	0	1.0	0
2	0	0	0
3	1.5	1.0	0
4	1.5	0	0
5	3.0	1.0	0
6	3.0	0	0
7	4.5	1.0	0
8	4.5	0	0

presents the required input for the PCTRAN Plus FEA software. Examining Figure 6.1 in detail we can work the problem and get a flavor for a typical PREP, ANALYSIS, and POST processing session. We begin by executing a batch file which puts us in the proper directory, then executes a program which turns a resident co-processor on, initializes graphics driver software, and sets up a virtual memory driver. These special software modules are particular to the ANSYS program and provide necessary environmental information to the PC for proper operation of the main FEA program. Most other vendors supply similar program modules to provide the user some flexibility as to the type of peripherals used by the FEA user. Following the environmental initialization the user would then execute the preprocessor program by typing PREP (in the case of ANSYS). Once in the preprocessor a series of statements can be entered which generate the model. In this example, we chose a two-dimensional thermal solid with four-nodal points per element. ANSYS will provide a short synopsis of the element type after entering the STIF,## command, where ## is the desired element number. Since we have decided to use element type 55 (two-dimensional isoparametric thermal solid) we enter the request into the key board: et,1,55. The following list provides the series of commands necessary to enter the two-dimensional conductivity problem:

```
\title,2-D conducting Bar
et,1,55                    Specify element type
kxx,1,50                   Enter the thermal conductivity
n,1,,1                     Define location of first node
n,2,,                      Second node
ngen,4,2,1,2,,1.5          Generate a set of nodes
e,1,2,4,3                  Specify first element
egen,3,2,1                 Generate a set of elements
```

nplot,1	Take a look at the nodal layout
eplot	Take a look at the element layout
nt,1,temp,100,2	Set nodal temperatures at nodes 1 and 2
nt,7,temp,350,8	Set nodal temperatures at nodes 7 and 8
afwrite	Write the analysis file
finish	Get out of the preprocessor

The next step in the process is to perform the analysis. This is accomplished in ANSYS by issuing the command: THERMAL <FILE27.DAT>. This command will read in the FILE27.DAT data and proceed with the analysis phase of the problem. Finally, the postprocessing phase is initiated by issuing the POST command. Within this module we may examine the result of the analysis phase. In this case, the nodal temperatures and a stress plot are provided for review and possibly further analysis. Comparing the results from the PCTRAN solution with that of the ANSYS solution we see that these programs produce identical results.

TWO-DIMENSIONAL HEAT CONDUCTION: ANSYS

```
ansys

C>cd thermal

C>8087 on

C>medium
ANSYS Medium Resolution Graphics Driver
(C)Copyright Swanson Analysis Systems,Inc. 1986

C>virtl
ANSYS Large Virtual Memory Driver
(C)Copyright Swanson Analysis Systems,Inc. 1986

C>
C>prep

ANSYS-PC/THERMAL PREP MODULE
Copyright(C)  1971 1978 1982 1985 1986
Swanson Analysis Systems, Inc.
As an Unpublished Work.
PROPRIETARY DATA - Unauthorized Use, Distribution
or Duplication is Prohibited.
All Rights Reserved.


    *** ANSYS REV 4.2 A3    ENSMINGER/DEMO   CP=      0.55 ***
    FOR SUPPORT CALL CUSTOMER SUPPORT  PHONE (412) 746-3304    TWX 510-690-8655


    TITLE
**ANSYS VERSION FOR DEMONSTRATION PURPOSES ONLY**

        ***** ANSYS ANALYSIS DEFINITION (PREP7) *****

ENTER  RESUME  TO RESUME EXISTING MODEL
ENTER  INFO     FOR PREP7 INFORMATION
ENTER  FINISH  TO LEAVE PREP7
THE VIRTUAL MEMORY SIZE IS 4194304 32 BIT WORDS

IMMEDIATE MODE IS AVAILABLE ON THIS GRAPHIC DEVICE
ENTER  /IMMED,YES  TO TURN ON  IMMEDIATE MODE
ENTER  /IMMED,NO   TO TURN OFF IMMEDIATE MODE
Immediate Mode Requires User Graphic Scaling
  Remember to Set /USER and Define /VIEW, /FOCUS and /DIST
PREP7 -INP=stif,55


***** DOCUMENTATION FOR  55     PARAMETR*****

TWO-DIMENSIONAL ISOPARAMETRIC THERMAL SOLID
NUMBER OF NODES= 4
DEGREES OF FREEDOM PER NODE= 1
REAL CONSTANTS    (0)
MATERIAL PROPERTIES=  KXX, KYY, DENS, C (GLOBAL)
CONVECTION FACES=  IJ,JK,KL,LI
KEYOPT  1 = 0 - FILM COEFF. AT FILM TEMP.
             1 - AT ELEMENT SURFACE
             2 - AT FLUID BULK
             3 - AT DIFFERENTIAL
```

TWO-DIMENSIONAL HEAT CONDUCTION: ANSYS

```
KEYOPT   3 = 0 - PLANE
         1 - AXISYMMETRIC
KEYOPT   5 = 1 - PRINT AT CENTROID
KEYOPT   6 = 1 - PRINT HEAT FLOW RATE
KEYOPT   7 = 0 - LUMPED SPECIFIC HEAT MATRIX
         = 1 - DISTRIBUTED, SPECIFIC HEAT MATRIX
PREP7 -INP=et,1,55


ELEMENT TYPE  1 USES STIF 55
 KEYOPT(1-9)=  0   0   0    0   0   0     0   0   0  INOTPR= 0
 NUMBER OF NODES=   4

QUAD. THERMAL SHELL

CURRENT NODAL DOF SET IS  TEMP
 TWO-DIMENSIONAL STRUCTURE
PREP7 -INP=Xtitle, 2-D conducting bar


***UNKNOWN OR NON-UNIQUE COMMAND(PREP7)= \TIT
PREP7 -INP=/title, 2-D Conducting barBar


NEW TITLE=  2-D CONDUCTING BAR


PREP7 -INP=kxx,1,50


 MATERIAL   1          COEFFICIENTS OF KXX  VS. TEMP EQUATION
  C0 =   50.00000

PROPERTY TABLE KXX   MAT=  1  NUM. POINTS=  2
       TEMPERATURE    DATA       TEMPERATURE     DATA
     0.00000E+00  50.000     2300.0      50.000
PREP7 -INP=n,1,,1


NODE   1 KCS= 0  X,Y,Z= 0.00000E+00  1.0000      0.00000E+00
PREP7 -INP=n,2,,


NODE   2 KCS= 0  X,Y,Z= 0.00000E+00 0.00000E+00 0.00000E+00
PREP7 -INP=ngen,4,2,1,2,,1.5


GENERATE   4 TOTAL SETS OF NODES WITH INCREMENT    2
   SET IS SELECTED NODES IN RANGE    1 TO   2 IN STEPS OF   1
   GEOMETRY INCREMENTS ARE  1.5000     0.00000E+00 0.00000E+00
PREP7 -INP=e,1,2,4,3

ELEMENT   1    1    2    4    3
PREP7 -INP=egen,3,2,1


GENERATE   3 TOTAL SETS OF ELEMENTS WITH NODE INCREMENT OF   2
   SET IS SELECTED ELEMENTS IN RANGE    1 TO   1 IN STEPS OF   1
NUMBER OF ELEMENTS=   3
```

nplot,1

ANSYS 4.2
APR 18 198
19:40:38
NODES
ZV=1
DIST=3.38
XF=2.25
YF=.5

1   3   5   7

Y
Σ X   4   6   8

2-D CONDUCTING BAR

eplot

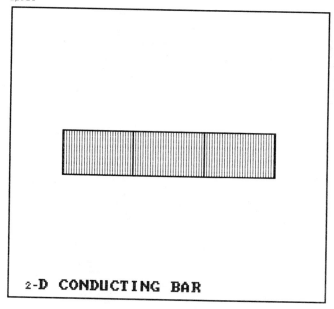

ANSYS 4.2
APR 18 198
19:43:23
ELEMENTS
ZV=1
DIST=3.38
XF=2.25
YF=.5

2-D CONDUCTING BAR

TWO-DIMENSIONAL HEAT CONDUCTION: ANSYS

```
PREP7 -INP=nt,1,temp,100,,2


SPECIFIED TEMP. DEFINITION FOR TEMP FOR SELECTED NODES IN RANGE      1 TO      2 B
Y    1
 VALUES=  100.00     ADDITIONAL DOFS=
PREP7 -INP=nt,7,temp,350,,8


SPECIFIED TEMP. DEFINITION FOR TEMP FOR SELECTED NODES IN RANGE      7 TO      8 B
Y    1
 VALUES=  350.00     ADDITIONAL DOFS=
PREP7 -INP=afwrit


***  NOTE  ***
NPRINT IS ZERO OR GREATER THAN NITTER. SOLUTION PRINTOUT
WILL BE SUPPRESSED UNLESS OTHER PRINT CONTROLS HAVE BEEN DEFINED.

***  NOTE  ***
DATA CHECKED - NO ERRORS FOUND

ANALYSIS DATA WRITTEN ON FILE27




ENTER  FINISH  TO LEAVE PREP7


PREP7 -INP=finish


ALL CURRENT PREP7 DATA WRITTEN TO FILE16
 FOR POSSIBLE RESUME FROM THIS POINT
Execution terminated : 0


C>thermal <file27.dat

ANSYS-PC/THERMAL SOLUTION MODULE
Copyright(C)  1971 1978 1982 1985 1986
Swanson Analysis Systems, Inc.
As an Unpublished Work.
PROPRIETARY DATA - Unauthorized Use, Distribution
 or Duplication is Prohibited.
All Rights Reserved.
```

```
post

ANSYS-PC/THERMAL POST MODULE
Copyright(C)  1971 1978 1982 1985 1986
Swanson Analysis Systems, Inc.
As an Unpublished Work.
PROPRIETARY DATA - Unauthorized Use, Distribution
 or Duplication is Prohibited.
All Rights Reserved.


   *** ANSYS REV 4.2 A3    ENSMINGER/DEMO   CP=      0.55 ***
   FOR SUPPORT CALL CUSTOMER SUPPORT  PHONE (412) 746-3304    TWX 510-690-8655


**ANSYS VERSION FOR DEMONSTRATION PURPOSES ONLY**

        ***** ANSYS RESULTS INTERPRETATION (POST1) *****
THE VIRTUAL MEMORY SIZE IS 4194304 32 BIT WORDS

ENTER   INFO  FOR POST1 DOCUMENTATION
ENTER   FINISH  TO LEAVE POST1
ENTER   /IMMED,YES  FOR MENU SYSTEM
POST1 -INP=set,1,1


USE LOAD STEP     1  ITERATION     1  SECTION   1  FOR LOAD CASE      1


GEOMETRY STORED FOR       8  NODES      3  ELEMENTS
  TITLE=  2-D CONDUCTING BAR


DISPLACEMENT STORED FOR      8 NODES

NODAL FORCES STORED FOR     3 ELEMENTS

REACTIONS STORED FOR     4 REACTIONS

FOR LOAD STEP=     1  ITERATION=     1  SECTION=    1
  TIME=  0.000000E+00   LOAD CASE=  1
  TITLE=  2-D CONDUCTING BAR

POST1 -INP=prtemp


PRINT NODAL TEMPERATURES

        ***** POST1 NODAL TEMPERATURE LISTING *****

LOAD STEP     1  ITERATION=     1  SECTION=  1
TIME=  0.00000E+00      LOAD CASE=  1

NODE     TEMP
   1    100.0
   2    100.0
   3    183.3
   4    183.3
```

TWO-DIMENSIONAL HEAT CONDUCTION: ANSYS

```
     5    266.7
     6    266.7
     7    350.0
     8    350.0

MAXIMUMS
NODE          7
VALUE    350.0
POST1 -INP=plns,tenp


PRODUCE STRESS PLOT,  LABEL= TENP

*** WARNING - ELEMENT STRESS ITEMS NOT AVAILABLE FOR PLOTTING.
POST1 -INP=plns,temp


PRODUCE STRESS PLOT,  LABEL= TEMP
```

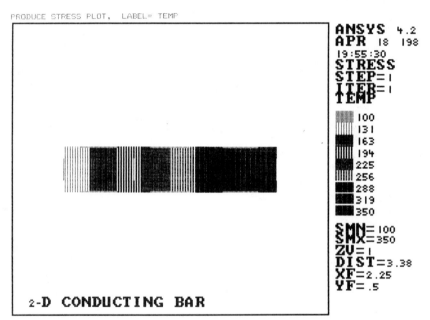

```
POST1 -INP=finish

Execution terminated : 0


C>
```

TWO-DIMENSIONAL HEAT CONDUCTION: PCTRAN PLUS

```
?PCTRAN

C:\PCT>echo off


P C T R A N    P l u s
(C) Copyright 1985, 1986    Brooks Scientific, Inc.
Version 3.2

ENTER JOB NAME: THERMAL




Portions of the PCTRAN menu system are copyrighted
by Alpha Software Corporation.

C:\PCT>PREP

PCTRAN Plus 3.3 PREPROCESSOR
(C) Copyright 1987 Brooks Scientific, Inc.

DEVICE DRIVER IS /PCTIBMG.DEV/
MODE =   1
      ... Please wait ...

New job: THERMAL

* ANALYSIS,1
Analysis type is  1 THERMAL
* ETYPE,1,29
Element type   1 is STIFF  29  2-D THERMAL SOLID (VER 2)
IEOPT =  0      4 Nodes; 1 DOF per node.
* KXX,1,50
Conductivity for material number   1 is   5.0000E+01
Material property "KXX " for material number   1 is   5.0000E+01
* R,1,1
Real set    1. value(s) =  1.0000E+00 0.0000E+00 0.0000E+00 0.0000E+00
0.0000E+00 0.0000E+00 0.0000E+00 0.0000E+00 0.0000E+00 0.0000E+00
* N,1,,1
Node:      1 =   0.0000E+00  1.0000E+00   0.0000E+00
* N,2,,
Node:      2 =   0.0000E+00  0.0000E+00   0.0000E+00
* NGEN,3,2,1,2,,1.5
Generate   3 sets of nodes with nodal increment    2
from node pattern    1 to    2 in steps of   1
DX =    1.5000 DY =      .0000 DZ =      .0000
* NLAB,ON
Plot label NLAB is: ON
*
```

TWO-DIMENSIONAL HEAT CONDUCTION: PCTRAN PLUS

```
 PCTRAN

 AUTO   NLAB
 VIEW:  0    0
 PLOT LIMITS
 1          3          5          7    X:   0.000E+00
                                           Y:   0.000E+00
 Y                                          Z:   0.000E+00
 |
 2___X      4          6          8

                                           THERMAL
                                           19:40:17
                                           10-08-87
```

```
* E,1,1,2,4,3
Element      1 =   1   1   1      1     2     4     3     0     0     0     0
* EGEN,2,2,1
Generate    2 sets of elements with nodal increment       2
from the element pattern       1 to       1 in steps of     1
* NT,1,TEMP,100,2
Define temperature from node       1 with label TEMP and value      1.0000E+02
to node      2 in steps of      1
* NT,7,TEMP,350,8
Define temperature from node       7 with label TEMP and value      3.5000E+02
to node      8 in steps of      1
* F,1,FX,0
Load case    1
Define force from node       1 with label FX    and value      0.0000E+00
to node      1 in steps of      1
* NLIST
List nodes from <N1> to <N2> in steps of <STEP>: 1,8,1
List nodes from       1 to       8 in steps of       1
Node       X            Y            Z            SN
------    ----------   ----------   ----------   -----   - - - - - -
   1    0.0000E+00   1.0000E+00   0.0000E+00       0     0 0 0 0 0 0
   2    0.0000E+00   0.0000E+00   0.0000E+00       0     0 0 0 0 0 0
   3    1.5000E+00   1.0000E+00   0.0000E+00       0     0 0 0 0 0 0
   4    1.5000E+00   0.0000E+00   0.0000E+00       0     0 0 0 0 0 0
   5    3.0000E+00   1.0000E+00   0.0000E+00       0     0 0 0 0 0 0
   6    3.0000E+00   0.0000E+00   0.0000E+00       0     0 0 0 0 0 0
   7    4.5000E+00   1.0000E+00   0.0000E+00       0     0 0 0 0 0 0
   8    4.5000E+00   0.0000E+00   0.0000E+00       0     0 0 0 0 0 0
* ELIST,1,3,1
List element connectivities from       1 to       3 in steps of       1
ENUM    TP MN RN    CONNECTIVITY

   1     1  1  1      1     2     4     3     0     0     0     0
   2     1  1  1      3     4     6     5     0     0     0     0
   3     1  1  1      5     6     8     7     0     0     0     0
* MPLIST
List Material properties
MNUM     EXX          NUXY         DENS         ALPX         KXX
   1   3.000E+07   3.000E-01   0.000E+00   0.000E+00   5.000E+01
* RLIST
List real numbers
RNUM    VALUE(S)
   1  1.0000E+00  0.0000E+00  0.0000E+00  0.0000E+00  0.0000E+00
      0.0000E+00  0.0000E+00  0.0000E+00  0.0000E+00  0.0000E+00
* ETLIST
NO.  STIFF   OPT        DESCRIPTION
   1    29     0      2-D THERMAL SOLID (VER 2)
* ELAB,ON
Plot label ELAB is: ON
* PLOT
```

TWO-DIMENSIONAL HEAT CONDUCTION: PCTRAN PLUS

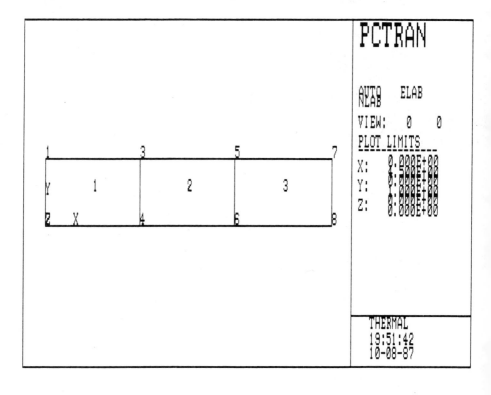

```
* SOLVER,6
Use solver number    6
* EXIT

    Statistics for job THERMAL
********************************

Analysis type:  1 THERMAL
Active coordinate system:    0   0.0000E+00   0.0000E+00   0.0000E+00
Number of load cases:  1
Max node number:       8
Number of system nodes:    0
Max element number:      3
Number of applied displacements:     4
Half bandwidth:      0
Input file echo: ON
Reorder load cases in ascending order
Load Case, # of Loads
      1       1
Reorder displacements in ascending order
Copying "PCTX" files back to job THERMAL
Exit preprocessor .....

C:\PCT>THERMAL
```

```
        THERMAL

C:\PCT>echo off

PCTRAN Plus 3.3 NODE RENUMBERING
(C) Copyright 1987 Brooks Scientific, Inc.

MAXN:        8
MAXE:        3
NSN:         8

NODE RENUMBERING [Y/N/(E to exit)]? N

NUMBER OF NODES        =      8
NUMBER OF SYSTEM NODES =      8
HALF BANDWIDTH         =      4
DEGREES OF FREEDOM     =      8

NODE RENUMBERING [Y/N/(E to exit)]? E

PCTRAN Plus 3.3 HEAT TRANSFER ELEMENT ASSEMBLY
(C) Copyright 1987 Brooks Scientific, Inc.

MAX NODE NUMBER:        8
MAX ELEMENT NUMBER:        3
HALF BANDWIDTH:        4
TOTAL DOF:        8
Assemble stiffness for use with solver        6
ASSEMBLING ELEMENT        1
ASSEMBLING ELEMENT        2
ASSEMBLING ELEMENT        3
EXIT CONDUCT . . .

PCTRAN Plus 3.3 LOAD CONVERSION
(C) Copyright 1987 Brooks Scientific, Inc.

NUMBER OF LOAD CASES:        1
# OF LOADS IN CASE  1:    1
NUMBER OF DISPLACEMENTS:        4

Applied displacements converted:        4

Load Case:   1
Forces converted:        1

Finished load conversion . . .

                QUICKSOLV (tm) Version 1.0
        Copyright (C) TEKTON ENGINEERING INC. 1987
                All rights reserved.

              44 Concord Ave., Suite 106
                Cambridge, MA 02138
                  (617) 576-3729

        Licensed to be used with PCTRAN (tm) Plus.
        Distributed by Brooks Scientific, Inc.

Number of Equations = 8
Matrix Size         = 24 elements
```

TWO-DIMENSIONAL HEAT CONDUCTION: PCTRAN PLUS

```
Half-Bandwidth      = 4
Mean Half-Bandwidth = 3
Memory allocated    = 1K

Assemble
Impose displacements
        4 specified displacements
Triangularize
Backsubstitute
        Load vector 1

Timing
        Assemble :              0.05 sec.
        Triangularize :         0.00 sec.
        Backsubstitute :        0.60 sec.
        Total :                 0.66 sec.

PCTRAN Plus 3.3 RESULT CONVERSION
(C) Copyright 1987 Brooks Scientific, Inc.

Deflections converted:      8
Reactions converted:        0

C:\PCT>POST1

C:\PCT>echo off

PCTRAN Plus 3.3 POSTPROCESSOR
(C) Copyright 1987 Brooks Scientific, Inc.

DEVICE DRIVER IS /PCTIBMG.DEV/
MODE =   1
        ... Please wait ...

JOBNAME: THERMAL

CALCULATE MIN,MAX POST DATA

   Statistics for THERMAL
***************************

Analysis type:  1 THERMAL
Active coordinate system:     0   0.0000E+00   0.0000E+00   0.0000E+00
Number of load cases:  1
Current load case:   1
Max node number:       8
Max element number:       3
Number of displacemenets:      4
Half bandwidth:     4
Reading POST codes from file "PCTRAN.CSD"
* PRTEMP
Print nodal temperatures for nodes NODE1 to NODE2 in steps of NINC 1,8,1
                Nodal Temperatures: Load Case   1

   NODE     TEMP
     1     1.0000E+02
     2     1.0000E+02
     3     1.8333E+02
     4     1.8333E+02
     5     2.6667E+02
```

```
   6     2.6667E+02
   7     3.5000E+02
   8     3.5000E+02

** MAX TEMPERATURE **
NODE:         7
          3.5000E+02
* EXIT
EXIT POSTPROCESSOR .....

C:\PCT>
```

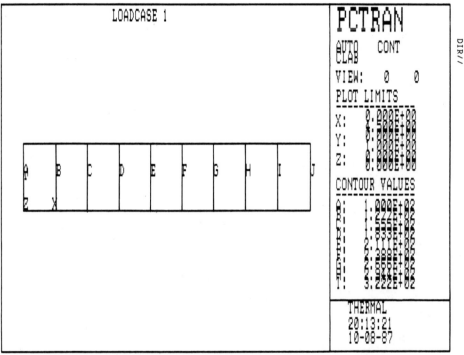

6.3.2 Transient Response of a Radiator Fin

The second problem considers a system in which a transient response is determined. The example used is taken directly from the ANSYS thermal manual which provides several examples of heat transfer analysis using FEA. The following printout displays the session input and resulting output.

THERMAL TRANSFER PROBLEM

```
    cd thermal

C>8087 on

C>medium
ANSYS Medium Resolution Graphics Driver
(C)Copyright Swanson Analysis Systems,Inc. 1986

C>virtl
ANSYS Large Virtual Memory Driver
(C)Copyright Swanson Analysis Systems,Inc. 1986

C>prep

ANSYS-PC/THERMAL PREP MODULE
Copyright(C)  1971 1978 1982 1985 1986
Swanson Analysis Systems, Inc.
As an Unpublished Work.
PROPRIETARY DATA - Unauthorized Use, Distribution
 or Duplication is Prohibited.
All Rights Reserved.


    *** ANSYS REV 4.2 A3    ENSMINGER/DEMO   CP=      0.61 ***
    FOR SUPPORT CALL CUSTOMER SUPPORT  PHONE (412) 746-3304    TWX 510-690-8655

    TITLE
**ANSYS VERSION FOR DEMONSTRATION PURPOSES ONLY**

        ***** ANSYS ANALYSIS DEFINITION (PREP7) *****
ENTER   RESUME   TO RESUME EXISTING MODEL
ENTER   INFO     FOR PREP7 INFORMATION
ENTER   FINISH   TO LEAVE PREP7
THE VIRTUAL MEMORY SIZE IS 4194304 32 BIT WORDS

IMMEDIATE MODE IS AVAILABLE ON THIS GRAPHIC DEVICE
  ENTER  /IMMED,YES  TO TURN ON  IMMEDIATE MODE
  ENTER  /IMMED,NO   TO TURN OFF IMMEDIATE MODE
  Immediate Mode Requires User Graphic Scaling
    Remember to Set /USER and Define /VIEW, /FOCUS and /DIST
PREP7 -INP=/title, radiator fin, transient response.


NEW TITLE=  RADIATOR FIN, TRANSIENT RESPONSE.

PREP7 -INP=et,1,70


ELEMENT TYPE  1 USES STIF 70
  KEYOPT(1-9)=   Ø    Ø   Ø    Ø    Ø   Ø     Ø   Ø    Ø   INOTPR= Ø
  NUMBER OF NODES=   8
```

THERMAL TRANSFER PROBLEM

```
ISOPAR. SOLID THERMAL

CURRENT NODAL DOF SET IS  TEMP
 THREE-DIMENSIONAL STRUCTURE
PREP7 -INP=et,2,57


ELEMENT TYPE  2 USES STIF 57
 KEYOPT(1-9)=   Ø   Ø   Ø     Ø   Ø   Ø     Ø   Ø   Ø   INOTPR= Ø
 NUMBER OF NODES=   4

ISOPAR. SOLID THERMAL

CURRENT NODAL DOF SET IS  TEMP
 THREE-DIMENSIONAL STRUCTURE
PREP7 -INP=kxx,1,Ø.25,-.ØØØØ3


MATERIAL   1          COEFFICIENTS OF KXX   VS. TEMP EQUATION
 CØ =  Ø.25ØØØØØ       C1 = -Ø.3ØØØØØØE-Ø4  C2 =  Ø.ØØØØØØØE+ØØ
 C3 =  Ø.ØØØØØØØE+ØØ   C4 =  Ø.ØØØØØØØE+ØØ

PROPERTY TABLE KXX   MAT=   1 NUM. POINTS=  2
     TEMPERATURE      DATA       TEMPERATURE      DATA
    Ø.ØØØØØE+ØØ Ø.25ØØØ       23ØØ.Ø      Ø.181ØØ
PREP7 -INP=dens,1,Ø.197


MATERIAL   1          COEFFICIENTS OF DENS VS. TEMP EQUATION
 CØ =  Ø.197ØØØØ

PROPERTY TABLE DENS  MAT=   1 NUM. POINTS=  2
     TEMPERATURE      DATA       TEMPERATURE      DATA
    Ø.ØØØØØE+ØØ Ø.197ØØ       23ØØ.Ø      Ø.197ØØ
PREP7 -INP=c,1,Ø.125,Ø.ØØØ1


MATERIAL   1          COEFFICIENTS OF C     VS. TEMP EQUATION
 CØ =  Ø.125ØØØØ       C1 =  Ø.1ØØØØØØE-Ø3  C2 =  Ø.ØØØØØØØE+ØØ
 C3 =  Ø.ØØØØØØØE+ØØ   C4 =  Ø.ØØØØØØØE+ØØ

PROPERTY TABLE C     MAT=   1 NUM. POINTS=  2
     TEMPERATURE      DATA       TEMPERATURE      DATA
    Ø.ØØØØØE+ØØ Ø.125ØØ       23ØØ.Ø      Ø.355ØØ
PREP7 -INP=kxx,2,1.5,-.ØØØ5


MATERIAL   2          COEFFICIENTS OF KXX   VS. TEMP EQUATION
 CØ =   1.5ØØØØØ       C1 = -Ø.5ØØØØØØE-Ø3  C2 =  Ø.ØØØØØØØE+ØØ
 C3 =  Ø.ØØØØØØØE+ØØ   C4 =  Ø.ØØØØØØØE+ØØ

PROPERTY TABLE KXX   MAT=   2 NUM. POINTS=  2
     TEMPERATURE      DATA       TEMPERATURE      DATA
```

THERMAL TRANSFER PROBLEM

```
      Ø.ØØØØØE+ØØ   1.5ØØØ        23ØØ.Ø      Ø.35ØØØ
PREP7 -INP=dens,2,.197


MATERIAL   2            COEFFICIENTS OF DENS VS. TEMP EQUATION
  CØ =  Ø.197ØØØØ

PROPERTY TABLE DENS  MAT=   2  NUM. POINTS=  2
      TEMPERATURE      DATA        TEMPERATURE        DATA
      Ø.ØØØØØE+ØØ Ø.197ØØ        23ØØ.Ø       Ø.197ØØ
PREP7 -INP=c,2,Ø.Ø5,Ø.ØØ1


MATERIAL   2            COEFFICIENTS OF C   VS. TEMP EQUATION
  CØ =  Ø.5ØØØØØØE-Ø1  C1 =  Ø.1ØØØØØØE-Ø2  C2 =  Ø.ØØØØØØØE+ØØ
  C3 =  Ø.ØØØØØØØE+ØØ  C4 =  Ø.ØØØØØØØE+ØØ

PROPERTY TABLE C    MAT=   2  NUM. POINTS=  2
      TEMPERATURE      DATA        TEMPERATURE        DATA
      Ø.ØØØØØE+ØØ Ø.5ØØØØE-Ø1  23ØØ.Ø        2.35ØØ
PREP7 -INP=r,1,.Ø755


REAL CONSTANT SET   1  ITEMS   1 TO   6
   Ø.755ØØE-Ø1  Ø.ØØØØØE+ØØ  Ø.ØØØØØE+ØØ  Ø.ØØØØØE+ØØ  Ø.ØØØØØE+ØØ  Ø.ØØØØØE+ØØ
PREP7 -INP=n,1,


NODE    1  KCS= Ø  X,Y,Z= Ø.ØØØØØE+ØØ Ø.ØØØØØE+ØØ Ø.ØØØØØE+ØØ
PREP7 -INP=n,3,,.12


NODE    3  KCS= Ø  X,Y,Z= Ø.ØØØØØE+ØØ Ø.12ØØØ       Ø.ØØØØØE+ØØ
PREP7 -INP=fill


FILL    1 POINTS BETWEEN NODE    1 AND NODE    3
   START WITH NODE    2 AND INCREMENT BY      1
PREP7 -INP=ngen,3,3,1,3,1,,,.Ø6


GENERATE    3 TOTAL SETS OF NODES WITH INCREMENT      3
   SET IS SELECTED NODES IN RANGE    1 TO   3 IN STEPS OF    1
   GEOMETRY INCREMENTS ARE Ø.ØØØØØE+ØØ Ø.ØØØØØE+ØØ Ø.6ØØØØE-Ø1
PREP7 -INP=ngen,2,9,1,9,1,.Ø6


GENERATE    2 TOTAL SETS OF NODES WITH INCREMENT      9
   SET IS SELECTED NODES IN RANGE    1 TO   9 IN STEPS OF    1
   GEOMETRY INCREMENTS ARE Ø.6ØØØØE-Ø1 Ø.ØØØØØE+ØØ Ø.ØØØØØE+ØØ
PREP7 -INP=ngen,5,1Ø,11,17,3,.Ø6




GENERATE    5 TOTAL SETS OF NODES WITH INCREMENT     1Ø
   SET IS SELECTED NODES IN RANGE   11 TO   17 IN STEPS OF    3
   GEOMETRY INCREMENTS ARE Ø.6ØØØØE-Ø1 Ø.ØØØØØE+ØØ Ø.ØØØØØE+ØØ
PREP7 -INP=nplot,1
```

```
                                              ANSYS 4.2
                                              OCT   4 198
                                              17:59: 3
                                              NODES
                                              ZV=1
                                              DIST= .225
                                              XF= .15
                                              YF= .06
                                              ZF= .06
    9      18

    8      17    27    37    47    57

   Y
   ZₓX    16



  RADIATOR FIN, TRANSIENT RESPONSE.
```

PREP7 -INP=view,1,1,1,1

***UNKNOWN OR NON-UNIQUE COMMAND(PREP7)= VIEW
PREP7 -INP=/view,1,1,1,1

VIEW POINT FOR WINDOW 1 1.0000   1.0000   1.0000
PREP7 -INP=nplot,1

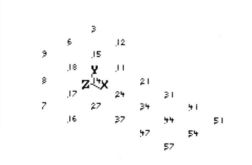

```
                                              ANSYS 4.2
                                              OCT   4 198
                                              18: 1:15
                                              NODES
                                              XV=1
                                              YV=1
                                              ZV=1
                                              DIST= .225
                                              XF= .15
                                              YF= .06
                                              ZF= .06

               3
         6       12
   9        15
      18   Y   11
   8    Z↳ₓX   21
      17    24      31
   7     27    34     41
      16    37    44     51
            47    54
               57



  RADIATOR FIN, TRANSIENT RESPONSE.
```

THERMAL TRANSFER PROBLEM

```
PREP7 -INP=e,1,2,5,4,10,11,14,13

ELEMENT    1      1     2     5     4    10    11    14    13

PREP7 -INP=egen,2,1,1


GENERATE    2 TOTAL SETS OF ELEMENTS WITH NODE INCREMENT OF     1
   SET IS SELECTED ELEMENTS IN RANGE    1 TO    1 IN STEPS OF     1
NUMBER OF ELEMENTS=     2
PREP7 -INP=egen,2,3,-2


GENERATE    2 TOTAL SETS OF ELEMENTS WITH NODE INCREMENT OF     3
   SET IS SELECTED ELEMENTS IN RANGE    1 TO    2 IN STEPS OF     1
NUMBER OF ELEMENTS=     4
PREP7 -INP=type,2


ELEMENT TYPE SET TO  2
PREP7 -INP=mat,2


MATERIAL NUMBER SET TO  2
PREP7 -INP=e,11,14,24,21

ELEMENT    5     11    14    24    21
PREP7 -INP=egen,2,3,-1


GENERATE    2 TOTAL SETS OF ELEMENTS WITH NODE INCREMENT OF     3
   SET IS SELECTED ELEMENTS IN RANGE    5 TO    5 IN STEPS OF     1
NUMBER OF ELEMENTS=     6
PREP7 -INP=egen,4,10,-2


GENERATE    4 TOTAL SETS OF ELEMENTS WITH NODE INCREMENT OF    10
   SET IS SELECTED ELEMENTS IN RANGE    5 TO    6 IN STEPS OF     1
NUMBER OF ELEMENTS=    12
PREP7 -INP=eplot
```

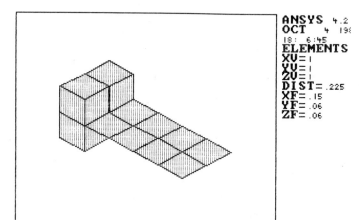

RADIATOR FIN. TRANSIENT RESPONSE.

```
PREP7 -INP=cvsf,0,1,0,.036,75
```

```
GENERATE CONVECTIONS ON SURFACE IN COORDINATE SYSTEM   0  NORMAL TO
   COMP.   X    AT  0.00000E+00  FILM COEF.=  0.36000E-01  TBULK=  75.000      TOL.=
NUMBER OF ELEMENT CONVECTIONS=    4
PREP7 -INP=cvsf,0,1,.06,.15,75
```

```
GENERATE CONVECTIONS ON SURFACE IN COORDINATE SYSTEM   0  NORMAL TO
   COMP.   X    AT  0.60000E-01  FILM COEF.=  0.15000     TBULK=  75.000      TOL.=
NUMBER OF ELEMENT CONVECTIONS=    8
PREP7 -INP=esel,type,2
```

```
ESEL  FOR LABEL= TYPE  FROM    2 TO    2 BY   1
     8 ELEMENTS (OF    12  DEFINED) SELECTED BY  ESEL   COMMAND.
PREP7 -INP=ec,all,1,.15,75
```

```
ELEM CONVECTION ON ALL SELECTED ELEMENTS   FACE  1
   HCOEF= 0.150000000      TBULK=  75.0000000
NUMBER OF ELEMENT CONVECTIONS=   16
PREP7 -INP=ec,all,2,.15,75
```

```
ELEM CONVECTION ON ALL SELECTED ELEMENTS   FACE  2
   HCOEF= 0.150000000      TBULK=  75.0000000
NUMBER OF ELEMENT CONVECTIONS=   24
PREP7 -INP=eall
```

```
    12 ELEMENTS (OF    12  DEFINED) SELECTED BY  EALL   COMMAND.
PREP7 -INP=lwrite
```

```
LOAD STEP    1  WRITTEN ON FILE23
PREP7 -INP=time,15
```

```
TIME=  15.000
PREP7 -INP=iter,-200,0
```

```
NITTER=   -200  NPRINT= 99900  NPOST=   200
USE CONVERGENCE AND/OR TIME STEP OPTIMIZATION
```

THERMAL TRANSFER PROBLEM

BOUNDARY CONDITION STEP OR RAMP DEPENDENT UPON KBC COMMAND.
PREP7 -INP=/com

PREP7 -INP=cvsf,0,1,0,.036,247

GENERATE CONVECTIONS ON SURFACE IN COORDINATE SYSTEM   0   NORMAL TO
  COMP.   X   AT 0.00000E+00 FILM COEF.= 0.36000E-01 TBULK=   247.00   TOL.=

NUMBER OF ELEMENT CONVECTIONS=   24
PREP7 -INP=cvsf,0,1,.06,.15,85

GENERATE CONVECTIONS ON SURFACE IN COORDINATE SYSTEM   0   NORMAL TO
  COMP.   X   AT 0.60000E-01 FILM COEF.= 0.15000   TBULK=   85.000   TOL.=

NUMBER OF ELEMENT CONVECTIONS=   24
PREP7 -INP=ersel,type,2

ERSE FOR LABEL= TYPE FROM   2 TO   2 BY   1

     8 ELEMENTS (OF   12   DEFINED) SELECTED BY   ERSE   COMMAND.
PREP7 -INP=ec,all,1,.15,85

ELEM CONVECTION ON ALL SELECTED ELEMENTS   FACE   1
  HCOEF= 0.150000000   TBULK= 85.0000000

NUMBER OF ELEMENT CONVECTIONS= 24
PREP7 -INP=ec,all,2,.15,85

ELEM CONVECTION ON ALL SELECTED ELEMENTS   FACE 2
  HCOEF= 0.150000000   TBULK= 85.0000000

NUMBER OF ELEMENT CONVECTIONS= 24
PREP7 -INP=eall

    12 ELEMENTS (OF   12   DEFINED) SELECTED BY   EALL   COMMAND.
PREP7 -INP=lwrite

LOAD STEP   2 WRITTEN ON FILE23
PREP7 -INP=afwrite

*** NOTE ***
NPRINT IS ZERO OR GREATER THAN NITTER. SOLUTION PRINTOUT
WILL BE SUPPRESSED UNLESS OTHER PRINT CONTROLS HAVE BEEN DEFINED.

```
*** NOTE ***
DATA CHECKED - NO ERRORS FOUND

LOADS DATA READ FROM FILE23

ANALYSIS DATA WRITTEN ON FILE27

ENTER FINISH TO LEAVE PREP7


PREP7 -INP=finish


ALL CURRENT PREP7 DATA WRITTEN TO FILE16
 FOR POSSIBLE RESUME FROM THIS POINT
Execution terminated : Ø

C>



C>thermal <file27.dat

ANSYS-PC/THERMAL SOLUTION MODULE
Copyright(C)  1971 1978 1982 1985 1986
Swanson Analysis Systems, Inc.
As an Unpublished Work.
PROPRIETARY DATA - Unauthorized Use, Distribution
 or Duplication is Prohibited.
All Rights Reserved.
```

THERMAL TRANSFER PROBLEM

```
***** NOTICE *****  THIS IS THE ANSYS-PC/THERMAL FINITE
ELEMENT PROGRAM.  NEITHER SWANSON ANALYSIS SYSTEMS, INC.
NOR THE DISTRIBUTOR SUPPLYING THIS PROGRAM ASSUME ANY
RESPONSIBILITY FOR THE VALIDITY, ACCURACY, OR APPLICABILITY
OF ANY RESULTS OBTAINED FROM THE ANSYS SYSTEM.  USERS
MUST VERIFY THEIR OWN RESULTS.


    *** ANSYS REV 4.2 A3    ENSMINGER/DEMO    CP=     3.35 ***
    FOR SUPPORT CALL CUSTOMER SUPPORT  PHONE (412) 746-3304    TWX 510-690-8655


RADIATOR FIN, TRANSIENT RESPONSE.
**ANSYS VERSION FOR DEMONSTRATION PURPOSES ONLY**
THE VIRTUAL MEMORY SIZE IS 4194304 32 BIT WORDS

NUMBER OF REAL CONSTANT SETS=   1

NUMBER OF ELEMENTS=    12

MAXIMUM NODE NUMBER=    57

    RANGE OF ELEMENT MAXIMUM CONDUCTIVITY IN GLOBAL COORDINATES
    MAXIMUM= 0.75770E-01  AT ELEMENT     12.
    MINIMUM= 0.51350E-02  AT ELEMENT      4.

    *** ELEMENT STIFFNESS FORMULATION TIMES
    TYPE  NUMBER  STIF  TOTAL CP   AVE CP

     1      4     70     3.78     0.945
     2      8     57     2.64     0.330
    TIME AT END OF ELEMENT STIFFNESS FORMULATION  CP=        48.560
    MAXIMUM WAVE FRONT ALLOWED=  200.
    MAXIMUM IN-CORE WAVE FRONT=    11.
        MATRIX SOLUTION TIMES
    READ IN ELEMENT STIFFNESSES     CP=           1.160
    NODAL COORD. TRANSFORMATION      CP=           0.000
    MATRIX TRIANGULARIZATION         CP=           1.760
    TIME AT END OF MATRIX TRIANGULARIZATION  CP=        57.020
    STRESS EVALUATION     ELEM=      7 L.S.=     1 ITER=      1  CP=     66.63

    *** ELEM. HT. FLOW CALC. TIMES
    TYPE  NUMBER  STIF  TOTAL CP   AVE CP

     1      4     70     0.28     0.070
     2      8     57     0.32     0.040

    *** NODAL HT. FLOW CALC. TIMES
    TYPE  NUMBER  STIF  TOTAL CP   AVE CP

     1      4     70     0.23     0.057
     2      8     57     0.34     0.042
```

```
*** LOAD STEP    1   ITER     1 COMPLETED.   TIME=   0.000000E+00
TIME INC=  0.000000E+00   CUM. ITER.=     1


 *** PROBLEM STATISTICS
NO. OF ACTIVE DEGREES OF FREEDOM =     30
R.M.S. WAVEFRONT SIZE =     6.5
TOTAL CP TIME=    73.600
*** LOAD STEP    2   ITER     1 COMPLETED.   TIME=   0.750000E-01
TIME INC=  0.750000E-01   CUM. ITER.=     2
ELEMENT FORMATION    ELEM=      1 L.S.=    2 ITER=      2  CP=     117.60
TRANSIENT OPTIMIZATION VALUE =   0.12696E-01 AT NODE     2
CRITERION =    10.000
*** LOAD STEP    2   ITER     2 COMPLETED.   TIME=   0.150000
TIME INC=  0.750000E-01   CUM. ITER.=     3
ELEMENT FORMATION    ELEM=      1 L.S.=    2 ITER=      6  CP=     150.39
TRANSIENT OPTIMIZATION VALUE =   0.10789E-01 AT NODE     8
CRITERION =    10.000
*** LOAD STEP    2   ITER     6 COMPLETED.   TIME=   0.450000
TIME INC=  0.300000   CUM. ITER.=     4
ELEMENT FORMATION    ELEM=      1 L.S.=    2 ITER=     22  CP=     183.51
TRANSIENT OPTIMIZATION VALUE =   0.54806E-02 AT NODE     5
CRITERION =    10.000
*** LOAD STEP    2   ITER    22 COMPLETED.   TIME=   1.65000
TIME INC=  1.20000   CUM. ITER.=     5
ELEMENT FORMATION    ELEM=      1 L.S.=    2 ITER=     86  CP=     216.63
TRANSIENT OPTIMIZATION VALUE =   0.81137E-03 AT NODE     5
CRITERION =    10.000
*** LOAD STEP    2   ITER    86 COMPLETED.   TIME=   6.45000
TIME INC=  4.80000   CUM. ITER.=     6
ELEMENT FORMATION    ELEM=      1 L.S.=    2 ITER=    200  CP=     250.24
TRANSIENT OPTIMIZATION VALUE =   0.20872E-02 AT NODE     8
CRITERION =    10.000
*** LOAD STEP    2   ITER   200 COMPLETED.   TIME=   15.0000
TIME INC=  8.55000   CUM. ITER.=     7

***** END OF INPUT ENCOUNTERED ON FILE 5.   FILE 5 REWOUND
Execution terminated : 0


C>
```

THERMAL TRANSFER PROBLEM

```
C>post
```

```
ANSYS-PC/THERMAL POST MODULE
Copyright(C)  1971 1978 1982 1985 1986
Swanson Analysis Systems, Inc.
As an Unpublished Work.
PROPRIETARY DATA - Unauthorized Use, Distribution
 or Duplication is Prohibited.
All Rights Reserved.
```

```
     *** ANSYS REV 4.2 A3    ENSMINGER/DEMO   CP=      0.66 ***
     FOR SUPPORT CALL CUSTOMER SUPPORT  PHONE (412) 746-3304    TWX 510-690-8655

**ANSYS VERSION FOR DEMONSTRATION PURPOSES ONLY**

          ***** ANSYS RESULTS INTERPRETATION (POST1) *****
     THE VIRTUAL MEMORY SIZE IS 4194304 32 BIT WORDS

ENTER   INFO   FOR POST1 DOCUMENTATION
ENTER   FINISH  TO LEAVE POST1
ENTER   /IMMED,YES   FOR MENU SYSTEM
POST1 -INP=set,2,


USE LOAD STEP     2  ITERATION     0  SECTION   1  FOR LOAD CASE     1


GEOMETRY STORED FOR     57  NODES      12  ELEMENTS
   TITLE=  RADIATOR FIN, TRANSIENT RESPONSE.

DISPLACEMENT STORED FOR   57 NODES

NODAL FORCES STORED FOR    12 ELEMENTS

 END OF FILE DETECTED ON FILE12

***NOTE- LOAD STEP    2  ITER  200  REACT. FORCES (GROUP 5) NOT ON FILE12

FOR LOAD STEP=     2  ITERATION=   200 SECTION=   1
 TIME=   15.0000       LOAD CASE=  1
 TITLE=  RADIATOR FIN, TRANSIENT RESPONSE.
POST1 -INP=prtemp


PRINT NODAL TEMPERATURES

         ***** POST1 NODAL TEMPERATURE LISTING *****

 LOAD STEP     2  ITERATION=  200 SECTION=  1
```

```
TIME=     15.000        LOAD CASE=  1

NODE      TEMP
   1     94.32
   2     94.57
   3     94.32
   4     94.32
   5     94.57
   6     94.32
   7     94.32
   8     94.57
   9     94.32
  10     93.74
  11     92.48
  12     93.74
  13     93.74
  14     92.48
MORE (YES,NO OR CONTINUOUS)=yes


          ***** POST1 NODAL TEMPERATURE LISTING *****

LOAD STEP     2  ITERATION=   200 SECTION=  1
TIME=     15.000        LOAD CASE=  1

NODE      TEMP
  15     93.74
  16     93.74
  17     92.48
  18     93.74
  21     92.23
  24     92.23
  27     92.23
  31     92.06
  34     92.06
  37     92.06
  41     91.96
  44     91.96
  47     91.96
  51     91.93
MORE (YES,NO OR CONTINUOUS)=yes


          ***** POST1 NODAL TEMPERATURE LISTING *****

LOAD STEP     2  ITERATION=   200 SECTION=  1
TIME=     15.000        LOAD CASE=  1

NODE      TEMP
  54     91.93
  57     91.93

MAXIMUMS
```

THERMAL TRANSFER PROBLEM

```
NODE        2
VALUE    94.57
POST1 -INP=/view,,1,2,3


VIEW POINT FOR WINDOW  1  1.0000      2.0000      3.0000
POST1 -INP=/type,,2


HIDDEN PLOT IN WINDOW  1
POST1 -INP=plnstr,temp


PRODUCE STRESS PLOT,  LABEL= TEMP
```

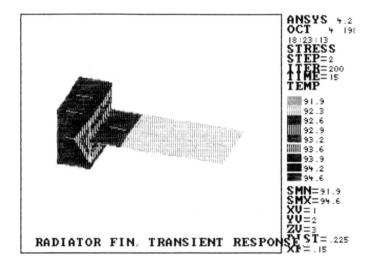

```
POST1 -INP=finish

Execution terminated : Ø


C>post26

ANSYS-PC/THERMAL POST26 MODULE
Copyright(C)  1971 1978 1982 1985 1986
Swanson Analysis Systems, Inc.
As an Unpublished Work.
PROPRIETARY DATA - Unauthorized Use, Distribution
 or Duplication is Prohibited.
All Rights Reserved.


   *** ANSYS REV 4.2 A3    ENSMINGER/DEMO    CP=     Ø.6Ø ***
   FOR SUPPORT CALL CUSTOMER SUPPORT  PHONE (412) 746-33Ø4    TWX 51Ø-69Ø-8655


**ANSYS VERSION FOR DEMONSTRATION PURPOSES ONLY**


          ***** GENERAL GRAPH POSTPROCESSOR (POST26) *****
THE VIRTUAL MEMORY SIZE IS 419434 32 BIT WORDS

ALL POST26 SPECIFICATIONS ARE RESET TO INITIAL DEFAULTS
POST26-INP=ylab,temp


Y AXIS LABEL= TEMP
POST26-INP=disp,2,14,temp,t14


VARIABLE  2 IS     14 TEMP  PHASE KEY= Ø
POST26-INP=plvar,2


STORAGE COMPLETE FOR    7 DATA POINTS

          SUMMARY OF VARIABLES STORED THIS STEP AND EXTREME VALUES
VARI TYPE  IDENTIFIERS  NAME    MINIMUM     AT TIME     MAXIMUM    AT TIME

  2 DISP     14 TEMP  14  T14    75.ØØ     Ø.ØØØØE+ØØ  92.48       15.ØØ
PLOT DEFINITION
CURVE   VARIABLE     NAME
    1      2    14  T14
```

THERMAL TRANSFER PROBLEM

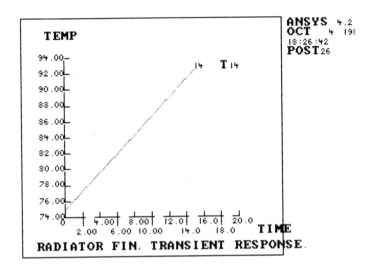

POST26-INP=finish

Execution terminated : 0

C>

6.3.3 Radiative Heat Transfer in a Solar Collector

This example demonstrates heat transfer by radiation. The problem is quite simple and straightforward. We being our problem by defining a conductive rod of some length and then the radiation link. The three-dimensional conductive bar used in this example will simulate a black, water-carrying pipe with an inside diameter of 0.25 inches. The radiation link will simulate the sun's rays. The following illustrates the problem.

HEAT TRANSFER BY RADIATION

```
    cd thermal

C>8087 on

C>medium
ANSYS Medium Resolution Graphics Driver
(C)Copyright Swanson Analysis Systems,Inc. 1986

C>virtl
ANSYS Large Virtual Memory Driver
(C)Copyright Swanson Analysis Systems,Inc. 1986

C>prep

ANSYS-PC/THERMAL PREP MODULE
Copyright(C) 1971 1978 1982 1985 1986
Swanson Analysis Systems, Inc.
As an Unpublished Work.
PROPRIETARY DATA - Unauthorized Use, Distribution
 or Duplication is Prohibited.
All Rights Reserved.

    *** ANSYS REV 4.2 A3    ENSMINGER/DEMO   CP=     0.60 ***
    FOR SUPPORT CALL CUSTOMER SUPPORT  PHONE (412) 746-3304    TWX 510-690-8655

    TITLE
**ANSYS VERSION FOR DEMONSTRATION PURPOSES ONLY**

       ***** ANSYS ANALYSIS DEFINITION (PREP7) *****

ENTER   RESUME   TO RESUME EXISTING MODEL
ENTER   INFO     FOR PREP7 INFORMATION
ENTER   FINISH   TO LEAVE PREP7
THE VIRTUAL MEMORY SIZE IS 4194304 32 BIT WORDS

IMMEDIATE MODE IS AVAILABLE ON THIS GRAPHIC DEVICE
 ENTER   /IMMED,YES  TO TURN ON   IMMEDIATE MODE
 ENTER   /IMMED,NO   TO TURN OFF IMMEDIATE MODE
 Immediate Mode Requires User Graphic Scaling
  Remember to Set /USER and Define /VIEW, /FOCUS and /DIST
PREP7 -INP=/title, solar collector, radiation link.

NEW TITLE= SOLAR COLLECTOR, RADIATION LINK.

PREP7 -INP=et,1,33

ELEMENT TYPE  1 USES STIF 33
  KEYOPT(1-9)=  0   0   0     0   0   0     0   0   0  INOTPR= 0
  NUMBER OF NODES=  2
```

HEAT TRANSFER BY RADIATION

CONDUCTING BAR, 3-D

CURRENT NODAL DOF SET IS   TEMP
 THREE-DIMENSIONAL STRUCTURE
 PREP7 -INP=et,2,31,,,,,,1

ELEMENT TYPE   2 USES STIF 31
 KEYOPT(1-9)=   Ø   Ø   Ø   Ø   Ø   1   Ø   Ø   Ø   INOTPR= Ø
 NUMBER OF NODES=   2

RADIATION LINK

CURRENT NODAL DOF SET IS   TEMP
 THREE-DIMENSIONAL STRUCTURE
 PREP7 -INP=toffst,46Ø

TEMPERATURE OFFSET FROM ABSOLUTE ZERO= 46Ø.ØØØ
 PREP7 -INP=r,1,.Ø49Ø9

 REAL CONSTANT SET   1 ITEMS   1 TO   6
  Ø.49Ø9ØE-Ø1 Ø.ØØØØØE+ØØ   Ø.ØØØØØE+ØØ   Ø.ØØØØØE+ØØ   Ø.ØØØØØE+ØØ   Ø.ØØØØØE+ØØ
 PREP7 -INP=r,2,2.5,.89,1.5

 REAL CONSTANT SET   2 ITEMS   1 TO   6
  2.5ØØØ   Ø.89ØØØ   1.5ØØØ   Ø.ØØØØØE+ØØ   Ø.ØØØØØE+ØØ   Ø.ØØØØØE+ØØ
 PREP7 -INP=/com

 PREP7 -INP=kxx,1,5.25

 MATERIAL   1   COEFFICIENTS OF KXX VS. TEMP EQUATION
  CØ =   5.25ØØØØ

 PROPERTY TABLE KXX   MAT=   1 NUM. POINTS=   2
       TEMPERATURE   DATA   TEMPERATURE   DATA
     Ø.ØØØØØE+ØØ 5.25ØØ   23ØØ.Ø   5.25ØØ
 PREP7 -INP=n,1,,1

 NODE   1 KCS= Ø   X,Y,Z= Ø.ØØØØØE+ØØ 1.ØØØØ   Ø.ØØØØØE+ØØ
 PREP7 -INP=n,6,5,1

 NODE   6 KCS= Ø   X,Y,Z= 5.ØØØØ   1.ØØØØ   Ø.ØØØØØE+ØØ
 PREP7 -INP=fill

 FILL   4 POINTS BETWEEN NODE   1 AND NODE   6

```
START WITH NODE    2 AND INCREMENT BY    1
PREP7 -INP=n,7,1,2


NODE    7 KCS= Ø X,Y,Z=  1.ØØØØ      2.ØØØØ      Ø.ØØØØØE+ØØ
PREP7 -INP=n,1Ø,4,2


NODE   1Ø KCS= Ø X,Y,Z=  4.ØØØØ      2.ØØØØ      Ø.ØØØØØE+ØØ
PREP7 -INP=fill


FILL    2 POINTS BETWEEN NODE    7 AND NODE   1Ø
 START WITH NODE    8 AND INCREMENT BY    1
PREP7 -INP=e,1,2

ELEMENT    1     1     2
PREP7 -INP=egen,5,1,1


GENERATE    5 TOTAL SETS OF ELEMENTS WITH NODE INCREMENT OF     1
  SET IS SELECTED ELEMENTS IN RANGE    1 TO    1 IN STEPS OF     1
NUMBER OF ELEMENTS=    5
PREP7 -INP=type,2


ELEMENT TYPE SET TO  2
PREP7 -INP=real,2


REAL CONSTANT NUMBER=  2
PREP7 -INP=e,2,7

ELEMENT    6     2     7
PREP7 -INP=egen,4,1,-1


GENERATE    4 TOTAL SETS OF ELEMENTS WITH NODE INCREMENT OF     1
  SET IS SELECTED ELEMENTS IN RANGE    6 TO    6 IN STEPS OF     1
NUMBER OF ELEMENTS=    9
PREP7 -INP=iter,-4Ø,4Ø


NITTER=   -4Ø NPRINT=   4Ø NPOST=   4Ø

USE CONVERGENCE AND/OR TIME STEP OPTIMIZATION
 BOUNDARY CONDITION STEP OR RAMP DEPENDENT UPON KBC COMMAND.
PREP7 -INP=/com


PREP7 -INP=kbc,1
```

HEAT TRANSFER BY RADIATION

```
STEP BOUNDARY CONDITION KEY= 1
PREP7 -INP=nt,1,temp,45


SPECIFIED TEMP. DEFINITION FOR TEMP FOR SELECTED NODES IN RANGE     1 TO     1 BY
  VALUES=  45.000      ADDITIONAL DOFS=
PREP7 -INP=nt,6,temp,195


SPECIFIED TEMP. DEFINITION FOR TEMP FOR SELECTED NODES IN RANGE     6 TO     6 BY
  VALUES=  195.00      ADDITIONAL DOFS=
PREP7 -INP=nrsel,y,2


NRSE  FOR LABEL= Y      BETWEEN  2.0000      AND  2.0000      KABS=  0.
  TOLERANCE=  0.100000E-01

      4 NODES (OF     10 DEFINED) SELECTED BY  NRSE  COMMAND.
PREP7 -INP=nt,all,emp,1500


SPECIFIED TEMP. DEFINITION FOR EMP  FOR ALL SELECTED NODES
  VALUES=  1500.0      ADDITIONAL DOFS=

***INVALID LABEL= EMP
PREP7 -INP=nall


     10 NODES (OF     10 DEFINED) SELECTED BY  NALL  COMMAND.
PREP7 -INP=afwrite


*** NOTE ***
DATA CHECKED - NO ERRORS FOUND

ANALYSIS DATA WRITTEN ON FILE27

ENTER  FINISH  TO LEAVE PREP7


PREP7 -INP=finish


ALL CURRENT PREP7 DATA WRITTEN TO FILE16
 FOR POSSIBLE RESUME FROM THIS POINT
Execution terminated : 0


C>
```

th

```
C>thermal <file27.dat
```

ANSYS-PC/THERMAL SOLUTION MODULE
Copyright(C) 1971 1978 1982 1985 1986
Swanson Analysis Systems, Inc.
As an Unpublished Work.
PROPRIETARY DATA - Unauthorized Use, Distribution
  or Duplication is Prohibited.
All Rights Reserved.

***** NOTICE ***** THIS IS THE ANSYS-PC/THERMAL FINITE
ELEMENT PROGRAM. NEITHER SWANSON ANALYSIS SYSTEMS, INC.
NOR THE DISTRIBUTOR SUPPLYING THIS PROGRAM ASSUME ANY
RESPONSIBILITY FOR THE VALIDITY, ACCURACY, OR APPLICABILITY
OF ANY RESULTS OBTAINED FROM THE ANSYS SYSTEM. USERS
MUST VERIFY THEIR OWN RESULTS.

*** ANSYS REV 4.2 A3   ENSMINGER/DEMO   CP=   3.52 ***
  FOR SUPPORT CALL CUSTOMER SUPPORT   PHONE (412) 746-3304   TWX 510-690-8655

 SOLAR COLLECTOR, RADIATION LINK.
ANSYS VERSION FOR DEMONSTRATION PURPOSES ONLY
THE VIRTUAL MEMORY SIZE IS 4194304 32 BIT WORDS

NUMBER OF REAL CONSTANT SETS=   2

NUMBER OF ELEMENTS=   9

MAXIMUM NODE NUMBER=   10

LOAD SUMMARY -   2 TEMPERATURES   Ø HEAT FLOWS   Ø CONVECTIONS

 RANGE OF ELEMENT MAXIMUM CONDUCTIVITY IN GLOBAL COORDINATES
MAXIMUM= Ø.257722E+ØØ AT ELEMENT   5.
MINIMUM= Ø.154633E-Ø1 AT ELEMENT   9.

 *** ELEMENT STIFFNESS FORMULATION TIMES
 TYPE NUMBER STIF TOTAL CP   AVE CP

   1   5   33   Ø.Ø6   Ø.Ø12
   2   4   31   Ø.Ø5   Ø.Ø12
TIME AT END OF ELEMENT STIFFNESS FORMULATION CP=   38.Ø1Ø
MAXIMUM WAVE FRONT ALLOWED= 2ØØ.
MAXIMUM IN-CORE WAVE FRONT=   5.
    MATRIX SOLUTION TIMES
 READ IN ELEMENT STIFFNESSES   CP=   Ø.59Ø
 NODAL COORD. TRANSFORMATION   CP=   Ø.ØØØ
 MATRIX TRIANGULARIZATION   CP=   Ø.35Ø
TIME AT END OF MATRIX TRIANGULARIZATION CP=   44.82Ø

 *** ELEM. HT. FLOW CALC. TIMES
 TYPE NUMBER STIF TOTAL CP   AVE CP

   2   4   31   Ø.Ø6   Ø.Ø15

 *** NODAL HT. FLOW CALC. TIMES
```
```

HEAT TRANSFER BY RADIATION

```
TYPE   NUMBER   STIF   TOTAL CP   AVE CP

 2       4       31     Ø.11      Ø.Ø27
*** LOAD STEP    1    ITER     1 COMPLETED.   TIME=   Ø.ØØØØØØE+ØØ
TIME INC=  Ø.ØØØØØØE+ØØ  CUM. ITER.=    1


*** PROBLEM STATISTICS
NO. OF ACTIVE DEGREES OF FREEDOM =      8
R.M.S. WAVEFRONT SIZE =     3.2
TOTAL CP TIME=    57.62Ø
TEMPERATURE INCREMENT =   195.ØØ     AT NODE     6   CRITERION =   1.ØØØØ
*** LOAD STEP    1    ITER     2 COMPLETED.   TIME=   Ø.ØØØØØØE+ØØ
TIME INC=  Ø.ØØØØØØE+ØØ  CUM. ITER.=    2
STRESS EVALUATION    ELEM=    1 L.S.=    1 ITER=    3 CP=     94.42
TEMPERATURE INCREMENT =  Ø.11369E-12 AT NODE     4   CRITERION =   1.ØØØØ
*** LOAD STEP    1    ITER     3 COMPLETED.   TIME=   Ø.ØØØØØØE+ØØ
TIME INC=  Ø.ØØØØØØE+ØØ  CUM. ITER.=    3

*** SOLUTION CONVERGED - LOAD STEP    1
CONVERGED AFTER ITERATION      3  CUM. ITER.=    3
NEXT ITERATION (     4Ø) SATISFIES PRINTOUT OR POST DATA REQUEST.


***** TEMPERATURE   SOLUTION *****      TIME =   Ø.ØØØØØE+ØØ
LOAD STEP=   1   ITERATION=    4Ø CUM. ITER.=    4
 NODE    TEMP      NODE    TEMP      NODE    TEMP      NODE    TEMP      NODE    TEMP
   1    45.ØØØ       2    75.ØØØ       3   1Ø5.ØØØ       4   135.ØØØ       5   165.ØØØ
   6   195.ØØØ       7    75.ØØØ       8   1Ø5.ØØØ       9   135.ØØØ      1Ø   165.ØØØ

    MAXIMUM TEMPERATURE=  195.ØØ     AT NODE     6
    MINIMUM TEMPERATURE=   45.ØØØ    AT NODE     1


***** ELEMENT HEAT FLOW RATES *****   TIME =   Ø.ØØØØØØE+ØØ
LOAD STEP=   1   ITER.=    4Ø  CUM. ITER.=    4

EL=    6 NODES=    2    7 MAT=   Ø  AREA=  2.5ØØØ
EMIS(I,J)=  1.5ØØ       1.5ØØ    TEMP(I,J)=   75.Ø    75.Ø HEAT REAT= Ø.ØØØØE+ØØ

EL=    7 NODES=    3    8 MAT=   Ø  AREA=  2.5ØØØ
EMIS(I,J)=  1.5ØØ       1.5ØØ    TEMP(I,J)=  1Ø5.Ø   1Ø5.Ø HEAT REAT= Ø.ØØØØE+ØØ

EL=    8 NODES=    4    9 MAT=   Ø  AREA=  2.5ØØØ
EMIS(I,J)=  1.5ØØ       1.5ØØ    TEMP(I,J)=  135.Ø   135.Ø HEAT REAT= Ø.ØØØØE+ØØ

EL=    9 NODES=    5   1Ø MAT=   Ø  AREA=  2.5ØØØ
EMIS(I,J)=  1.5ØØ       1.5ØØ    TEMP(I,J)=  165.Ø   165.Ø HEAT REAT= Ø.ØØØØE+ØØ

    ***** HEAT FLOW RATES INTO NODES *****  TIME= Ø.ØØØØØE+ØØ
LOAD STEP=    1   ITERATION=    4Ø  CUM. ITER.=    4
   NODE   HEAT
     1  -7.7317
```

```
     6  7.7317


TOTAL Ø.15099E-13
  TEMPERATURE INCREMENT =  Ø.11369E-12 AT NODE    4   CRITERION =   1.ØØØØ
*** LOAD STEP    1   ITER    4Ø  COMPLETED.   TIME=  Ø.ØØØØØØE+ØØ
  TIME INC=  Ø.ØØØØØØE+ØØ  CUM. ITER.=    4

***** END OF INPUT ENCOUNTERED ON FILE 5.  FILE 5 REWOUND
Execution terminated : Ø


C>



   prep

ANSYS-PC/THERMAL PREP MODULE
Copyright(C)  1971 1978 1982 1985 1986
Swanson Analysis Systems, Inc.
As an Unpublished Work.
PROPRIETARY DATA - Unauthorized Use, Distribution
  or Duplication is Prohibited.
All Rights Reserved.


    *** ANSYS REV 4.2 A3    ENSMINGER/DEMO   CP=     Ø.6Ø ***
    FOR SUPPORT CALL CUSTOMER SUPPORT  PHONE (412) 746-33Ø4    TWX 51Ø-69Ø-8655

    TITLE
**ANSYS VERSION FOR DEMONSTRATION PURPOSES ONLY**

      ***** ANSYS ANALYSIS DEFINITION (PREP7) *****

ENTER  RESUME  TO RESUME EXISTING MODEL
ENTER  INFO    FOR PREP7 INFORMATION
ENTER  FINISH  TO LEAVE PREP7
THE VIRTUAL MEMORY SIZE IS 41943Ø4 32 BIT WORDS

IMMEDIATE MODE IS AVAILABLE ON THIS GRAPHIC DEVICE
ENTER  /IMMED,YES  TO TURN ON  IMMEDIATE MODE
ENTER  /IMMED,NO   TO TURN OFF IMMEDIATE MODE
  Immediate Mode Requires User Graphic Scaling
   Remember to Set /USER and Define /VIEW, /FOCUS and /DIST
PREP7 -INP=resume


RESUME PREP7 DATA FROM FILE16

TITLE= SOLAR COLLECTOR, RADIATION LINK.
NUMBER OF ELEMENTS=   9
NUMBER OF NODES=   1Ø
PREP7 -INP=eplot,1
```

HEAT TRANSFER BY RADIATION

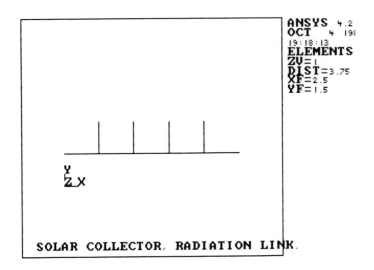

```
Write fault error writing device PRN
Abort, Retry, Ignore :r

PREP7 -INP=nplot,1
```

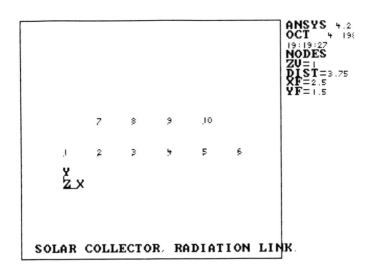

```
PREP7 -INP=finish


ALL CURRENT PREP7 DATA WRITTEN TO FILE16
 FOR POSSIBLE RESUME FROM THIS POINT
Execution terminated : 0


C>


C>post

ANSYS-PC/THERMAL POST MODULE
Copyright(C)  1971 1978 1982 1985 1986
Swanson Analysis Systems, Inc.
As an Unpublished Work.
PROPRIETARY DATA - Unauthorized Use, Distribution
 or Duplication is Prohibited.
All Rights Reserved.


    *** ANSYS REV 4.2 A3    ENSMINGER/DEMO    CP=    0.66 ***
    FOR SUPPORT CALL CUSTOMER SUPPORT  PHONE (412) 746-3304    TWX 510-690-8655


**ANSYS VERSION FOR DEMONSTRATION PURPOSES ONLY**

        ***** ANSYS RESULTS INTERPRETATION (POST1) *****
THE VIRTUAL MEMORY SIZE IS 4194304 32 BIT WORDS

ENTER  INFO  FOR POST1 DOCUMENTATION
ENTER  FINISH  TO LEAVE POST1
ENTER  /IMMED,YES  FOR MENU SYSTEM
POST1 -INP=set,1,


USE LOAD STEP    1 ITERATION    0 SECTION   1 FOR LOAD CASE    1


GEOMETRY STORED FOR    10 NODES    9 ELEMENTS
 TITLE=  SOLAR COLLECTOR, RADIATION LINK.

DISPLACEMENT STORED FOR   10 NODES

NODAL FORCES STORED FOR   9 ELEMENTS

REACTIONS STORED FOR    2 REACTIONS

FOR LOAD STEP=    1 ITERATION=    40 SECTION=   1
 TIME=  0.000000E+00   LOAD CASE=  1
 TITLE=  SOLAR COLLECTOR, RADIATION LINK.
POST1 -INP=prrfor


PRINT REACTION FORCES PER NODE

        ***** POST1 REACTION FORCE LISTING *****

LOAD STEP    1 ITERATION=    40 SECTION=  1
 TIME=  0.00000E+00    LOAD CASE=  1
```

HEAT TRANSFER BY RADIATION

```
THE FOLLOWING X,Y,Z FORCES ARE IN NODAL COORDINATES

NODE      HEAT
   1    -7.732
   6     7.732

TOTAL    0.1510E-13
POST1 -INP=prtemp


PRINT NODAL TEMPERATURES

          ***** POST1 NODAL TEMPERATURE LISTING *****

  LOAD STEP      1  ITERATION=      40  SECTION=   1
  TIME=    0.00000E+00      LOAD CASE=   1

  NODE       TEMP
     1     45.00
     2     75.00
     3    105.0
     4    135.0
     5    165.0
     6    195.0
     7     75.00
     8    105.0
     9    135.0
    10    165.0

MAXIMUMS
NODE         6
VALUE     195.0
POST1 -INP=finish

Execution terminated : 0


C>
```

6.4 WHAT TO LOOK FOR IN AN FEA THERMAL PROGRAM

The most important elements one generally looks for in any computer program are speed and accuracy. But as with most programs one must always make a trade-off of one form or another. Chapter 3 discusses these trade-offs.

Several FEA program vendors offer good thermal analysis programs. A list of the programs used for comparison is given below.

COSMOS/M. Good documentation, excellent program offering a wide range of solution types (i.e., steady state, non-linear, and transient). One fault is that the documentation is not well organized and lacks an adequate index.

LIBRA. Good program offering steady state and transient solution types. Well-organized documentation but model building is hampered by the program's method of data input. Also, LIBRA makes it difficult to build a model by not providing a quick visual method of review of the data.

ANSYS. An excellent program and well-documented, which provides professional capabilities. Only drawback is the cost to performance ratio. The program is only available on a lease basis. Recent releases (v. 4.3) now allow the user to purchase the program. The user should check with his ANSYS representative.

PCTRAN. This program has good documentation, provides an ANSYS style of input, but lacks the ability to do transient thermal problems. PCTRAN Plus is no longer in production. Swenson Analysis Systems (ANSYS) purchased the assets of Broohs Scientific, Inc. (PCTRAN Plus).

7
Fluid Models

7.1 INTRODUCTION

The objective of this chapter is to introduce the FEA user to the application of finite element analysis to fluid mechanics by example. The application and use of fluid FEA is less than its structural and thermal counterparts, and therefore there are fewer fluid FEA programs in general use and only one (at this writing) for PC based systems. The program and examples used in this chapter were written by Dr. David Roylance of MIT.

There are two approaches to solving fluid problems with the finite element concept. The first method is somewhat restrictive in the type of fluid problems which may be solved, since the user utilizes an FEA thermal program, and using substitution of variables, creates the model.

Since the simplest type of flow to model is potential flow, this method works well. Potential flow models are typically irrotational, invisid (viscous effects are ignored), and incompressible. Fluid flow of this type is analogous to steady state thermal conduction. The substitution of variables is straightforward, with the thermal conductivity matrix replaced with a unity matrix. The calculated nodal temperature values are analogous to the stream function.

The flow potential is calculated at each node. From this information, the velocity components at the centroid of each element can be determined by calculating the flow potential derivatives over the element. Many simple fluid problems may solved using a thermal program since ideal fluid problems may be solved using the same governing equations as thermal problems.

The second method uses a finite element program specifically designed for fluid flow. Currently only one program provides this capability for PCs. The program is one written by Dr. David Roylance of the Department of Materials Science and Engineering at the Massachusetts Institute of Technology. Dr. Roylance's program was written as an adjunct module to the PCTran Plus FEA program. The program will address simple ideal fluid models as well as problems involving viscous fluids in which nonisothermal conditions exist. The examples provided will touch on a few of the principal features of this program.

7.2 FLUIDS

A fluid is a continuous material composed of discrete molecules each able to act independently when acted upon by an external force. In general, though, a fluid is treated as a bulk material with forces applied over a large area with respect to the individual molecules of which it is composed. Fluids are liquids and gases. Because of the very nature of fluids it is difficult at best to develop a program which will run efficiently on a PC.

For these problems addressed herein, two common simplifying boundary conditions are considered. The first condition states that when a fluid is at a solid boundary the velocity of the fluid at that boundary is zero. The second condition states that when two immiscible liquids are in contact they will act as if under tension at the boundary of the two fluids. As with all simulations one should use good engineering judgement and check as best one can as to the validity of the results.

7.3 EXAMPLE PROBLEM 1: COUETTE/POISEUILLE FLOW

The first model is a classical fluid problem. The problem is one of simple drag/pressure flow here an incompressible fluid is contained between two boundaries of infinite extent. The upper boundary is set into motion in the x direction thereby creating a drag at that surface. Additionally, a pressure is applied in the negative x direction providing a resistance to the plane flow. The analytical solution for this problem is well known and is given by the following equation,

$$U \quad u(y) = \frac{y}{H} \, U$$

where u(y) is the velocity as a function of the height y and U is the velocity of the upper boundary at H. Figure 7.1 displays the element mesh for this prob-

flow demo1 — Couette/Poiseuille channel flow

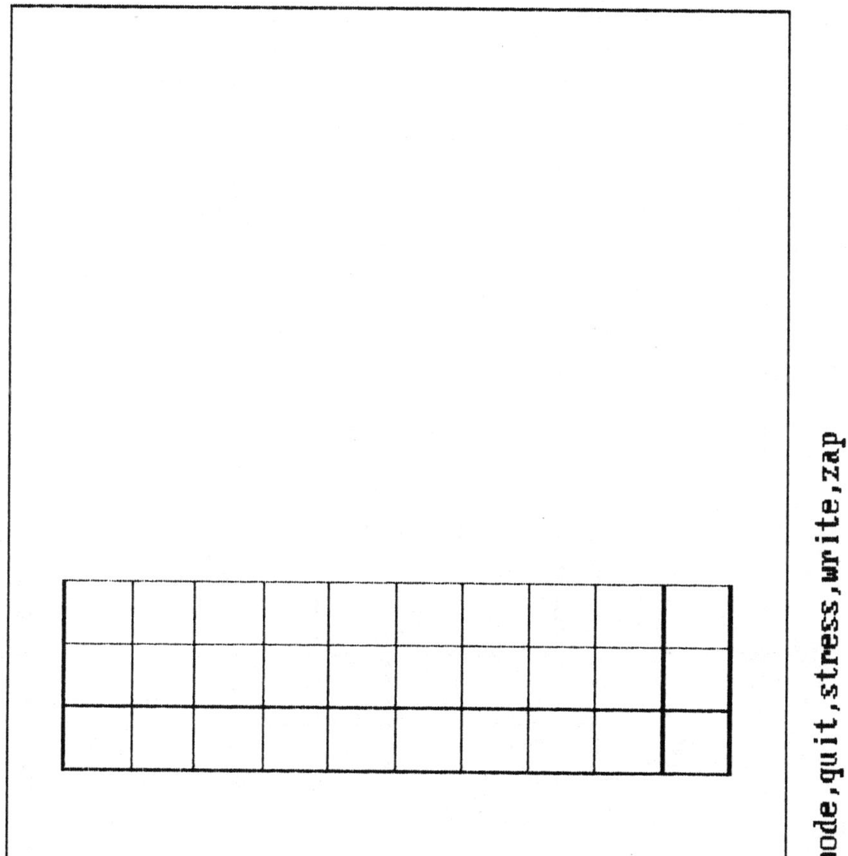

options: boun,disp,elem,mesh,node,quit,stress,write,zap

===>

Figure 7.1 Example 1: Input Geometry.

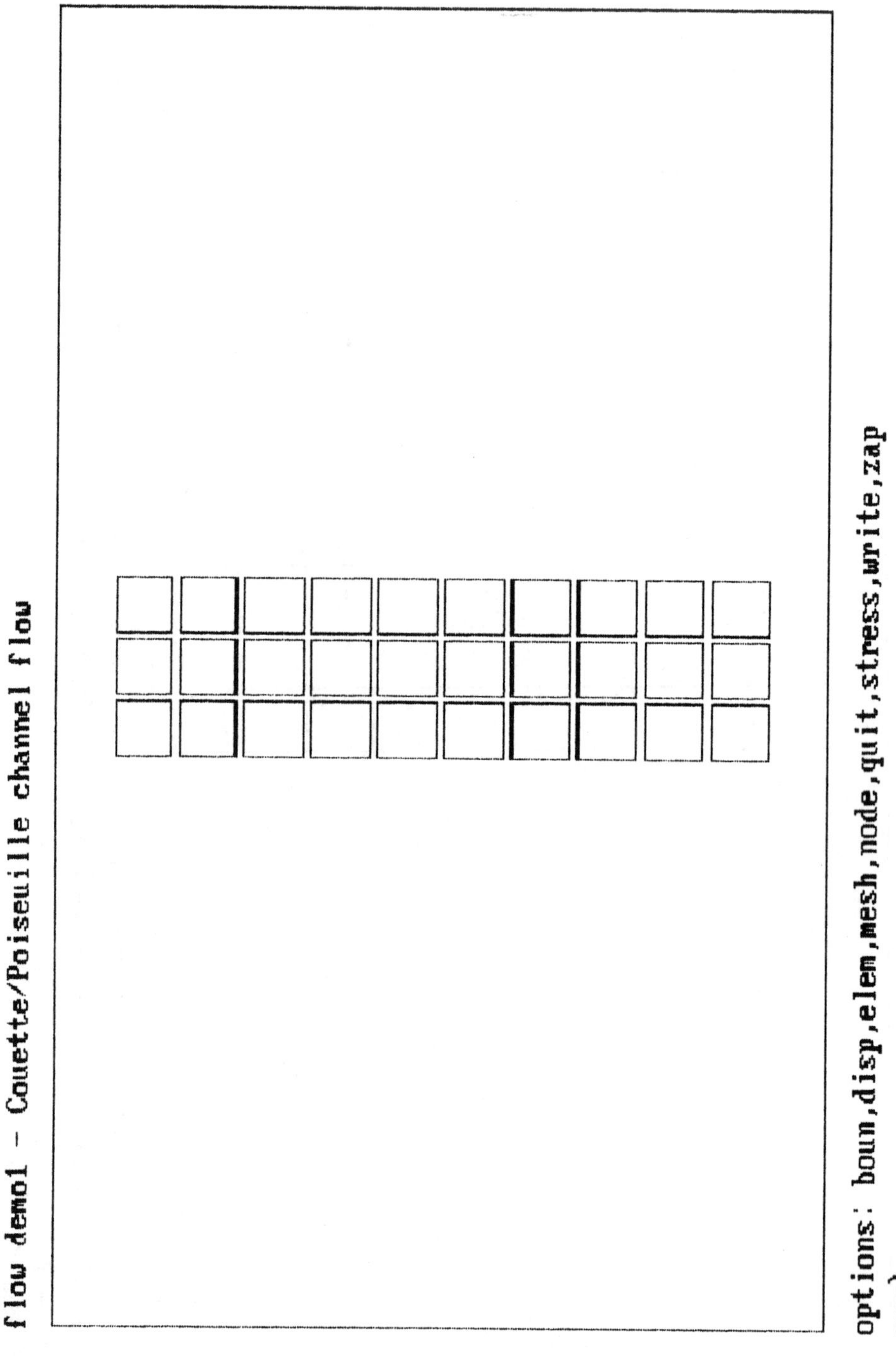

flow demo1 – Couette/Poiseuille channel flow

options: boun,disp,elem,mesh,node,quit,stress,write,zap
===>

Figure 7.2 Example 1: Chruelon Mesh

lem as displayed by the flow program. At the time of this writing a preprocessor interface was not complete, therefore the nodal and elemental data input for the problems in this chapter are not available. Instead the nodal, elemental, and material data are presented to the program in a form compatible with the PCTRAN input format. Node 1 is located at the lower left corner of the mesh and node 44 is located at the upper right corner. Figure 7.2 displays the shrunken mesh for this problem. The shrunken mesh can be used to help define the individual elements in more complicated geometries.

The section entitled "Input Data List" (below) displays the Couette/Poiseuille flow input data in the PCTRAN format. A slight variation in the format is introduced in the material section and at the end of the input data. The material section provides all the pertinent information about the fluid being used. At the end of the input section are a series of macro commands used by the fluid program to implement the various options available. This list of macros is located immediately following the "forc" input section. The first "end" is used to signify the end of the numerical input. The "macr" tells the program to start processing the following commands until the next "end" is encountered. Each macro command performs a primary function for the determination of the solution set. The "tang" command assembles the stiffness matrix which is also known as the "tangent" stiffness matrix. The "form" command sets up the right hand side of the matrix equation (the unbalanced force vector). The "solv" command is obvious, it solves the assembled set of equations for a change necessary to eliminate the unbalanced force. For further information on the specific details of this method read Zienkiewicz's textbook, "The Finite Element Method". The above three macros followed by the "disp" macro, which sends the computed displacements to an output file, macro are required for basic linear problems. Likewise, the "stre" command outputs the stress data to a file. To address more complicated nonlinear problems one may wish to use the "loop" and "next" for iterative techniques which may compute the stiffness matrix several times in order to reach a preset tolerance or to vary a parameter such as the viscosity of a material. Again, the above listed reference is suggested for a more detailed look into the use of these macros. The section "Output Data File" (below) lists the output data as a result of the flow program calculations. The output reiterates the input parameters such as the number of nodes, elements, degrees of freedom per node, nodal coordinates, elemental connectivities, and material parameters, to name a few. Finally the nodal displacements are presented in Figure 7.3. Using the analytical solution given above one sees immediately that an exact correlation exist.

```
flow demo1 - Couette/Poiseuille channel flow

nodal displacements      time  0.00000E+00

node      1 coord      2 coord      1 displ      2 displ
   1   0.0000E+00   0.0000E+00   0.0000E+00   0.0000E+00
   2   0.1000       0.0000E+00   0.0000E+00   0.0000E+00
   3   0.2000       0.0000E+00   0.0000E+00   0.0000E+00
   4   0.3000       0.0000E+00   0.0000E+00   0.0000E+00
   5   0.0000E+00   0.1000       0.5500E-01   0.0000E+00
   6   0.1000       0.1000       0.5500E-01   0.0000E+00
   7   0.2000       0.1000       0.5500E-01   0.0000E+00
   8   0.3000       0.1000       0.5500E-01   0.0000E+00
   9   0.0000E+00   0.2000       0.1200       0.0000E+00
  10   0.1000       0.2000       0.1200       0.0000E+00
  11   0.2000       0.2000       0.1200       0.0000E+00
  12   0.3000       0.2000       0.1200       0.0000E+00
  13   0.0000E+00   0.3000       0.1950       0.0000E+00
  14   0.1000       0.3000       0.1950       0.0000E+00
  15   0.2000       0.3000       0.1950       0.0000E+00
  16   0.3000       0.3000       0.1950       0.0000E+00
  17   0.0000E+00   0.4000       0.2800       0.0000E+00
  18   0.1000       0.4000       0.2800       0.0000E+00
  19   0.2000       0.4000       0.2800       0.0000E+00
  20   0.3000       0.4000       0.2800       0.0000E+00
  21   0.0000E+00   0.5000       0.3750       0.0000E+00
  22   0.1000       0.5000       0.3750       0.0000E+00
  23   0.2000       0.5000       0.3750       0.0000E+00
  24   0.3000       0.5000       0.3750       0.0000E+00
  25   0.0000E+00   0.6000       0.4800       0.0000E+00
  26   0.1000       0.6000       0.4800       0.0000E+00
  27   0.2000       0.6000       0.4800       0.0000E+00
  28   0.3000       0.6000       0.4800       0.0000E+00
  29   0.0000E+00   0.7000       0.5950       0.0000E+00
  30   0.1000       0.7000       0.5950       0.0000E+00
  31   0.2000       0.7000       0.5950       0.0000E+00
  32   0.3000       0.7000       0.5950       0.0000E+00
  33   0.0000E+00   0.8000       0.7200       0.0000E+00
  34   0.1000       0.8000       0.7200       0.0000E+00
  35   0.2000       0.8000       0.7200       0.0000E+00
  36   0.3000       0.8000       0.7200       0.0000E+00
  37   0.0000E+00   0.9000       0.8550       0.0000E+00
  38   0.1000       0.9000       0.8550       0.0000E+00
  39   0.2000       0.9000       0.8550       0.0000E+00
  40   0.3000       0.9000       0.8550       0.0000E+00
  41   0.0000E+00   1.000        1.000        0.0000E+00
  42   0.1000       1.000        1.000        0.0000E+00
  43   0.2000       1.000        1.000        0.0000E+00
  44   0.3000       1.000        1.000        0.0000E+00

**macro instruction disp completed. time used this step =  3.24 total time = 36
.85
**macro instruction   5 initiated**  stre      v1 =    0.0000E+00 , v2 =    0
.0000E+00
```

Figure 7.3   Example 1: Nodal Displacement Output.

INPUT DATA LIST

```
flow demo1 - Couette/Poiseuille channel flow
    44    30    1    2    2    4    0
coor
     1    0   .000000   .000000
     2    0   .100000   .000000
     3    0   .200000   .000000
     4    0   .300000   .000000
     5    0   .000000   .100000
     6    0   .100000   .100000
     7    0   .200000   .100000
     8    0   .300000   .100000
     9    0   .000000   .200000
    10    0   .100000   .200000
    11    0   .200000   .200000
    12    0   .300000   .200000
    13    0   .000000   .300000
    14    0   .100000   .300000
    15    0   .200000   .300000
    16    0   .300000   .300000
    17    0   .000000   .400000
    18    0   .100000   .400000
    19    0   .200000   .400000
    20    0   .300000   .400000
    21    0   .000000   .500000
    22    0   .100000   .500000
    23    0   .200000   .500000
    24    0   .300000   .500000
    25    0   .000000   .600000
    26    0   .100000   .600000
    27    0   .200000   .600000
    28    0   .300000   .600000
    29    0   .000000   .700000
    30    0   .100000   .700000
    31    0   .200000   .700000
    32    0   .300000   .700000
    33    0   .000000   .800000
    34    0   .100000   .800000
    35    0   .200000   .800000
    36    0   .300000   .800000
    37    0   .000000   .900000
    38    0   .100000   .900000
    39    0   .200000   .900000
    40    0   .300000   .900000
    41    0   .000000  1.000000
    42    0   .100000  1.000000
    43    0   .200000  1.000000
    44    0   .300000  1.000000

elem
     1    1    1    2    6    5    1
     2    1    2    3    7    6    1
     3    1    3    4    8    7    1
     4    1    5    6   10    9    1
     5    1    6    7   11   10    1
     6    1    7    8   12   11    1
     7    1    9   10   14   13    1
     8    1   10   11   15   14    1
     9    1   11   12   16   15    1
    10    1   13   14   18   17    1
    11    1   14   15   19   18    1
```

INPUT DATA LIST

```
12   1   15   16   20   19   1
13   1   17   18   22   21   1
14   1   18   19   23   22   1
15   1   19   20   24   23   1
16   1   21   22   26   25   1
17   1   22   23   27   26   1
18   1   23   24   28   27   1
19   1   25   26   30   29   1
20   1   26   27   31   30   1
21   1   27   28   32   31   1
22   1   29   30   34   33   1
23   1   30   31   35   34   1
24   1   31   32   36   35   1
25   1   33   34   38   37   1
26   1   34   35   39   38   1
27   1   35   36   40   39   1
28   1   37   38   42   41   1
29   1   38   39   43   42   1
30   1   39   40   44   43   1
```

mate
```
        1        3!  material number, element type number
                 2!  regular integration order
                 0!  flag for axisymmetry (0 for no)
                 0!  flag for time stepping (0 for no)
        0.000E+00!  theta factor in transient algorithm
                 1!  degree of freedom number for velocities
                 1!  penalty integration order
                 0!  momentum convection integration order:
        1.000E+07!  penalty factor
        0.000E+00!  density
        1.000E+00!  viscosity coefficient
        0.000E+00!  lambda factor in carreau model
        1.000E+00!  power-law exponent
        0.000E+00!  viscosity activation energy
        0.000E+00!  reference temperature
        0.000E+00!  thermal expansion coefficient
                 0!  degree of freedom number for temperature
                 0!  thermal convection integration order
        0.000E+00!  thermal conductivity
        0.000E+00!  specific heat
        0.000E+00!  thermomechanical dissipation factor
                 0!  degree of freedom number for reaction
                 0!  species convection integration order
        0.000E+00!  species diffusivity
        1.000E+00!  kinetic order of reaction
        0.000E+00!  rate constant preexponential factor
        0.000E+00!  activation energy for reaction / gas constant
        0.000E+00!  heat of reaction
                 0!  degree of freedom number for stream function
```

boun
```
        1    0    1    1
        2    0    1    1
        3    0    1    1
        4    0    1    1
        5    0    0    1
        6    0    0    1
        7    0    0    1
        8    0    0    1
```

```
   9     0     0     1
  10     0     0     1
  11     0     0     1
  12     0     0     1
  13     0     0     1
  14     0     0     1
  15     0     0     1
  16     0     0     1
  17     0     0     1
  18     0     0     1
  19     0     0     1
  20     0     0     1
  21     0     0     1
  22     0     0     1
  23     0     0     1
  24     0     0     1
  25     0     0     1
  26     0     0     1
  27     0     0     1
  28     0     0     1
  29     0     0     1
  30     0     0     1
  31     0     0     1
  32     0     0     1
  33     0     0     1
  34     0     0     1
  35     0     0     1
  36     0     0     1
  37     0     0     1
  38     0     0     1
  39     0     0     1
  40     0     0     1
  41     0     1     1
  42     0     1     1
  43     0     1     1
  44     0     1     1

forc
   8     0   -.0300    .0000
  12     0   -.0300    .0000
  16     0   -.0300    .0000
  20     0   -.0300    .0000
  24     0   -.0300    .0000
  28     0   -.0300    .0000
  32     0   -.0300    .0000
  36     0   -.0300    .0000
  40     0   -.0300    .0000
  41     0  1.0000    .0000
  42     0  1.0000    .0000
  43     0  1.0000    .0000
  44     0  1.0000    .0000

end
macr
tang
form
solv
disp
stre
end
```

OUTPUT DATA FILE

```
flow demo1 - Couette/Poiseuille channel flow

    number of nodal points        =     44
    number of elements            =     30
    number of material sets       =      1
    dimension of coordinate space=      2
    degrees of freedom/node       =      2
    nodes per element (maximum)   =      4
    extra d.o.f. to element       =      0


flow demo1 - Couette/Poiseuille channel flow

    nodal coordinates

    node      1 coord       2 coord
       1    0.0000E+00    0.0000E+00
       2    0.1000       0.0000E+00
       3    0.2000       0.0000E+00
       4    0.3000       0.0000E+00
       5    0.0000E+00    0.1000
       6    0.1000       0.1000
       7    0.2000       0.1000
       8    0.3000       0.1000
       9    0.0000E+00    0.2000
      10    0.1000       0.2000
      11    0.2000       0.2000
      12    0.3000       0.2000
      13    0.0000E+00    0.3000
      14    0.1000       0.3000
      15    0.2000       0.3000
      16    0.3000       0.3000
      17    0.0000E+00    0.4000
      18    0.1000       0.4000
      19    0.2000       0.4000
      20    0.3000       0.4000
      21    0.0000E+00    0.5000
      22    0.1000       0.5000
      23    0.2000       0.5000
      24    0.3000       0.5000
      25    0.0000E+00    0.6000
      26    0.1000       0.6000
      27    0.2000       0.6000
      28    0.3000       0.6000
      29    0.0000E+00    0.7000
      30    0.1000       0.7000
      31    0.2000       0.7000
      32    0.3000       0.7000
      33    0.0000E+00    0.8000
      34    0.1000       0.8000
      35    0.2000       0.8000
      36    0.3000       0.8000
      37    0.0000E+00    0.9000
      38    0.1000       0.9000
      39    0.2000       0.9000
      40    0.3000       0.9000
      41    0.0000E+00    1.000
      42    0.1000       1.000
      43    0.2000       1.000
      44    0.3000       1.000
```

```
flow demo1 - Couette/Poiseuille channel flow

    elements
```

element	material	1 node	2 node	3 node	4 node
1	1	1	2	6	5
2	1	2	3	7	6
3	1	3	4	8	7
4	1	5	6	10	9
5	1	6	7	11	10
6	1	7	8	12	11
7	1	9	10	14	13
8	1	10	11	15	14
9	1	11	12	16	15
10	1	13	14	18	17
11	1	14	15	19	18
12	1	15	16	20	19
13	1	17	18	22	21
14	1	18	19	23	22
15	1	19	20	24	23
16	1	21	22	26	25
17	1	22	23	27	26
18	1	23	24	28	27
19	1	25	26	30	29
20	1	26	27	31	30
21	1	27	28	32	31
22	1	29	30	34	33
23	1	30	31	35	34
24	1	31	32	36	35
25	1	33	34	38	37
26	1	34	35	39	38
27	1	35	36	40	39
28	1	37	38	42	41
29	1	38	39	43	42
30	1	39	40	44	43

OUTPUT DATA FILE

flow demo1 - Couette/Poiseuille channel flow

    material properties

    material set 1 for element type 3   : material number, element type numb
er

    two-dimensional transport element

        regular integration order = 2
        flag for axisymmetry (0 for no) = 0
        flag for time stepping (0 for no) = 0
        transient algorithm theta value = 0.0000E+00

   momentum parameters:

        velocity degree of freedom = 1
        penalty integration order = 1
        momentum convection integration order = 0

        penalty multiplier = 0.1000E+08
        density = 0.0000E+00
        viscosity coefficient =   1.000
        Carreau lambda factor = 0.0000E+00
        power-law exponent =   1.000
        viscosity activation energy = 0.0000E+00
        reference temperature = 0.0000E+00
        thermal expansion coefficient = 0.0000E+00

   thermal parameters:

        temperature degree of freedom = 0
        thermal convection integration order = 0
        coefficient of thermal conduction = 0.0000E+00
        specific heat = 0.0000E+00
        dissipation factor = 0.0000E+00

   species conversion parameters:

        conversion degree of freedom = 0
        species convection integration order = 0
        species diffusivity = 0.0000E+00
        kinetic reaction order =   1.000
        preexponential rate factor = 0.0000E+00
        reaction activation energy = 0.0000E+00
        heat of reaction = 0.0000E+00

        stream function degree of freedom = 0

```
flow demo1 - Couette/Poiseuille channel flow

    nodal b.c.

    node      1 b.c.      2 b.c.
     1           1           1
     2           1           1
     3           1           1
     4           1           1
     5           0           1
     6           0           1
     7           0           1
     8           0           1
     9           0           1
    10           0           1
    11           0           1
    12           0           1
    13           0           1
    14           0           1
    15           0           1
    16           0           1
    17           0           1
    18           0           1
    19           0           1
    20           0           1
    21           0           1
    22           0           1
    23           0           1
    24           0           1
    25           0           1
    26           0           1
    27           0           1
    28           0           1
    29           0           1
    30           0           1
    31           0           1
    32           0           1
    33           0           1
    34           0           1
    35           0           1
    36           0           1
    37           0           1
    38           0           1
    39           0           1
    40           0           1
    41           1           1
    42           1           1
    43           1           1
    44           1           1
```

OUTPUT DATA FILE

```
flow demo1 - Couette/Poiseuille channel flow

    nodal force/displ

    node        1 force         2 force
      1      0.0000E+00      0.0000E+00
      2      0.0000E+00      0.0000E+00
      3      0.0000E+00      0.0000E+00
      4      0.0000E+00      0.0000E+00
      5      0.0000E+00      0.0000E+00
      6      0.0000E+00      0.0000E+00
      7      0.0000E+00      0.0000E+00
      8     -0.3000E-01      0.0000E+00
      9      0.0000E+00      0.0000E+00
     10      0.0000E+00      0.0000E+00
     11      0.0000E+00      0.0000E+00
     12     -0.3000E-01      0.0000E+00
     13      0.0000E+00      0.0000E+00
     14      0.0000E+00      0.0000E+00
     15      0.0000E+00      0.0000E+00
     16     -0.3000E-01      0.0000E+00
     17      0.0000E+00      0.0000E+00
     18      0.0000E+00      0.0000E+00
     19      0.0000E+00      0.0000E+00
     20     -0.3000E-01      0.0000E+00
     21      0.0000E+00      0.0000E+00
     22      0.0000E+00      0.0000E+00
     23      0.0000E+00      0.0000E+00
     24     -0.3000E-01      0.0000E+00
     25      0.0000E+00      0.0000E+00
     26      0.0000E+00      0.0000E+00
     27      0.0000E+00      0.0000E+00
     28     -0.3000E-01      0.0000E+00
     29      0.0000E+00      0.0000E+00
     30      0.0000E+00      0.0000E+00
     31      0.0000E+00      0.0000E+00
     32     -0.3000E-01      0.0000E+00
     33      0.0000E+00      0.0000E+00
     34      0.0000E+00      0.0000E+00
     35      0.0000E+00      0.0000E+00
     36     -0.3000E-01      0.0000E+00
     37      0.0000E+00      0.0000E+00
     38      0.0000E+00      0.0000E+00
     39      0.0000E+00      0.0000E+00
     40     -0.3000E-01      0.0000E+00
     41      1.000           0.0000E+00
     42      1.000           0.0000E+00
     43      1.000           0.0000E+00
     44      1.000           0.0000E+00
```

```
flow demo1 - Couette/Poiseuille channel flow

   macro instructions

   macro statement     variable 1        variable 2
        tang          0.00000E+00       0.00000E+00
        form          0.00000E+00       0.00000E+00
        solv          0.00000E+00       0.00000E+00
        disp          0.00000E+00       0.00000E+00
        stre          0.00000E+00       0.00000E+00
        end           0.00000E+00       0.00000E+00


**macro instruction loop completed. time used this step =  1.16 total time = 19
.50
**macro instruction  1 initiated**  tang        v1 =    0.0000E+00 , v2 =     0
.0000E+00


**macro instruction tang completed. time used this step =  8.40 total time = 27
.90
**macro instruction  2 initiated**  form        v1 =    0.0000E+00 , v2 =     0
.0000E+00
     force convergence test
        rnmax =     1.5742        rm    =    1.5742        tol   =    0.1000
0E-08


**macro instruction form completed. time used this step =  5.11 total time = 33
.01
**macro instruction  3 initiated**  solv        v1 =    0.0000E+00 , v2 =     0
.0000E+00


**macro instruction solv completed. time used this step =  0.60 total time = 33
.61
**macro instruction  4 initiated**  disp        v1 =    0.0000E+00 , v2 =     0
.0000E+00

     time  0.00000E+00

     proportional load   1.0000
```

7.4 EXAMPLE PROBLEM 2: 4:1 ENTRY FLOW

The second example, 4:1 entry flow, is found in many areas of fluid dynamics. Here a viscous fluid moving through a channel with a fully developed parabolic flow is suddenly met with a constriction one quarter the original diameter. In this problem, one may take advantage of symmetry and produce only the upper half of the problem geometry. Figure 7.4 displays the input geometry while the Example 2: Input Data List (below) lists the input data. There are 128 nodes and 100 elements with the centerline boundary being set to zero. The "forc" section sets the parabolic flow at the right side of the model. In this example the "loop" and "next" macros are used since in this problem we are going to compute the temperature and stream function for this flow. Graphical output for the streamfunction and temperature variables are shown in Figures 7.5 and 7.6, respectively. As in the first problem, the input data are reiterated in the Example 2 Output Data List and the macro instruction execution log is displayed. Finally, Example 2 Nodal Displacement List (below) lists the output displacements by coordinate. The displacement columns represent the x-velocity, y-velocity, stream-function and nodal temperatures for this problem.

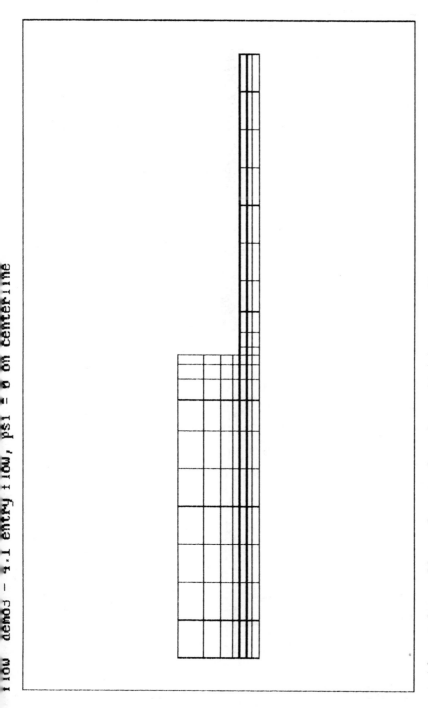

flow demo3 — 4.1 entry flow, psi = 0 on centerline

options: boun,disp,elem,mesh,node,quit,stress,write,zap

===}

Figure 7.4

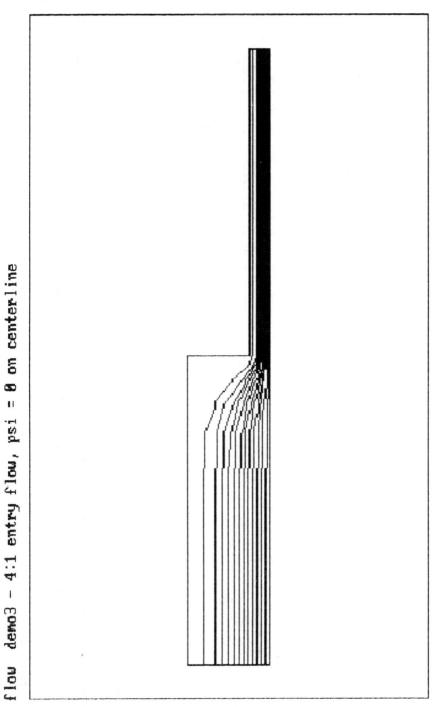

flow demo3 — 4:1 entry flow, psi = 0 on centerline

options: boun,disp,elem,mesh,node,quit,stress,write,zap
===>

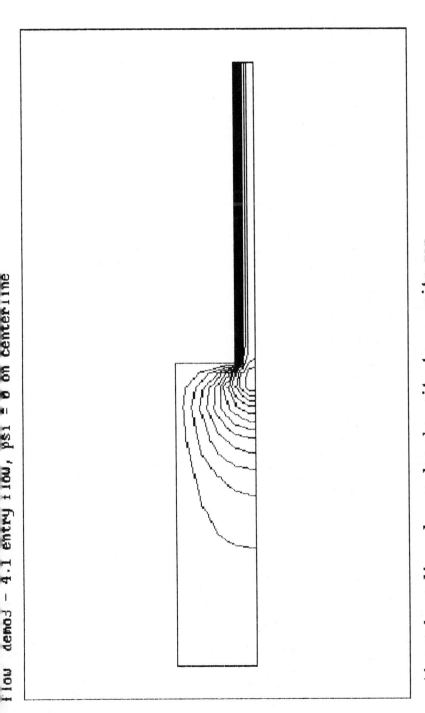

Figure 7.6

EXAMPLE 2: INPUT DATA LIST

```
flow   demo3 - 4:1 entry flow, psi = 0 on centerline
   128  100    1    2    4    4    0
coor
     1     0 -.4000+01  .0000
     2     0 -.4000+01  .8333-01
     3     0 -.4000+01  .1667+00
     4     0 -.4000+01  .2500+00
     5     0 -.4000+01  .3423+00
     6     0 -.4000+01  .4808+00
     7     0 -.4000+01  .6885+00
     8     0 -.4000+01  .1000+01
     9     0 -.3500+01  .0000
    10     0 -.3500+01  .8333-01
    11     0 -.3500+01  .1667+00
    12     0 -.3500+01  .2500+00
    13     0 -.3500+01  .3423+00
    14     0 -.3500+01  .4808+00
    15     0 -.3500+01  .6885+00
    16     0 -.3500+01  .1000+01
    17     0 -.3000+01  .0000
    18     0 -.3000+01  .8333-01
    19     0 -.3000+01  .1667+00
    20     0 -.3000+01  .2500+00
    21     0 -.3000+01  .3423+00
    22     0 -.3000+01  .4808+00
    23     0 -.3000+01  .6885+00
    24     0 -.3000+01  .1000+01
    25     0 -.2500+01  .0000
    26     0 -.2500+01  .8333-01
    27     0 -.2500+01  .1667+00
    28     0 -.2500+01  .2500+00
    29     0 -.2500+01  .3423+00
    30     0 -.2500+01  .4808+00
    31     0 -.2500+01  .6885+00
    32     0 -.2500+01  .1000+01
    33     0 -.2000+01  .0000
    34     0 -.2000+01  .8333-01
    35     0 -.2000+01  .1667+00
    36     0 -.2000+01  .2500+00
    37     0 -.2000+01  .3423+00
    38     0 -.2000+01  .4808+00
    39     0 -.2000+01  .6885+00
    40     0 -.2000+01  .1000+01
    41     0 -.1500+01  .0000
    42     0 -.1500+01  .8333-01
    43     0 -.1500+01  .1667+00
    44     0 -.1500+01  .2500+00
    45     0 -.1500+01  .3423+00
    46     0 -.1500+01  .4808+00
    47     0 -.1500+01  .6885+00
    48     0 -.1500+01  .1000+01
    49     0 -.1000+01  .2794-08
    50     0 -.1000+01  .8333-01
    51     0 -.1000+01  .1667+00
    52     0 -.1000+01  .2500+00
    53     0 -.1000+01  .3423+00
    54     0 -.1000+01  .4808+00
    55     0 -.1000+01  .6885+00
    56     0 -.1000+01  .1000+01
    57     0 -.5846+00  .0000
```

```
 58   0 -.5846+00   .8333-01
 59   0 -.5846+00   .1667+00
 60   0 -.5846+00   .2500+00
 61   0 -.5846+00   .3423+00
 62   0 -.5846+00   .4808+00
 63   0 -.5846+00   .6885+00
 64   0 -.5846+00   .1000+01
 65   0 -.3077+00   .0000
 66   0 -.3077+00   .8333-01
 67   0 -.3077+00   .1667+00
 68   0 -.3077+00   .2500+00
 69   0 -.3077+00   .3423+00
 70   0 -.3077+00   .4808+00
 71   0 -.3077+00   .6885+00
 72   0 -.3077+00   .1000+01
 73   0 -.1231+00   .0000
 74   0 -.1231+00   .8333-01
 75   0 -.1231+00   .1667+00
 76   0 -.1231+00   .2500+00
 77   0 -.1231+00   .3423+00
 78   0 -.1231+00   .4808+00
 79   0 -.1231+00   .6885+00
 80   0 -.1231+00   .1000+01
 81   0   0.        .1863-08
 82   0   0.        .8333-01
 83   0   0.        .1667+00
 84   0   0.        .2500+00
 85   0   0.        .3423+00
 86   0   0.        .4808+00
 87   0   0.        .6885+00
 88   0   0.        .1000+01
 89   0  .1231+00   .0000
 90   0  .1231+00   .8333-01
 91   0  .1231+00   .1667+00
 92   0  .1231+00   .2500+00
 93   0  .3077+00   .0000
 94   0  .3077+00   .8333-01
 95   0  .3077+00   .1667+00
 96   0  .3077+00   .2500+00
 97   0  .5846+00   .0000
 98   0  .5846+00   .8333-01
 99   0  .5846+00   .1667+00
100   0  .5846+00   .2500+00
101   0  .1000+01   .1863-08
102   0  .1000+01   .8333-01
103   0  .1000+01   .1667+00
104   0  .1000+01   .2500+00
105   0  .1500+01   .0000
106   0  .1500+01   .8333-01
107   0  .1500+01   .1667+00
108   0  .1500+01   .2500+00
109   0  .2000+01   .0000
110   0  .2000+01   .8333-01
111   0  .2000+01   .1667+00
112   0  .2000+01   .2500+00
113   0  .2500+01   .0000
114   0  .2500+01   .8333-01
115   0  .2500+01   .1667+00
116   0  .2500+01   .2500+00
117   0  .3000+01   .0000
```

EXAMPLE 2: INPUT DATA LIST

118	0	.3000+01	.8333-01
119	0	.3000+01	.1667+00
120	0	.3000+01	.2500+00
121	0	.3500+01	.0000
122	0	.3500+01	.8333-01
123	0	.3500+01	.1667+00
124	0	.3500+01	.2500+00
125	0	.4000+01	.0000
126	0	.4000+01	.8333-01
127	0	.4000+01	.1667+00
128	0	.4000+01	.2500+00

elem

1	1	1	9	10	2	0
2	1	2	10	11	3	0
3	1	3	11	12	4	0
4	1	4	12	13	5	0
5	1	5	13	14	6	0
6	1	6	14	15	7	0
7	1	7	15	16	8	0
8	1	9	17	18	10	0
9	1	10	18	19	11	0
10	1	11	19	20	12	0
11	1	12	20	21	13	0
12	1	13	21	22	14	0
13	1	14	22	23	15	0
14	1	15	23	24	16	0
15	1	17	25	26	18	0
16	1	18	26	27	19	0
17	1	19	27	28	20	0
18	1	20	28	29	21	0
19	1	21	29	30	22	0
20	1	22	30	31	23	0
21	1	23	31	32	24	0
22	1	25	33	34	26	0
23	1	26	34	35	27	0
24	1	27	35	36	28	0
25	1	28	36	37	29	0
26	1	29	37	38	30	0
27	1	30	38	39	31	0
28	1	31	39	40	32	0
29	1	33	41	42	34	0
30	1	34	42	43	35	0
31	1	35	43	44	36	0
32	1	36	44	45	37	0
33	1	37	45	46	38	0
34	1	38	46	47	39	0
35	1	39	47	48	40	0
36	1	41	49	50	42	0
37	1	42	50	51	43	0
38	1	43	51	52	44	0
39	1	44	52	53	45	0
40	1	45	53	54	46	0
41	1	46	54	55	47	0
42	1	47	55	56	48	0
43	1	49	57	58	50	0
44	1	50	58	59	51	0
45	1	51	59	60	52	0
46	1	52	60	61	53	0
47	1	53	61	62	54	0

```
 48    1    54    62    63    55     0
 49    1    55    63    64    56     0
 50    1    57    65    66    58     0
 51    1    58    66    67    59     0
 52    1    59    67    68    60     0
 53    1    60    68    69    61     0
 54    1    61    69    70    62     0
 55    1    62    70    71    63     0
 56    1    63    71    72    64     0
 57    1    65    73    74    66     0
 58    1    66    74    75    67     0
 59    1    67    75    76    68     0
 60    1    68    76    77    69     0
 61    1    69    77    78    70     0
 62    1    70    78    79    71     0
 63    1    71    79    80    72     0
 64    1    73    81    82    74     0
 65    1    74    82    83    75     0
 66    1    75    83    84    76     0
 67    1    76    84    85    77     0
 68    1    77    85    86    78     0
 69    1    78    86    87    79     0
 70    1    79    87    88    80     0
 71    1    81    89    90    82     0
 72    1    82    90    91    83     0
 73    1    83    91    92    84     0
 74    1    89    93    94    90     0
 75    1    90    94    95    91     0
 76    1    91    95    96    92     0
 77    1    93    97    98    94     0
 78    1    94    98    99    95     0
 79    1    95    99   100    96     0
 80    1    97   101   102    98     0
 81    1    98   102   103    99     0
 82    1    99   103   104   100     0
 83    1   101   105   106   102     0
 84    1   102   106   107   103     0
 85    1   103   107   108   104     0
 86    1   105   109   110   106     0
 87    1   106   110   111   107     0
 88    1   107   111   112   108     0
 89    1   109   113   114   110     0
 90    1   110   114   115   111     0
 91    1   111   115   116   112     0
 92    1   113   117   118   114     0
 93    1   114   118   119   115     0
 94    1   115   119   120   116     0
 95    1   117   121   122   118     0
 96    1   118   122   123   119     0
 97    1   119   123   124   120     0
 98    1   121   125   126   122     0
 99    1   122   126   127   123     0
100    1   123   127   128   124     0

mate
    1    3
         2! regular integration order
         0! flag for axisymmetry (0 for no)
         0! flag for time stepping (0 for no)
       0.! theta factor in transient algorithm
```

EXAMPLE 2: INPUT DATA LIST

```
   1¦ degree of freedom number for velocities
   1¦ penalty integration order
   0¦ momentum convection integration order:
1.d7¦ penalty factor
  0.¦ density
  1.¦ viscosity coefficient
  0.¦ lambda factor in carreau model
  1.¦ power-law exponent
  0.¦ viscosity activation energy
  0.¦ reference temperature
  0.¦ thermal expansion coefficient
   3¦ degree of freedom number for temperature
   0¦ thermal convection integration order
  1.¦ thermal conductivity
  0.¦ specific heat
  1.¦ thermomechanical dissipation factor
   0¦ degree of freedom number for reaction
   0¦ species convection integration order
  0.¦ species diffusivity
  1.¦ kinetic order of reaction
  0.¦ rate constant preexponential factor
  0.¦ activation energy for reaction / gas constant
  0.¦ heat of reaction
   4¦ degree of freedom number for stream function
```

```
boun
    1    0    1    1    1    1
    2    0    1    1    1    0
    3    0    1    1    1    0
    4    0    1    1    1    0
    5    0    1    1    1    0
    6    0    1    1    1    0
    7    0    1    1    1    0
    8    0    1    1    1    0
    9    1    0   -1    0    1
   16    0    1    1    1    0
   17    0    0    1    0    1
   24    0    1    1    1    0
   25    0    0    1    0    1
   32    0    1    1    1    0
   33    0    0    1    0    1
   40    0    1    1    1    0
   41    0    0    1    0    1
   48    0    1    1    1    0
   49    0    0    1    0    1
   56    0    1    1    1    0
   57    0    0    1    0    1
   64    0    1    1    1    0
   65    0    0    1    0    1
   72    0    1    1    1    0
   73    0    0    1    0    1
   80    0    1    1    1    0
   81    0    0    1    0    1
   84    0    1    1    1    0
   85    0    1    1    1    0
   86    0    1    1    1    0
   87    0    1    1    1    0
   88    0    1    1    1    0
   89    0    0    1    0    1
   92    0    1    1    1    0
```

```
 93    0    0    1    0    1
 96    0    1    1    1    0
 97    0    0    1    0    1
100    0    1    1    1    0
101    0    0    1    0    1
104    0    1    1    1    0
105    0    0    1    0    1
108    0    1    1    1    0
109    0    0    1    0    1
112    0    1    1    1    0
113    0    0    1    0    1
116    0    1    1    1    0
117    0    0    1    0    1
120    0    1    1    1    0
121    0    0    1    0    1
124    0    1    1    1    0
125    1    0   -1    0    1
128    0    1    1    1    0

forc
  1    0  .1000+01  .0000    0.       0.
  2    0  .9931+00  .0000    0.       0.
  3    0  .9722+00  .0000    0.       0.
  4    0  .9375+00  .0000    0.       0.
  5    0  .8828+00  .0000    0.       0.
  6    0  .7688+00  .0000    0.       0.
  7    0  .5260+00  .0000    0.       0.

end
macr
tang
loop          2
form
solv
next
disp
end
stop
```

EXAMPLE 2: OUTPUT DATA LIST

```
flow   demo3 - 4:1 entry flow, psi = 0 on centerline

        number of nodal points      =     128
        number of elements          =     100
        number of material sets     =       1
        dimension of coordinate space=      2
        degrees of freedom/node     =       4
        nodes per element (maximum) =       4
        extra d.o.f. to element     =       0
```

EXAMPLE 2: OUTPUT DATA LIST

flow demo3 - 4:1 entry flow, psi = 0 on centerline

nodal coordinates

node	1 coord	2 coord
1	-4.000	0.0000E+00
2	-4.000	0.8333E-01
3	-4.000	0.1667
4	-4.000	0.2500
5	-4.000	0.3423
6	-4.000	0.4808
7	-4.000	0.6885
8	-4.000	1.000
9	-3.500	0.0000E+00
10	-3.500	0.8333E-01
11	-3.500	0.1667
12	-3.500	0.2500
13	-3.500	0.3423
14	-3.500	0.4808
15	-3.500	0.6885
16	-3.500	1.000
17	-3.000	0.0000E+00
18	-3.000	0.8333E-01
19	-3.000	0.1667
20	-3.000	0.2500
21	-3.000	0.3423
22	-3.000	0.4808
23	-3.000	0.6885
24	-3.000	1.000
25	-2.500	0.0000E+00
26	-2.500	0.8333E-01
27	-2.500	0.1667
28	-2.500	0.2500
29	-2.500	0.3423
30	-2.500	0.4808
31	-2.500	0.6885
32	-2.500	1.000
33	-2.000	0.0000E+00
34	-2.000	0.8333E-01
35	-2.000	0.1667
36	-2.000	0.2500
37	-2.000	0.3423
38	-2.000	0.4808
39	-2.000	0.6885
40	-2.000	1.000
41	-1.500	0.0000E+00
42	-1.500	0.8333E-01
43	-1.500	0.1667
44	-1.500	0.2500
45	-1.500	0.3423
46	-1.500	0.4808
47	-1.500	0.6885
48	-1.500	1.000
49	-1.000	0.2794E-08
50	-1.000	0.8333E-01

```
flow  demo3 - 4:1 entry flow, psi = 0 on centerline

    nodal coordinates

    node      1 coord       2 coord
      51     -1.000        0.1667
      52     -1.000        0.2500
      53     -1.000        0.3423
      54     -1.000        0.4808
      55     -1.000        0.6885
      56     -1.000        1.000
      57     -0.5846       0.0000E+00
      58     -0.5846       0.8333E-01
      59     -0.5846       0.1667
      60     -0.5846       0.2500
      61     -0.5846       0.3423
      62     -0.5846       0.4808
      63     -0.5846       0.6885
      64     -0.5846       1.000
      65     -0.3077       0.0000E+00
      66     -0.3077       0.8333E-01
      67     -0.3077       0.1667
      68     -0.3077       0.2500
      69     -0.3077       0.3423
      70     -0.3077       0.4808
      71     -0.3077       0.6885
      72     -0.3077       1.000
      73     -0.1231       0.0000E+00
      74     -0.1231       0.8333E-01
      75     -0.1231       0.1667
      76     -0.1231       0.2500
      77     -0.1231       0.3423
      78     -0.1231       0.4808
      79     -0.1231       0.6885
      80     -0.1231       1.000
      81      0.0000E+00   0.1863E-08
      82      0.0000E+00   0.8333E-01
      83      0.0000E+00   0.1667
      84      0.0000E+00   0.2500
      85      0.0000E+00   0.3423
      86      0.0000E+00   0.4808
      87      0.0000E+00   0.6885
      88      0.0000E+00   1.000
      89      0.1231       0.0000E+00
      90      0.1231       0.8333E-01
      91      0.1231       0.1667
      92      0.1231       0.2500
      93      0.3077       0.0000E+00
      94      0.3077       0.8333E-01
      95      0.3077       0.1667
      96      0.3077       0.2500
      97      0.5846       0.0000E+00
      98      0.5846       0.8333E-01
      99      0.5846       0.1667
     100      0.5846       0.2500
```

EXAMPLE 2: OUTPUT DATA LIST

flow   demo3 — 4:1 entry flow, psi = 0 on centerline

    nodal coordinates

node	1 coord	2 coord
101	1.000	0.1863E-08
102	1.000	0.8333E-01
103	1.000	0.1667
104	1.000	0.2500
105	1.500	0.0000E+00
106	1.500	0.8333E-01
107	1.500	0.1667
108	1.500	0.2500
109	2.000	0.0000E+00
110	2.000	0.8333E-01
111	2.000	0.1667
112	2.000	0.2500
113	2.500	0.0000E+00
114	2.500	0.8333E-01
115	2.500	0.1667
116	2.500	0.2500
117	3.000	0.0000E+00
118	3.000	0.8333E-01
119	3.000	0.1667
120	3.000	0.2500
121	3.500	0.0000E+00
122	3.500	0.8333E-01
123	3.500	0.1667
124	3.500	0.2500
125	4.000	0.0000E+00
126	4.000	0.8333E-01
127	4.000	0.1667
128	4.000	0.2500

```
flow  demo3 - 4:1 entry flow, psi = 0 on centerline

     elements
```

element	material	1 node	2 node	3 node	4 node
1	1	1	9	10	2
2	1	2	10	11	3
3	1	3	11	12	4
4	1	4	12	13	5
5	1	5	13	14	6
6	1	6	14	15	7
7	1	7	15	16	8
8	1	9	17	18	10
9	1	10	18	19	11
10	1	11	19	20	12
11	1	12	20	21	13
12	1	13	21	22	14
13	1	14	22	23	15
14	1	15	23	24	16
15	1	17	25	26	18
16	1	18	26	27	19
17	1	19	27	28	20
18	1	20	28	29	21
19	1	21	29	30	22
20	1	22	30	31	23
21	1	23	31	32	24
22	1	25	33	34	26
23	1	26	34	35	27
24	1	27	35	36	28
25	1	28	36	37	29
26	1	29	37	38	30
27	1	30	38	39	31
28	1	31	39	40	32
29	1	33	41	42	34
30	1	34	42	43	35
31	1	35	43	44	36
32	1	36	44	45	37
33	1	37	45	46	38
34	1	38	46	47	39
35	1	39	47	48	40
36	1	41	49	50	42
37	1	42	50	51	43
38	1	43	51	52	44
39	1	44	52	53	45
40	1	45	53	54	46
41	1	46	54	55	47
42	1	47	55	56	48
43	1	49	57	58	50
44	1	50	58	59	51
45	1	51	59	60	52
46	1	52	60	61	53
47	1	53	61	62	54
48	1	54	62	63	55
49	1	55	63	64	56
50	1	57	65	66	58

EXAMPLE 2: OUTPUT DATA LIST

```
flow  demo3 - 4:1 entry flow, psi = 0 on centerline

    elements
```

element	material	1 node	2 node	3 node	4 node
51	1	58	66	67	59
52	1	59	67	68	60
53	1	60	68	69	61
54	1	61	69	70	62
55	1	62	70	71	63
56	1	63	71	72	64
57	1	65	73	74	66
58	1	66	74	75	67
59	1	67	75	76	68
60	1	68	76	77	69
61	1	69	77	78	70
62	1	70	78	79	71
63	1	71	79	80	72
64	1	73	81	82	74
65	1	74	82	83	75
66	1	75	83	84	76
67	1	76	84	85	77
68	1	77	85	86	78
69	1	78	86	87	79
70	1	79	87	88	80
71	1	81	89	90	82
72	1	82	90	91	83
73	1	83	91	92	84
74	1	89	93	94	90
75	1	90	94	95	91
76	1	91	95	96	92
77	1	93	97	98	94
78	1	94	98	99	95
79	1	95	99	100	96
80	1	97	101	102	98
81	1	98	102	103	99
82	1	99	103	104	100
83	1	101	105	106	102
84	1	102	106	107	103
85	1	103	107	108	104
86	1	105	109	110	106
87	1	106	110	111	107
88	1	107	111	112	108
89	1	109	113	114	110
90	1	110	114	115	111
91	1	111	115	116	112
92	1	113	117	118	114
93	1	114	118	119	115
94	1	115	119	120	116
95	1	117	121	122	118
96	1	118	122	123	119
97	1	119	123	124	120
98	1	121	125	126	122
99	1	122	126	127	123
100	1	123	127	128	124

```
flow  demo3 - 4:1 entry flow, psi = 0 on centerline

    material properties

    material set  1 for element type  3


    two-dimensional transport element

        regular integration order = 2
        flag for axisymmetry (0 for no) = 0
        flag for time stepping (0 for no) = 0
        transient algorithm theta value =  0.0000E+00

    momentum parameters:

        velocity degree of freedom = 1
        penalty integration order = 1
        momentum convection integration order = 0

        penalty multiplier =  0.1000E+08
        density =  0.0000E+00
        viscosity coefficient =   1.000
        Carreau lambda factor =  0.0000E+00
        power-law exponent =   1.000
        viscosity activation energy =  0.0000E+00
        reference temperature =  0.0000E+00
        thermal expansion coefficient =  0.0000E+00

    thermal parameters:

        temperature degree of freedom = 3
        thermal convection integration order = 0
        coefficient of thermal conduction =   1.000
        specific heat =  0.0000E+00
        dissipation factor =   1.000

    species conversion parameters:

        conversion degree of freedom = 0
        species convection integration order = 0
        species diffusivity =  0.0000E+00
        kinetic reaction order =   1.000
        preexponential rate factor =  0.0000E+00
        reaction activation energy =  0.0000E+00
        heat of reaction =  0.0000E+00


    stream function degree of freedom = 4
```

EXAMPLE 2: OUTPUT DATA LIST

```
flow   demo3 - 4:1 entry flow, psi = 0 on centerline

    nodal b.c.
```

node	1 b.c.	2 b.c.	3 b.c.	4 b.c.
1	1	1	1	1
2	1	1	1	0
3	1	1	1	0
4	1	1	1	0
5	1	1	1	0
6	1	1	1	0
7	1	1	1	0
8	1	1	1	0
9	0	-1	0	1
10	0	-1	0	0
11	0	-1	0	0
12	0	-1	0	0
13	0	-1	0	0
14	0	-1	0	0
15	0	-1	0	0
16	1	1	1	0
17	0	1	0	1
24	1	1	1	0
25	0	1	0	1
32	1	1	1	0
33	0	1	0	1
40	1	1	1	0
41	0	1	0	1
48	1	1	1	0
49	0	1	0	1
56	1	1	1	0
57	0	1	0	1
64	1	1	1	0
65	0	1	0	1
72	1	1	1	0
73	0	1	0	1
80	1	1	1	0
81	0	1	0	1
84	1	1	1	0
85	1	1	1	0
86	1	1	1	0
87	1	1	1	0
88	1	1	1	0
89	0	1	0	1
92	1	1	1	0
93	0	1	0	1
96	1	1	1	0
97	0	1	0	1
100	1	1	1	0
101	0	1	0	1
104	1	1	1	0
105	0	1	0	1
108	1	1	1	0
109	0	1	0	1
112	1	1	1	0
113	0	1	0	1
116	1	1	1	0
117	0	1	0	1
120	1	1	1	0
121	0	1	0	1

124	1	1	1	0
125	0	−1	0	1
126	0	−1	0	0
127	0	−1	0	0
128	1	1	1	0

flow   demo3 — 4:1 entry flow, psi = 0 on centerline

    nodal force/displ

node	1 force	2 force	3 force	4 force
1	1.000	0.0000E+00	0.0000E+00	0.0000E+00
2	0.9931	0.0000E+00	0.0000E+00	0.0000E+00
3	0.9722	0.0000E+00	0.0000E+00	0.0000E+00
4	0.9375	0.0000E+00	0.0000E+00	0.0000E+00
5	0.8828	0.0000E+00	0.0000E+00	0.0000E+00
6	0.7688	0.0000E+00	0.0000E+00	0.0000E+00
7	0.5260	0.0000E+00	0.0000E+00	0.0000E+00
8	0.0000E+00	0.0000E+00	0.0000E+00	0.0000E+00
9	0.0000E+00	0.0000E+00	0.0000E+00	0.0000E+00
10	0.0000E+00	0.0000E+00	0.0000E+00	0.0000E+00
11	0.0000E+00	0.0000E+00	0.0000E+00	0.0000E+00
12	0.0000E+00	0.0000E+00	0.0000E+00	0.0000E+00
13	0.0000E+00	0.0000E+00	0.0000E+00	0.0000E+00
14	0.0000E+00	0.0000E+00	0.0000E+00	0.0000E+00
15	0.0000E+00	0.0000E+00	0.0000E+00	0.0000E+00
16	0.0000E+00	0.0000E+00	0.0000E+00	0.0000E+00
17	0.0000E+00	0.0000E+00	0.0000E+00	0.0000E+00
18	0.0000E+00	0.0000E+00	0.0000E+00	0.0000E+00
19	0.0000E+00	0.0000E+00	0.0000E+00	0.0000E+00
20	0.0000E+00	0.0000E+00	0.0000E+00	0.0000E+00
21	0.0000E+00	0.0000E+00	0.0000E+00	0.0000E+00
22	0.0000E+00	0.0000E+00	0.0000E+00	0.0000E+00
23	0.0000E+00	0.0000E+00	0.0000E+00	0.0000E+00
24	0.0000E+00	0.0000E+00	0.0000E+00	0.0000E+00
25	0.0000E+00	0.0000E+00	0.0000E+00	0.0000E+00
26	0.0000E+00	0.0000E+00	0.0000E+00	0.0000E+00
27	0.0000E+00	0.0000E+00	0.0000E+00	0.0000E+00
28	0.0000E+00	0.0000E+00	0.0000E+00	0.0000E+00
29	0.0000E+00	0.0000E+00	0.0000E+00	0.0000E+00
30	0.0000E+00	0.0000E+00	0.0000E+00	0.0000E+00
31	0.0000E+00	0.0000E+00	0.0000E+00	0.0000E+00
32	0.0000E+00	0.0000E+00	0.0000E+00	0.0000E+00
33	0.0000E+00	0.0000E+00	0.0000E+00	0.0000E+00
34	0.0000E+00	0.0000E+00	0.0000E+00	0.0000E+00
35	0.0000E+00	0.0000E+00	0.0000E+00	0.0000E+00
36	0.0000E+00	0.0000E+00	0.0000E+00	0.0000E+00
37	0.0000E+00	0.0000E+00	0.0000E+00	0.0000E+00
38	0.0000E+00	0.0000E+00	0.0000E+00	0.0000E+00
39	0.0000E+00	0.0000E+00	0.0000E+00	0.0000E+00
40	0.0000E+00	0.0000E+00	0.0000E+00	0.0000E+00
41	0.0000E+00	0.0000E+00	0.0000E+00	0.0000E+00
42	0.0000E+00	0.0000E+00	0.0000E+00	0.0000E+00
43	0.0000E+00	0.0000E+00	0.0000E+00	0.0000E+00
44	0.0000E+00	0.0000E+00	0.0000E+00	0.0000E+00
45	0.0000E+00	0.0000E+00	0.0000E+00	0.0000E+00
46	0.0000E+00	0.0000E+00	0.0000E+00	0.0000E+00
47	0.0000E+00	0.0000E+00	0.0000E+00	0.0000E+00
48	0.0000E+00	0.0000E+00	0.0000E+00	0.0000E+00
49	0.0000E+00	0.0000E+00	0.0000E+00	0.0000E+00
50	0.0000E+00	0.0000E+00	0.0000E+00	0.0000E+00

EXAMPLE 2: OUTPUT DATA LIST

flow demo3 - 4:1 entry flow, psi = 0 on centerline

    nodal force/displ

node	1 force	2 force	3 force	4 force
51	0.0000E+00	0.0000E+00	0.0000E+00	0.0000E+00
52	0.0000E+00	0.0000E+00	0.0000E+00	0.0000E+00
53	0.0000E+00	0.0000E+00	0.0000E+00	0.0000E+00
54	0.0000E+00	0.0000E+00	0.0000E+00	0.0000E+00
55	0.0000E+00	0.0000E+00	0.0000E+00	0.0000E+00
56	0.0000E+00	0.0000E+00	0.0000E+00	0.0000E+00
57	0.0000E+00	0.0000E+00	0.0000E+00	0.0000E+00
58	0.0000E+00	0.0000E+00	0.0000E+00	0.0000E+00
59	0.0000E+00	0.0000E+00	0.0000E+00	0.0000E+00
60	0.0000E+00	0.0000E+00	0.0000E+00	0.0000E+00
61	0.0000E+00	0.0000E+00	0.0000E+00	0.0000E+00
62	0.0000E+00	0.0000E+00	0.0000E+00	0.0000E+00
63	0.0000E+00	0.0000E+00	0.0000E+00	0.0000E+00
64	0.0000E+00	0.0000E+00	0.0000E+00	0.0000E+00
65	0.0000E+00	0.0000E+00	0.0000E+00	0.0000E+00
66	0.0000E+00	0.0000E+00	0.0000E+00	0.0000E+00
67	0.0000E+00	0.0000E+00	0.0000E+00	0.0000E+00
68	0.0000E+00	0.0000E+00	0.0000E+00	0.0000E+00
69	0.0000E+00	0.0000E+00	0.0000E+00	0.0000E+00
70	0.0000E+00	0.0000E+00	0.0000E+00	0.0000E+00
71	0.0000E+00	0.0000E+00	0.0000E+00	0.0000E+00
72	0.0000E+00	0.0000E+00	0.0000E+00	0.0000E+00
73	0.0000E+00	0.0000E+00	0.0000E+00	0.0000E+00
74	0.0000E+00	0.0000E+00	0.0000E+00	0.0000E+00
75	0.0000E+00	0.0000E+00	0.0000E+00	0.0000E+00
76	0.0000E+00	0.0000E+00	0.0000E+00	0.0000E+00
77	0.0000E+00	0.0000E+00	0.0000E+00	0.0000E+00
78	0.0000E+00	0.0000E+00	0.0000E+00	0.0000E+00
79	0.0000E+00	0.0000E+00	0.0000E+00	0.0000E+00
80	0.0000E+00	0.0000E+00	0.0000E+00	0.0000E+00
81	0.0000E+00	0.0000E+00	0.0000E+00	0.0000E+00
82	0.0000E+00	0.0000E+00	0.0000E+00	0.0000E+00
83	0.0000E+00	0.0000E+00	0.0000E+00	0.0000E+00
84	0.0000E+00	0.0000E+00	0.0000E+00	0.0000E+00
85	0.0000E+00	0.0000E+00	0.0000E+00	0.0000E+00
86	0.0000E+00	0.0000E+00	0.0000E+00	0.0000E+00
87	0.0000E+00	0.0000E+00	0.0000E+00	0.0000E+00
88	0.0000E+00	0.0000E+00	0.0000E+00	0.0000E+00
89	0.0000E+00	0.0000E+00	0.0000E+00	0.0000E+00
90	0.0000E+00	0.0000E+00	0.0000E+00	0.0000E+00
91	0.0000E+00	0.0000E+00	0.0000E+00	0.0000E+00
92	0.0000E+00	0.0000E+00	0.0000E+00	0.0000E+00
93	0.0000E+00	0.0000E+00	0.0000E+00	0.0000E+00
94	0.0000E+00	0.0000E+00	0.0000E+00	0.0000E+00
95	0.0000E+00	0.0000E+00	0.0000E+00	0.0000E+00
96	0.0000E+00	0.0000E+00	0.0000E+00	0.0000E+00
97	0.0000E+00	0.0000E+00	0.0000E+00	0.0000E+00
98	0.0000E+00	0.0000E+00	0.0000E+00	0.0000E+00
99	0.0000E+00	0.0000E+00	0.0000E+00	0.0000E+00
100	0.0000E+00	0.0000E+00	0.0000E+00	0.0000E+00

```
flow  demo3 - 4:1 entry flow, psi = 0 on centerline

  nodal force/displ

  node      1 force       2 force       3 force       4 force
   101    0.0000E+00    0.0000E+00    0.0000E+00    0.0000E+00
   102    0.0000E+00    0.0000E+00    0.0000E+00    0.0000E+00
   103    0.0000E+00    0.0000E+00    0.0000E+00    0.0000E+00
   104    0.0000E+00    0.0000E+00    0.0000E+00    0.0000E+00
   105    0.0000E+00    0.0000E+00    0.0000E+00    0.0000E+00
   106    0.0000E+00    0.0000E+00    0.0000E+00    0.0000E+00
   107    0.0000E+00    0.0000E+00    0.0000E+00    0.0000E+00
   108    0.0000E+00    0.0000E+00    0.0000E+00    0.0000E+00
   109    0.0000E+00    0.0000E+00    0.0000E+00    0.0000E+00
   110    0.0000E+00    0.0000E+00    0.0000E+00    0.0000E+00
   111    0.0000E+00    0.0000E+00    0.0000E+00    0.0000E+00
   112    0.0000E+00    0.0000E+00    0.0000E+00    0.0000E+00
   113    0.0000E+00    0.0000E+00    0.0000E+00    0.0000E+00
   114    0.0000E+00    0.0000E+00    0.0000E+00    0.0000E+00
   115    0.0000E+00    0.0000E+00    0.0000E+00    0.0000E+00
   116    0.0000E+00    0.0000E+00    0.0000E+00    0.0000E+00
   117    0.0000E+00    0.0000E+00    0.0000E+00    0.0000E+00
   118    0.0000E+00    0.0000E+00    0.0000E+00    0.0000E+00
   119    0.0000E+00    0.0000E+00    0.0000E+00    0.0000E+00
   120    0.0000E+00    0.0000E+00    0.0000E+00    0.0000E+00
   121    0.0000E+00    0.0000E+00    0.0000E+00    0.0000E+00
   122    0.0000E+00    0.0000E+00    0.0000E+00    0.0000E+00
   123    0.0000E+00    0.0000E+00    0.0000E+00    0.0000E+00
   124    0.0000E+00    0.0000E+00    0.0000E+00    0.0000E+00
   125    0.0000E+00    0.0000E+00    0.0000E+00    0.0000E+00
   126    0.0000E+00    0.0000E+00    0.0000E+00    0.0000E+00
   127    0.0000E+00    0.0000E+00    0.0000E+00    0.0000E+00
   128    0.0000E+00    0.0000E+00    0.0000E+00    0.0000E+00
```

EXAMPLE 2: MACRO INSTRUCTION EXECUTION LOG

```
flow   demo3 - 4:1 entry flow, psi = 0 on centerline

      macro instructions

      macro statement        variable 1          variable 2
              tang         0.00000E+00          0.00000E+00
              loop         2.0000               0.00000E+00
              form         0.00000E+00          0.00000E+00
              solv         0.00000E+00          0.00000E+00
              next         0.00000E+00          0.00000E+00
              disp         0.00000E+00          0.00000E+00
              end          0.00000E+00          0.00000E+00


  **macro instruction loop completed. time used this step =  7.80 total time = 52
.40
  **macro instruction   1 initiated**  tang       v1 =    0.0000E+00 , v2 =    0
.0000E+00


  **macro instruction tang completed. time used this step = 41.85 total time = 94
.25
  **macro instruction   2 initiated**  loop       v1 =    2.000      , v2 =
6.000


  **macro instruction loop completed. time used this step =  0.28 total time = 94
.53
  **macro instruction   3 initiated**  form       v1 =    0.0000E+00 , v2 =    0
.0000E+00
      force convergence test
          rnmax =    0.48480E+07     rm   =   0.48480E+07      tol   =    0.1000
0E-08


  **macro instruction form completed. time used this step = 21.91 total time =116
.44
  **macro instruction   4 initiated** · solv       v1 =    0.0000E+00 , v2 =    0
.0000E+00


  **macro instruction solv completed. time used this step = 32.90 total time =149
.34
  **macro instruction   5 initiated**  next       v1 =    1.000      , v2 =
3.000


  **macro instruction loop completed. time used this step =  0.28 total time =149
.62
  **macro instruction   3 initiated**  form       v1 =    0.0000E+00 , v2 =    0
.0000E+00
      force convergence test
          rnmax =    0.48480E+07     rm   =   57.366            tol   =    0.1000
0E-08


  **macro instruction form completed. time used this step = 23.40 total time =173
.02
  **macro instruction   4 initiated**  solv       v1 =    0.0000E+00 , v2 =    0
.0000E+00
```

```
**macro instruction solv completed. time used this step =  4.55 total time =177
.57
**macro instruction   5 initiated**  next         v1 =     2.000      , v2 =
3.000

**macro instruction next completed. time used this step =  0.28 total time =177
.85
**macro instruction   6 initiated**  disp         v1 =     0.0000E+00 , v2 =      0
.0000E+00

    time   0.00000E+00

    proportional load    1.0000
```

EXAMPLE 2: NODAL DISPLACEMENT LIST

flow demo3 — 4:1 entry flow, psi = 0 on centerline

nodal displacements   time 0.00000E+00

node	1 coord	2 coord	1 displ	2 displ	3 displ	4 displ
1	-4.000	0.0000E+00	1.000	0.0000E+00	0.0000E+00	0.0000E+00
2	-4.000	0.8333E-01	0.9931	0.0000E+00	0.0000E+00	0.8304E-01
3	-4.000	0.1667	0.9722	0.0000E+00	0.0000E+00	0.1650
4	-4.000	0.2500	0.9375	0.0000E+00	0.0000E+00	0.2445
5	-4.000	0.3423	0.8828	0.0000E+00	0.0000E+00	0.3285
6	-4.000	0.4808	0.7688	0.0000E+00	0.0000E+00	0.4429
7	-4.000	0.6885	0.5260	0.0000E+00	0.0000E+00	0.5773
8	-4.000	1.000	0.0000E+00	0.0000E+00	0.0000E+00	0.6592
9	-3.500	0.0000E+00	0.9999	0.0000E+00	0.2074	0.0000E+00
10	-3.500	0.8333E-01	0.9931	0.0000E+00	0.2084	0.8304E-01
11	-3.500	0.1667	0.9721	0.0000E+00	0.2111	0.1650
12	-3.500	0.2500	0.9375	0.0000E+00	0.2152	0.2445
13	-3.500	0.3423	0.8827	0.0000E+00	0.2202	0.3285
14	-3.500	0.4808	0.7688	0.0000E+00	0.2249	0.4429
15	-3.500	0.6885	0.5259	0.0000E+00	0.2064	0.5773
16	-3.500	1.000	0.0000E+00	0.0000E+00	0.0000E+00	0.6592
17	-3.000	0.0000E+00	0.9932	0.0000E+00	0.3678	0.0000E+00
18	-3.000	0.8333E-01	1.003	-0.4893E-03	0.3678	0.8316E-01
19	-3.000	0.1667	0.9578	0.2754E-03	0.3669	0.1649
20	-3.000	0.2500	0.9519	0.2494E-03	0.3653	0.2444
21	-3.000	0.3423	0.8733	-0.6737E-03	0.3607	0.3287
22	-3.000	0.4808	0.7735	0.6193E-03	0.3442	0.4427
23	-3.000	0.6885	0.5246	-0.8094E-03	0.2794	0.5775
24	-3.000	1.000	0.0000E+00	0.0000E+00	0.0000E+00	0.6592
25	-2.500	0.0000E+00	1.008	0.0000E+00	0.5350	0.0000E+00
26	-2.500	0.8333E-01	0.9809	0.1652E-02	0.5333	0.8287E-01
27	-2.500	0.1667	0.9906	-0.9399E-03	0.5285	0.1650
28	-2.500	0.2500	0.9187	-0.8568E-03	0.5193	0.2446
29	-2.500	0.3423	0.8947	0.2238E-02	0.5034	0.3283
30	-2.500	0.4808	0.7628	-0.2021E-02	0.4653	0.4430
31	-2.500	0.6885	0.5275	0.2640E-02	0.3577	0.5770
32	-2.500	1.000	0.0000E+00	0.0000E+00	0.0000E+00	0.6592
33	-2.000	0.0000E+00	0.9894	0.0000E+00	0.8206	0.0000E+00
34	-2.000	0.8333E-01	1.003	-0.2172E-02	0.8168	0.8299E-01
35	-2.000	0.1667	0.9490	0.3736E-02	0.8043	0.1643
36	-2.000	0.2500	0.9571	0.4180E-02	0.7832	0.2437
37	-2.000	0.3423	0.8697	-0.1413E-02	0.7487	0.3280
38	-2.000	0.4808	0.7762	0.6036E-02	0.6724	0.4420
39	-2.000	0.6885	0.5259	-0.3566E-02	0.4918	0.5772
40	-2.000	1.000	0.0000E+00	0.0000E+00	0.0000E+00	0.6592
41	-1.500	0.0000E+00	1.000	0.0000E+00	1.420	0.0000E+00
42	-1.500	0.8333E-01	0.9753	0.4879E-02	1.410	0.8232E-01
43	-1.500	0.1667	0.9920	-0.3654E-02	1.383	0.1643
44	-1.500	0.2500	0.9081	-0.3101E-02	1.338	0.2435
45	-1.500	0.3423	0.8935	0.7134E-02	1.265	0.3266
46	-1.500	0.4808	0.7627	-0.3192E-02	1.113	0.4413
47	-1.500	0.6885	0.5340	0.8637E-02	0.7772	0.5760
48	-1.500	1.000	0.0000E+00	0.0000E+00	0.0000E+00	0.6591
49	-1.000	0.2794E-08	1.065	0.0000E+00	2.717	0.0000E+00
50	-1.000	0.8333E-01	1.071	-0.3167E-01	2.699	0.8901E-01

flow demo3 - 4:1 entry flow, psi = 0 on centerline

nodal displacements   time 0.00000E+00

node	1 coord	2 coord	1 displ	2 displ	3 displ	4 displ
51	-1.000	0.1667	0.9902	-0.3881E-01	2.641	0.1749
52	-1.000	0.2500	1.009	-0.5590E-01	2.536	0.2582
53	-1.000	0.3423	0.8792	-0.8214E-01	2.373	0.3453
54	-1.000	0.4808	0.7531	-0.6520E-01	2.034	0.4584
55	-1.000	0.6885	0.4719	-0.4729E-01	1.346	0.5856
56	-1.000	1.000	0.0000E+00	0.0000E+00	0.0000E+00	0.6591
57	-0.5846	0.0000E+00	1.409	0.0000E+00	4.973	0.0000E+00
58	-0.5846	0.8333E-01	1.364	-0.9602E-01	4.899	0.1155
59	-0.5846	0.1667	1.352	-0.2202	4.693	0.2287
60	-0.5846	0.2500	1.080	-0.2899	4.387	0.3300
61	-0.5846	0.3423	0.9655	-0.2985	3.926	0.4244
62	-0.5846	0.4808	0.6070	-0.2955	3.109	0.5333
63	-0.5846	0.6885	0.2418	-0.1253	1.772	0.6214
64	-0.5846	1.000	0.0000E+00	0.0000E+00	0.0000E+00	0.6591
65	-0.3077	0.0000E+00	2.257	0.0000E+00	7.178	0.0000E+00
66	-0.3077	0.8333E-01	2.104	-0.3821	7.022	0.1817
67	-0.3077	0.1667	1.580	-0.5496	6.525	0.3353
68	-0.3077	0.2500	1.359	-0.6328	5.696	0.4578
69	-0.3077	0.3423	0.6870	-0.6246	4.650	0.5522
70	-0.3077	0.4808	0.2622	-0.3159	2.998	0.6179
71	-0.3077	0.6885	0.5361E-01	-0.8638E-01	1.390	0.6507
72	-0.3077	1.000	0.0000E+00	0.0000E+00	0.0000E+00	0.6591
73	-0.1231	0.0000E+00	3.301	0.0000E+00	7.358	0.0000E+00
74	-0.1231	0.8333E-01	3.017	-0.5012	7.118	0.2633
75	-0.1231	0.1667	2.234	-1.041	6.316	0.4822
76	-0.1231	0.2500	0.6406	-0.9288	5.162	0.6019
77	-0.1231	0.3423	0.1714	-0.3200	3.044	0.6394
78	-0.1231	0.4808	0.4379E-01	-0.7792E-01	1.538	0.6543
79	-0.1231	0.6885	0.9694E-03	-0.2419E-02	0.6396	0.6589
80	-0.1231	1.000	0.0000E+00	0.0000E+00	0.0000E+00	0.6591
81	0.0000E+00	0.1863E-08	3.878	0.0000E+00	6.536	0.0000E+00
82	0.0000E+00	0.8333E-01	3.526	-0.2334	6.212	0.3085
83	0.0000E+00	0.1667	2.443	-0.1796	4.964	0.5573
84	0.0000E+00	0.2500	0.0000E+00	0.0000E+00	0.0000E+00	0.6591
85	0.0000E+00	0.3423	0.0000E+00	0.0000E+00	0.0000E+00	0.6591
86	0.0000E+00	0.4808	0.0000E+00	0.0000E+00	0.0000E+00	0.6591
87	0.0000E+00	0.6885	0.0000E+00	0.0000E+00	0.0000E+00	0.6591
88	0.0000E+00	1.000	0.0000E+00	0.0000E+00	0.0000E+00	0.6591
89	0.1231	0.0000E+00	4.071	0.0000E+00	6.021	0.0000E+00
90	0.1231	0.8333E-01	3.654	0.1515E-01	5.881	0.3219
91	0.1231	0.1667	2.218	0.2702E-01	4.638	0.5667
92	0.1231	0.2500	0.0000E+00	0.0000E+00	0.0000E+00	0.6591
93	0.3077	0.0000E+00	4.085	0.0000E+00	5.897	0.0000E+00
94	0.3077	0.8333E-01	3.600	0.3323E-02	5.783	0.3202
95	0.3077	0.1667	2.265	-0.5576E-02	4.637	0.5647
96	0.3077	0.2500	0.0000E+00	0.0000E+00	0.0000E+00	0.6590
97	0.5846	0.0000E+00	4.057	0.0000E+00	5.825	0.0000E+00
98	0.5846	0.8333E-01	3.623	-0.1904E-02	5.726	0.3200
99	0.5846	0.1667	2.257	0.2903E-02	4.599	0.5651
100	0.5846	0.2500	0.0000E+00	0.0000E+00	0.0000E+00	0.6590

EXAMPLE 2: NODAL DISPLACEMENT LIST

flow demo3 - 4:1 entry flow, psi = 0 on centerline

nodal displacements   time 0.00000E+00

node	1 coord	2 coord	1 displ	2 displ	3 displ	4 displ
101	1.000	0.1863E-08	4.073	0.0000E+00	5.823	0.0000E+00
102	1.000	0.8333E-01	3.611	0.1090E-02	5.717	0.3202
103	1.000	0.1667	2.260	-0.2159E-02	4.592	0.5649
104	1.000	0.2500	0.0000E+00	0.0000E+00	0.0000E+00	0.6590
105	1.500	0.0000E+00	4.063	0.0000E+00	5.819	0.0000E+00
106	1.500	0.8333E-01	3.618	-0.7040E-03	5.719	0.3201
107	1.500	0.1667	2.258	0.1750E-02	4.595	0.5650
108	1.500	0.2500	0.0000E+00	0.0000E+00	0.0000E+00	0.6590
109	2.000	0.0000E+00	4.070	0.0000E+00	5.822	0.0000E+00
110	2.000	0.8333E-01	3.613	0.4711E-03	5.717	0.3201
111	2.000	0.1667	2.260	-0.1385E-02	4.592	0.5649
112	2.000	0.2500	0.0000E+00	0.0000E+00	0.0000E+00	0.6590
113	2.500	0.0000E+00	4.064	0.0000E+00	5.819	0.0000E+00
114	2.500	0.8333E-01	3.618	-0.3087E-03	5.719	0.3201
115	2.500	0.1667	2.258	0.1037E-02	4.595	0.5650
116	2.500	0.2500	0.0000E+00	0.0000E+00	0.0000E+00	0.6590
117	3.000	0.0000E+00	4.070	0.0000E+00	5.821	0.0000E+00
118	3.000	0.8333E-01	3.613	0.1895E-03	5.717	0.3201
119	3.000	0.1667	2.260	-0.6875E-03	4.592	0.5649
120	3.000	0.2500	0.0000E+00	0.0000E+00	0.0000E+00	0.6590
121	3.500	0.0000E+00	4.065	0.0000E+00	5.819	0.0000E+00
122	3.500	0.8333E-01	3.618	-0.8958E-04	5.719	0.3201
123	3.500	0.1667	2.258	0.3435E-03	4.595	0.5650
124	3.500	0.2500	0.0000E+00	0.0000E+00	0.0000E+00	0.6590
125	4.000	0.0000E+00	4.070	0.0000E+00	5.821	0.0000E+00
126	4.000	0.8333E-01	3.613	0.0000E+00	5.717	0.3201
127	4.000	0.1667	2.260	0.0000E+00	4.592	0.5649
128	4.000	0.2500	0.0000E+00	0.0000E+00	0.0000E+00	0.6590

**macro instruction disp completed. time used this step = 12.63 total time =195.92

7.5 EXAMPLE PROBLEM 3: THERMALLY DRIVEN BUOYANT FLOW

The last problem is interesting since the fluid is initially not flowing. Instead, the temperatures at the two opposing vertical boundaries are set to different levels. Intuitively, this temperature gradient gives rise to a vertical force upward on the "warm" wall since the fluid is less dense than the cooler fluid on the opposing wall. Likewise, the cooler heavier fluid will move down thereby creating a clockwise circulation. Figure 7.7 displays the problem input geometry while the section "Example 3 Input Data" (below) provides the input data for this problem. Notice that in the force section of the input data, column four is set to unity for these specific nodes. These are the temperature values for the indicated nodes. Note that the reference temperature is set to zero in the material block of the input data. Again, the macro file at the end of the input data directs the flow program through a series of computations which result in the solution. Figure 7.8 displays the displaced mesh. Here we clearly see the clockwise flow. The output data is listed in the section "Example 3 Output Data" and the macro execution log is also given. Notice that the macro execution log provides the execution times for each step in the program. The Nodal Displacement List gives the nodal displacements for this problem, giving the coordinates and the associated displacements. These displacements are x-velocity, y-velocity, nodal temperatures, and the streamfunction values. Finally, Figure 7.9 plots the streamfunction for this example by using the plot command in the post processor.

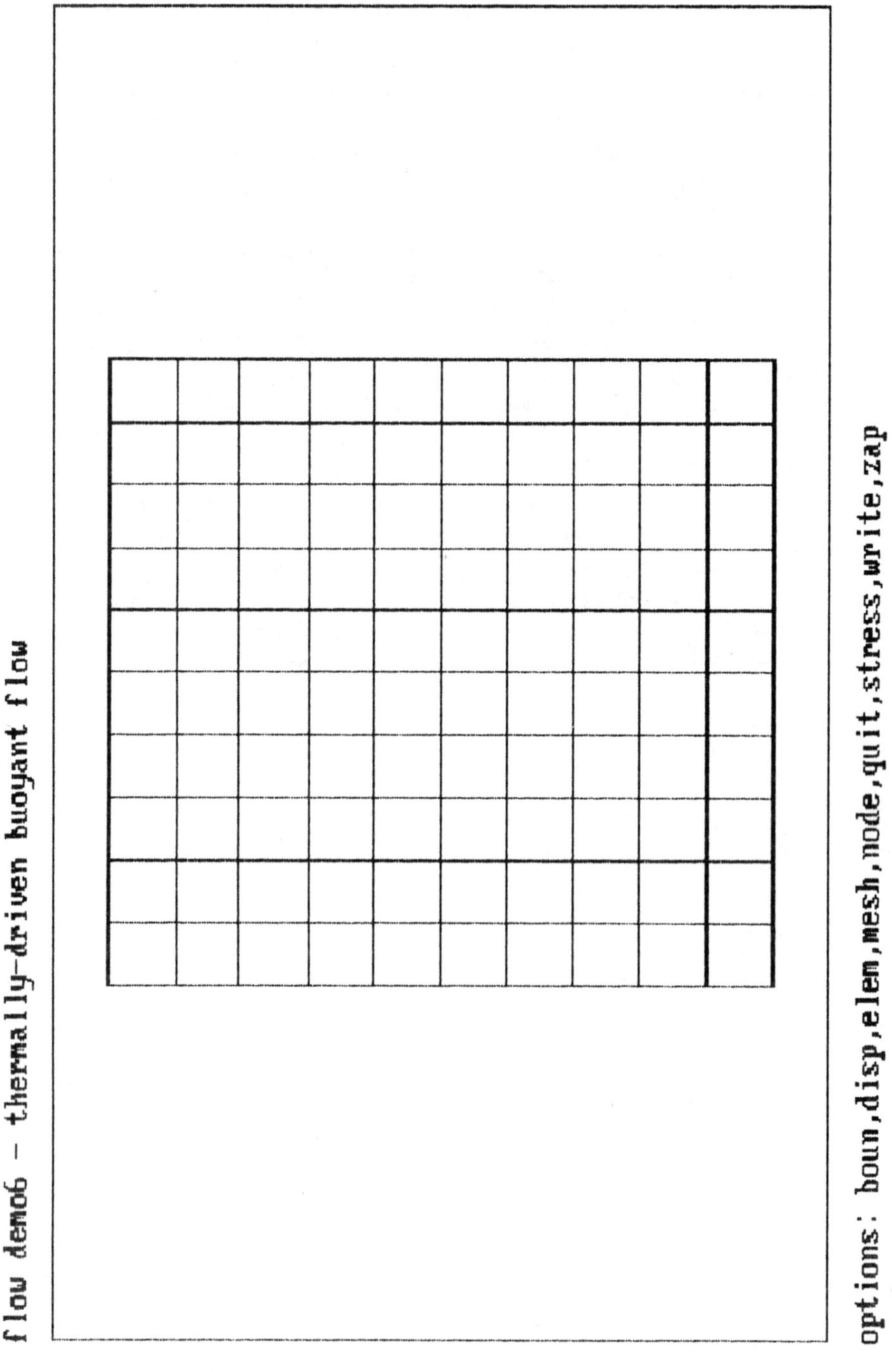

flow demo6 - thermally-driven buoyant flow

options: boun,disp,elem,mesh,node,quit,stress,write,zap

===>

Figure 7.7

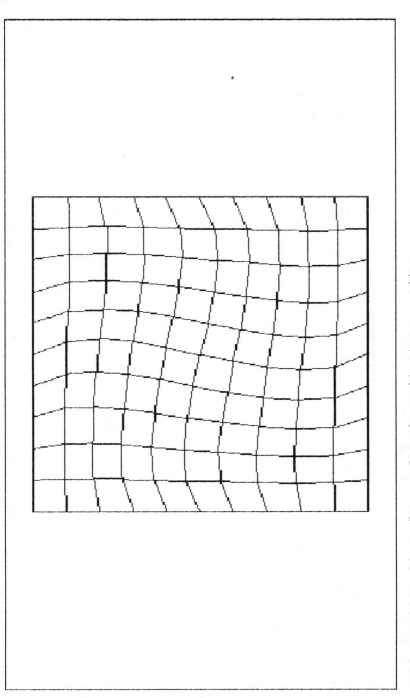

flow demo6 – thermally-driven buoyant flow

options: boun,disp,elem,mesh,node,quit,stress,write,zap
===>
Figure 7.8

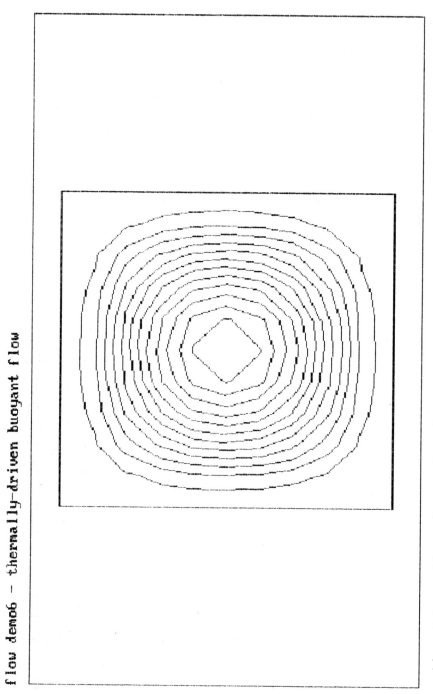

flow demo6 - thermally-driven buoyant flow

options: boun,disp,elem,mesh,node,quit,stress,write,zap
===>

Figure 7.9

EXAMPLE 3: INPUT DATA LIST

```
flow demo6 - thermally-driven buoyant flow
   121   100     1     2     4     4     0
coor
     1     0    .000000   1.000000
     2     0    .100000   1.000000
     3     0    .200000   1.000000
     4     0    .300000   1.000000
     5     0    .400000   1.000000
     6     0    .500000   1.000000
     7     0    .600000   1.000000
     8     0    .700000   1.000000
     9     0    .800000   1.000000
    10     0    .900000   1.000000
    11     0   1.000000   1.000000
    12     0    .000000    .900000
    13     0    .100000    .900000
    14     0    .200000    .900000
    15     0    .300000    .900000
    16     0    .400000    .900000
    17     0    .500000    .900000
    18     0    .600000    .900000
    19     0    .700000    .900000
    20     0    .800000    .900000
    21     0    .900000    .900000
    22     0   1.000000    .900000
    23     0    .000000    .800000
    24     0    .100000    .800000
    25     0    .200000    .800000
    26     0    .300000    .800000
    27     0    .400000    .800000
    28     0    .500000    .800000
    29     0    .600000    .800000
    30     0    .700000    .800000
    31     0    .800000    .800000
    32     0    .900000    .800000
    33     0   1.000000    .800000
    34     0    .000000    .700000
    35     0    .100000    .700000
    36     0    .200000    .700000
    37     0    .300000    .700000
    38     0    .400000    .700000
    39     0    .500000    .700000
    40     0    .600000    .700000
    41     0    .700000    .700000
    42     0    .800000    .700000
    43     0    .900000    .700000
    44     0   1.000000    .700000
    45     0    .000000    .600000
    46     0    .100000    .600000
    47     0    .200000    .600000
    48     0    .300000    .600000
    49     0    .400000    .600000
    50     0    .500000    .600000
    51     0    .600000    .600000
    52     0    .700000    .600000
    53     0    .800000    .600000
    54     0    .900000    .600000
    55     0   1.000000    .600000
    56     0    .000000    .500000
    57     0    .100000    .500000
```

EXAMPLE 3: INPUT DATA LIST

```
 58    0   .200000   .500000
 59    0   .300000   .500000
 60    0   .400000   .500000
 61    0   .500000   .500000
 62    0   .600000   .500000
 63    0   .700000   .500000
 64    0   .800000   .500000
 65    0   .900000   .500000
 66    0  1.000000   .500000
 67    0   .000000   .400000
 68    0   .100000   .400000
 69    0   .200000   .400000
 70    0   .300000   .400000
 71    0   .400000   .400000
 72    0   .500000   .400000
 73    0   .600000   .400000
 74    0   .700000   .400000
 75    0   .800000   .400000
 76    0   .900000   .400000
 77    0  1.000000   .400000
 78    0   .000000   .300000
 79    0   .100000   .300000
 80    0   .200000   .300000
 81    0   .300000   .300000
 82    0   .400000   .300000
 83    0   .500000   .300000
 84    0   .600000   .300000
 85    0   .700000   .300000
 86    0   .800000   .300000
 87    0   .900000   .300000
 88    0  1.000000   .300000
 89    0   .000000   .200000
 90    0   .100000   .200000
 91    0   .200000   .200000
 92    0   .300000   .200000
 93    0   .400000   .200000
 94    0   .500000   .200000
 95    0   .600000   .200000
 96    0   .700000   .200000
 97    0   .800000   .200000
 98    0   .900000   .200000
 99    0  1.000000   .200000
100    0   .000000   .100000
101    0   .100000   .100000
102    0   .200000   .100000
103    0   .300000   .100000
104    0   .400000   .100000
105    0   .500000   .100000
106    0   .600000   .100000
107    0   .700000   .100000
108    0   .800000   .100000
109    0   .900000   .100000
110    0  1.000000   .100000
111    0   .000000   .000000
112    0   .100000   .000000
113    0   .200000   .000000
114    0   .300000   .000000
115    0   .400000   .000000
116    0   .500000   .000000
117    0   .600000   .000000
```

```
118    0   .700000    .000000
119    0   .800000    .000000
120    0   .900000    .000000
121    0  1.000000    .000000

elem
  1    1    1    2   13   12   1
  2    1    2    3   14   13   1
  3    1    3    4   15   14   1
  4    1    4    5   16   15   1
  5    1    5    6   17   16   1
  6    1    6    7   18   17   1
  7    1    7    8   19   18   1
  8    1    8    9   20   19   1
  9    1    9   10   21   20   1
 10    1   10   11   22   21   1
 11    1   12   13   24   23   1
 12    1   13   14   25   24   1
 13    1   14   15   26   25   1
 14    1   15   16   27   26   1
 15    1   16   17   28   27   1
 16    1   17   18   29   28   1
 17    1   18   19   30   29   1
 18    1   19   20   31   30   1
 19    1   20   21   32   31   1
 20    1   21   22   33   32   1
 21    1   23   24   35   34   1
 22    1   24   25   36   35   1
 23    1   25   26   37   36   1
 24    1   26   27   38   37   1
 25    1   27   28   39   38   1
 26    1   28   29   40   39   1
 27    1   29   30   41   40   1
 28    1   30   31   42   41   1
 29    1   31   32   43   42   1
 30    1   32   33   44   43   1
 31    1   34   35   46   45   1
 32    1   35   36   47   46   1
 33    1   36   37   48   47   1
 34    1   37   38   49   48   1
 35    1   38   39   50   49   1
 36    1   39   40   51   50   1
 37    1   40   41   52   51   1
 38    1   41   42   53   52   1
 39    1   42   43   54   53   1
 40    1   43   44   55   54   1
 41    1   45   46   57   56   1
 42    1   46   47   58   57   1
 43    1   47   48   59   58   1
 44    1   48   49   60   59   1
 45    1   49   50   61   60   1
 46    1   50   51   62   61   1
 47    1   51   52   63   62   1
 48    1   52   53   64   63   1
 49    1   53   54   65   64   1
 50    1   54   55   66   65   1
 51    1   56   57   68   67   1
 52    1   57   58   69   68   1
 53    1   58   59   70   69   1
 54    1   59   60   71   70   1
```

EXAMPLE 3: INPUT DATA LIST

```
 55    1    60    61    72    71    1
 56    1    61    62    73    72    1
 57    1    62    63    74    73    1
 58    1    63    64    75    74    1
 59    1    64    65    76    75    1
 60    1    65    66    77    76    1
 61    1    67    68    79    78    1
 62    1    68    69    80    79    1
 63    1    69    70    81    80    1
 64    1    70    71    82    81    1
 65    1    71    72    83    82    1
 66    1    72    73    84    83    1
 67    1    73    74    85    84    1
 68    1    74    75    86    85    1
 69    1    75    76    87    86    1
 70    1    76    77    88    87    1
 71    1    78    79    90    89    1
 72    1    79    80    91    90    1
 73    1    80    81    92    91    1
 74    1    81    82    93    92    1
 75    1    82    83    94    93    1
 76    1    83    84    95    94    1
 77    1    84    85    96    95    1
 78    1    85    86    97    96    1
 79    1    86    87    98    97    1
 80    1    87    88    99    98    1
 81    1    89    90   101   100    1
 82    1    90    91   102   101    1
 83    1    91    92   103   102    1
 84    1    92    93   104   103    1
 85    1    93    94   105   104    1
 86    1    94    95   106   105    1
 87    1    95    96   107   106    1
 88    1    96    97   108   107    1
 89    1    97    98   108   108    1
 90    1    98    99   110   109    1
 91    1   100   101   112   111    1
 92    1   101   102   113   112    1
 93    1   102   103   114   113    1
 94    1   103   104   115   114    1
 95    1   104   105   116   115    1
 96    1   105   106   117   116    1
 97    1   106   107   118   117    1
 98    1   107   108   119   118    1
 99    1   108   109   120   119    1
100    1   109   110   121   120    1
```

```
mate
      1      3! material number, element type number
             2! regular integration order
             0! flag for axisymmetry (0 for no)
             0! flag for time stepping (0 for no)
   0.000E+00! theta factor in transient algorithm
             1! degree of freedom number for velocities
             1! penalty integration order
             0! momentum convection integration order:
   1.000E+07! penalty factor
   1.000E+00! density
   1.000E+00! viscosity coefficient
   0.000E+00! lambda factor in carreau model
```

```
1.000E+00! power-law exponent
0.000E+00! viscosity activation energy
0.000E+00! reference temperature
1.000E+00! thermal expansion coefficient
        3! degree of freedom number for temperature
        0! thermal convection integration order
1.000E+00! thermal conductivity
1.000E+00! specific heat
0.000E+00! thermomechanical dissipation factor
        0! degree of freedom number for reaction
        0! species convection integration order
0.000E+00! species diffusivity
1.000E+00! kinetic order of reaction
0.000E+00! rate constant preexponential factor
0.000E+00! activation energy for reaction / gas constant
0.000E+00! heat of reaction
        4! degree of freedom number for stream function
```

boun

```
  1    0    1    1    1    1
  2    0    1    1    0    0
  3    0    1    1    0    0
  4    0    1    1    0    0
  5    0    1    1    0    0
  6    0    1    1    0    0
  7    0    1    1    0    0
  8    0    1    1    0    0
  9    0    1    1    0    0
 10    0    1    1    0    0
 11    0    1    1    1    0
 12    0    1    1    1    1
 22    0    1    1    1    0
 23    0    1    1    1    1
 33    0    1    1    1    0
 34    0    1    1    1    1
 44    0    1    1    1    0
 45    0    1    1    1    1
 55    0    1    1    1    0
 56    0    1    1    1    1
 66    0    1    1    1    0
 67    0    1    1    1    1
 77    0    1    1    1    0
 78    0    1    1    1    1
 88    0    1    1    1    0
 89    0    1    1    1    1
 99    0    1    1    1    0
100    0    1    1    1    1
110    0    1    1    1    0
111    0    1    1    1    1
112    0    1    1    0    1
113    0    1    1    0    1
114    0    1    1    0    1
115    0    1    1    0    1
116    0    1    1    0    1
117    0    1    1    0    1
118    0    1    1    0    1
119    0    1    1    0    1
120    0    1    1    0    1
121    0    1    1    0    1
```

EXAMPLE 3: INPUT DATA LIST

```
forc
      1      0      .0000      .0000      1.0000
     12      0      .0000      .0000      1.0000
     23      0      .0000      .0000      1.0000
     34      0      .0000      .0000      1.0000
     45      0      .0000      .0000      1.0000
     56      0      .0000      .0000      1.0000
     67      0      .0000      .0000      1.0000
     78      0      .0000      .0000      1.0000
     89      0      .0000      .0000      1.0000
    100      0      .0000      .0000      1.0000
    111      0      .0000      .0000      1.0000

end
macr
tang
loop              3
form
solv
next
disp
end
stop
```

EXAMPLE 3: OUTPUT DATA LIST

```
flow demo6 — thermally-driven buoyant flow

      number of nodal points        =    121
      number of elements            =    100
      number of material sets       =      1
      dimension of coordinate space=      2
      degrees of freedom/node       =      4
      nodes per element (maximum)   =      4
      extra d.o.f. to element       =      0
```

```
flow demo6 - thermally-driven buoyant flow

    nodal coordinates

    node        1 coord         2 coord
       1     0.0000E+00        1.000
       2     0.1000            1.000
       3     0.2000            1.000
       4     0.3000            1.000
       5     0.4000            1.000
       6     0.5000            1.000
       7     0.6000            1.000
       8     0.7000            1.000
       9     0.8000            1.000
      10     0.9000            1.000
      11     1.000             1.000
      12     0.0000E+00        0.9000
      13     0.1000            0.9000
      14     0.2000            0.9000
      15     0.3000            0.9000
      16     0.4000            0.9000
      17     0.5000            0.9000
      18     0.6000            0.9000
      19     0.7000            0.9000
      20     0.8000            0.9000
      21     0.9000            0.9000
      22     1.000             0.9000
      23     0.0000E+00        0.8000
      24     0.1000            0.8000
      25     0.2000            0.8000
      26     0.3000            0.8000
      27     0.4000            0.8000
      28     0.5000            0.8000
      29     0.6000            0.8000
      30     0.7000            0.8000
      31     0.8000            0.8000
      32     0.9000            0.8000
      33     1.000             0.8000
      34     0.0000E+00        0.7000
      35     0.1000            0.7000
      36     0.2000            0.7000
      37     0.3000            0.7000
      38     0.4000            0.7000
      39     0.5000            0.7000
      40     0.6000            0.7000
      41     0.7000            0.7000
      42     0.8000            0.7000
      43     0.9000            0.7000
      44     1.000             0.7000
      45     0.0000E+00        0.6000
      46     0.1000            0.6000
      47     0.2000            0.6000
      48     0.3000            0.6000
      49     0.4000            0.6000
      50     0.5000            0.6000
```

EXAMPLE 3: OUTPUT DATA LIST

```
flow demo6 - thermally-driven buoyant flow

    nodal coordinates

      node        1 coord        2 coord
        51      0.6000         0.6000
        52      0.7000         0.6000
        53      0.8000         0.6000
        54      0.9000         0.6000
        55       1.000         0.6000
        56      0.0000E+00     0.5000
        57      0.1000         0.5000
        58      0.2000         0.5000
        59      0.3000         0.5000
        60      0.4000         0.5000
        61      0.5000         0.5000
        62      0.6000         0.5000
        63      0.7000         0.5000
        64      0.8000         0.5000
        65      0.9000         0.5000
        66       1.000         0.5000
        67      0.0000E+00     0.4000
        68      0.1000         0.4000
        69      0.2000         0.4000
        70      0.3000         0.4000
        71      0.4000         0.4000
        72      0.5000         0.4000
        73      0.6000         0.4000
        74      0.7000         0.4000
        75      0.8000         0.4000
        76      0.9000         0.4000
        77       1.000         0.4000
        78      0.0000E+00     0.3000
        79      0.1000         0.3000
        80      0.2000         0.3000
        81      0.3000         0.3000
        82      0.4000         0.3000
        83      0.5000         0.3000
        84      0.6000         0.3000
        85      0.7000         0.3000
        86      0.8000         0.3000
        87      0.9000         0.3000
        88       1.000         0.3000
        89      0.0000E+00     0.2000
        90      0.1000         0.2000
        91      0.2000         0.2000
        92      0.3000         0.2000
        93      0.4000         0.2000
        94      0.5000         0.2000
        95      0.6000         0.2000
        96      0.7000         0.2000
        97      0.8000         0.2000
        98      0.9000         0.2000
        99       1.000         0.2000
       100      0.0000E+00     0.1000
```

```
flow demo6 - thermally-driven buoyant flow

     nodal coordinates

     node      1 coord       2 coord
      101     0.1000        0.1000
      102     0.2000        0.1000
      103     0.3000        0.1000
      104     0.4000        0.1000
      105     0.5000        0.1000
      106     0.6000        0.1000
      107     0.7000        0.1000
      108     0.8000        0.1000
      109     0.9000        0.1000
      110      1.000        0.1000
      111     0.0000E+00    0.0000E+00
      112     0.1000        0.0000E+00
      113     0.2000        0.0000E+00
      114     0.3000        0.0000E+00
      115     0.4000        0.0000E+00
      116     0.5000        0.0000E+00
      117     0.6000        0.0000E+00
      118     0.7000        0.0000E+00
      119     0.8000        0.0000E+00
      120     0.9000        0.0000E+00
      121      1.000        0.0000E+00
```

EXAMPLE 3: OUTPUT DATA LIST

```
flow demo6 - thermally-driven buoyant flow

    elements

    element   material   1 node   2 node   3 node   4 node
          1          1        1        2       13       12
          2          1        2        3       14       13
          3          1        3        4       15       14
          4          1        4        5       16       15
          5          1        5        6       17       16
          6          1        6        7       18       17
          7          1        7        8       19       18
          8          1        8        9       20       19
          9          1        9       10       21       20
         10          1       10       11       22       21
         11          1       12       13       24       23
         12          1       13       14       25       24
         13          1       14       15       26       25
         14          1       15       16       27       26
         15          1       16       17       28       27
         16          1       17       18       29       28
         17          1       18       19       30       29
         18          1       19       20       31       30
         19          1       20       21       32       31
         20          1       21       22       33       32
         21          1       23       24       35       34
         22          1       24       25       36       35
         23          1       25       26       37       36
         24          1       26       27       38       37
         25          1       27       28       39       38
         26          1       28       29       40       39
         27          1       29       30       41       40
         28          1       30       31       42       41
         29          1       31       32       43       42
         30          1       32       33       44       43
         31          1       34       35       46       45
         32          1       35       36       47       46
         33          1       36       37       48       47
         34          1       37       38       49       48
         35          1       38       39       50       49
         36          1       39       40       51       50
         37          1       40       41       52       51
         38          1       41       42       53       52
         39          1       42       43       54       53
         40          1       43       44       55       54
         41          1       45       46       57       56
         42          1       46       47       58       57
         43          1       47       48       59       58
         44          1       48       49       60       59
         45          1       49       50       61       60
         46          1       50       51       62       61
         47          1       51       52       63       62
         48          1       52       53       64       63
         49          1       53       54       65       64
         50          1       54       55       66       65
```

flow demo6 - thermally-driven buoyant flow

    elements

element	material	1 node	2 node	3 node	4 node
51	1	56	57	68	67
52	1	57	58	69	68
53	1	58	59	70	69
54	1	59	60	71	70
55	1	60	61	72	71
56	1	61	62	73	72
57	1	62	63	74	73
58	1	63	64	75	74
59	1	64	65	76	75
60	1	65	66	77	76
61	1	67	68	79	78
62	1	68	69	80	79
63	1	69	70	81	80
64	1	70	71	82	81
65	1	71	72	83	82
66	1	72	73	84	83
67	1	73	74	85	84
68	1	74	75	86	85
69	1	75	76	87	86
70	1	76	77	88	87
71	1	78	79	90	89
72	1	79	80	91	90
73	1	80	81	92	91
74	1	81	82	93	92
75	1	82	83	94	93
76	1	83	84	95	94
77	1	84	85	96	95
78	1	85	86	97	96
79	1	86	87	98	97
80	1	87	88	99	98
81	1	89	90	101	100
82	1	90	91	102	101
83	1	91	92	103	102
84	1	92	93	104	103
85	1	93	94	105	104
86	1	94	95	106	105
87	1	95	96	107	106
88	1	96	97	108	107
89	1	97	98	109	108
90	1	98	99	110	109
91	1	100	101	112	111
92	1	101	102	113	112
93	1	102	103	114	113
94	1	103	104	115	114
95	1	104	105	116	115
96	1	105	106	117	116
97	1	106	107	118	117
98	1	107	108	119	118
99	1	108	109	120	119
100	1	109	110	121	120

EXAMPLE 3: OUTPUT DATA LIST

```
flow demo6 - thermally-driven buoyant flow

    material properties

    material set   1 for element type   3 ! material number, element type numb
er


    two-dimensional transport element

        regular integration order = 2
        flag for axisymmetry (0 for no) = 0
        flag for time stepping (0 for no) = 0
        transient algorithm theta value =   0.0000E+00

    momentum parameters:

        velocity degree of freedom = 1
        penalty integration order = 1
        momentum convection integration order = 0

        penalty multiplier =   0.1000E+08
        density =    1.000
        viscosity coefficient =    1.000
        Carreau lambda factor =   0.0000E+00
        power-law exponent =    1.000
        viscosity activation energy =   0.0000E+00
        reference temperature =   0.0000E+00
        thermal expansion coefficient =    1.000

    thermal parameters:

        temperature degree of freedom = 3
        thermal convection integration order = 0
        coefficient of thermal conduction =    1.000
        specific heat =    1.000
        dissipation factor =   0.0000E+00

    species conversion parameters:

        conversion degree of freedom = 0
        species convection integration order = 0
        species diffusivity =   0.0000E+00
        kinetic reaction order =    1.000
        preexponential rate factor =   0.0000E+00
        reaction activation energy =   0.0000E+00
        heat of reaction =   0.0000E+00


    stream function degree of freedom = 4
```

flow demo6 - thermally-driven buoyant flow

nodal b.c.

node	1 b.c.	2 b.c.	3 b.c.	4 b.c.
1	1	1	1	1
2	1	1	0	0
3	1	1	0	0
4	1	1	0	0
5	1	1	0	0
6	1	1	0	0
7	1	1	0	0
8	1	1	0	0
9	1	1	0	0
10	1	1	0	0
11	1	1	1	0
12	1	1	1	1
22	1	1	1	0
23	1	1	1	1
33	1	1	1	0
34	1	1	1	1
44	1	1	1	0
45	1	1	1	1
55	1	1	1	0
56	1	1	1	1
66	1	1	1	0
67	1	1	1	1
77	1	1	1	0
78	1	1	1	1
88	1	1	1	0
89	1	1	1	1
99	1	1	1	0
100	1	1	1	1
110	1	1	1	0
111	1	1	1	1
112	1	1	0	1
113	1	1	0	1
114	1	1	0	1
115	1	1	0	1
116	1	1	0	1
117	1	1	0	1
118	1	1	0	1
119	1	1	0	1
120	1	1	0	1
121	1	1	0	1

EXAMPLE 3: OUTPUT DATA LIST

```
flow demo6 - thermally-driven buoyant flow

    nodal force/displ

    node      1 force      2 force      3 force      4 force
       1   0.0000E+00   0.0000E+00    1.000      0.0000E+00
       2   0.0000E+00   0.0000E+00   0.0000E+00   0.0000E+00
       3   0.0000E+00   0.0000E+00   0.0000E+00   0.0000E+00
       4   0.0000E+00   0.0000E+00   0.0000E+00   0.0000E+00
       5   0.0000E+00   0.0000E+00   0.0000E+00   0.0000E+00
       6   0.0000E+00   0.0000E+00   0.0000E+00   0.0000E+00
       7   0.0000E+00   0.0000E+00   0.0000E+00   0.0000E+00
       8   0.0000E+00   0.0000E+00   0.0000E+00   0.0000E+00
       9   0.0000E+00   0.0000E+00   0.0000E+00   0.0000E+00
      10   0.0000E+00   0.0000E+00   0.0000E+00   0.0000E+00
      11   0.0000E+00   0.0000E+00   0.0000E+00   0.0000E+00
      12   0.0000E+00   0.0000E+00    1.000      0.0000E+00
      13   0.0000E+00   0.0000E+00   0.0000E+00   0.0000E+00
      14   0.0000E+00   0.0000E+00   0.0000E+00   0.0000E+00
      15   0.0000E+00   0.0000E+00   0.0000E+00   0.0000E+00
      16   0.0000E+00   0.0000E+00   0.0000E+00   0.0000E+00
      17   0.0000E+00   0.0000E+00   0.0000E+00   0.0000E+00
      18   0.0000E+00   0.0000E+00   0.0000E+00   0.0000E+00
      19   0.0000E+00   0.0000E+00   0.0000E+00   0.0000E+00
      20   0.0000E+00   0.0000E+00   0.0000E+00   0.0000E+00
      21   0.0000E+00   0.0000E+00   0.0000E+00   0.0000E+00
      22   0.0000E+00   0.0000E+00   0.0000E+00   0.0000E+00
      23   0.0000E+00   0.0000E+00    1.000      0.0000E+00
      24   0.0000E+00   0.0000E+00   0.0000E+00   0.0000E+00
      25   0.0000E+00   0.0000E+00   0.0000E+00   0.0000E+00
      26   0.0000E+00   0.0000E+00   0.0000E+00   0.0000E+00
      27   0.0000E+00   0.0000E+00   0.0000E+00   0.0000E+00
      28   0.0000E+00   0.0000E+00   0.0000E+00   0.0000E+00
      29   0.0000E+00   0.0000E+00   0.0000E+00   0.0000E+00
      30   0.0000E+00   0.0000E+00   0.0000E+00   0.0000E+00
      31   0.0000E+00   0.0000E+00   0.0000E+00   0.0000E+00
      32   0.0000E+00   0.0000E+00   0.0000E+00   0.0000E+00
      33   0.0000E+00   0.0000E+00   0.0000E+00   0.0000E+00
      34   0.0000E+00   0.0000E+00    1.000      0.0000E+00
      35   0.0000E+00   0.0000E+00   0.0000E+00   0.0000E+00
      36   0.0000E+00   0.0000E+00   0.0000E+00   0.0000E+00
      37   0.0000E+00   0.0000E+00   0.0000E+00   0.0000E+00
      38   0.0000E+00   0.0000E+00   0.0000E+00   0.0000E+00
      39   0.0000E+00   0.0000E+00   0.0000E+00   0.0000E+00
      40   0.0000E+00   0.0000E+00   0.0000E+00   0.0000E+00
      41   0.0000E+00   0.0000E+00   0.0000E+00   0.0000E+00
      42   0.0000E+00   0.0000E+00   0.0000E+00   0.0000E+00
      43   0.0000E+00   0.0000E+00   0.0000E+00   0.0000E+00
      44   0.0000E+00   0.0000E+00   0.0000E+00   0.0000E+00
      45   0.0000E+00   0.0000E+00    1.000      0.0000E+00
      46   0.0000E+00   0.0000E+00   0.0000E+00   0.0000E+00
      47   0.0000E+00   0.0000E+00   0.0000E+00   0.0000E+00
      48   0.0000E+00   0.0000E+00   0.0000E+00   0.0000E+00
      49   0.0000E+00   0.0000E+00   0.0000E+00   0.0000E+00
      50   0.0000E+00   0.0000E+00   0.0000E+00   0.0000E+00
```

```
flow demo6 - thermally-driven buoyant flow

    nodal force/displ

    node      1 force         2 force         3 force         4 force
      51    0.0000E+00      0.0000E+00      0.0000E+00      0.0000E+00
      52    0.0000E+00      0.0000E+00      0.0000E+00      0.0000E+00
      53    0.0000E+00      0.0000E+00      0.0000E+00      0.0000E+00
      54    0.0000E+00      0.0000E+00      0.0000E+00      0.0000E+00
      55    0.0000E+00      0.0000E+00      0.0000E+00      0.0000E+00
      56    0.0000E+00      0.0000E+00      1.000           0.0000E+00
      57    0.0000E+00      0.0000E+00      0.0000E+00      0.0000E+00
      58    0.0000E+00      0.0000E+00      0.0000E+00      0.0000E+00
      59    0.0000E+00      0.0000E+00      0.0000E+00      0.0000E+00
      60    0.0000E+00      0.0000E+00      0.0000E+00      0.0000E+00
      61    0.0000E+00      0.0000E+00      0.0000E+00      0.0000E+00
      62    0.0000E+00      0.0000E+00      0.0000E+00      0.0000E+00
      63    0.0000E+00      0.0000E+00      0.0000E+00      0.0000E+00
      64    0.0000E+00      0.0000E+00      0.0000E+00      0.0000E+00
      65    0.0000E+00      0.0000E+00      0.0000E+00      0.0000E+00
      66    0.0000E+00      0.0000E+00      0.0000E+00      0.0000E+00
      67    0.0000E+00      0.0000E+00      1.000           0.0000E+00
      68    0.0000E+00      0.0000E+00      0.0000E+00      0.0000E+00
      69    0.0000E+00      0.0000E+00      0.0000E+00      0.0000E+00
      70    0.0000E+00      0.0000E+00      0.0000E+00      0.0000E+00
      71    0.0000E+00      0.0000E+00      0.0000E+00      0.0000E+00
      72    0.0000E+00      0.0000E+00      0.0000E+00      0.0000E+00
      73    0.0000E+00      0.0000E+00      0.0000E+00      0.0000E+00
      74    0.0000E+00      0.0000E+00      0.0000E+00      0.0000E+00
      75    0.0000E+00      0.0000E+00      0.0000E+00      0.0000E+00
      76    0.0000E+00      0.0000E+00      0.0000E+00      0.0000E+00
      77    0.0000E+00      0.0000E+00      0.0000E+00      0.0000E+00
      78    0.0000E+00      0.0000E+00      1.000           0.0000E+00
      79    0.0000E+00      0.0000E+00      0.0000E+00      0.0000E+00
      80    0.0000E+00      0.0000E+00      0.0000E+00      0.0000E+00
      81    0.0000E+00      0.0000E+00      0.0000E+00      0.0000E+00
      82    0.0000E+00      0.0000E+00      0.0000E+00      0.0000E+00
      83    0.0000E+00      0.0000E+00      0.0000E+00      0.0000E+00
      84    0.0000E+00      0.0000E+00      0.0000E+00      0.0000E+00
      85    0.0000E+00      0.0000E+00      0.0000E+00      0.0000E+00
      86    0.0000E+00      0.0000E+00      0.0000E+00      0.0000E+00
      87    0.0000E+00      0.0000E+00      0.0000E+00      0.0000E+00
      88    0.0000E+00      0.0000E+00      0.0000E+00      0.0000E+00
      89    0.0000E+00      0.0000E+00      1.000           0.0000E+00
      90    0.0000E+00      0.0000E+00      0.0000E+00      0.0000E+00
      91    0.0000E+00      0.0000E+00      0.0000E+00      0.0000E+00
      92    0.0000E+00      0.0000E+00      0.0000E+00      0.0000E+00
      93    0.0000E+00      0.0000E+00      0.0000E+00      0.0000E+00
      94    0.0000E+00      0.0000E+00      0.0000E+00      0.0000E+00
      95    0.0000E+00      0.0000E+00      0.0000E+00      0.0000E+00
      96    0.0000E+00      0.0000E+00      0.0000E+00      0.0000E+00
      97    0.0000E+00      0.0000E+00      0.0000E+00      0.0000E+00
      98    0.0000E+00      0.0000E+00      0.0000E+00      0.0000E+00
      99    0.0000E+00      0.0000E+00      0.0000E+00      0.0000E+00
     100    0.0000E+00      0.0000E+00      1.000           0.0000E+00
```

EXAMPLE 3: OUTPUT DATA LIST

```
flow demo6 - thermally-driven buoyant flow

    nodal force/displ
        node      1 force       2 force       3 force       4 force
        101     0.0000E+00    0.0000E+00    0.0000E+00    0.0000E+00
        102     0.0000E+00    0.0000E+00    0.0000E+00    0.0000E+00
        103     0.0000E+00    0.0000E+00    0.0000E+00    0.0000E+00
        104     0.0000E+00    0.0000E+00    0.0000E+00    0.0000E+00
        105     0.0000E+00    0.0000E+00    0.0000E+00    0.0000E+00
        106     0.0000E+00    0.0000E+00    0.0000E+00    0.0000E+00
        107     0.0000E+00    0.0000E+00    0.0000E+00    0.0000E+00
        108     0.0000E+00    0.0000E+00    0.0000E+00    0.0000E+00
        109     0.0000E+00    0.0000E+00    0.0000E+00    0.0000E+00
        110     0.0000E+00    0.0000E+00    0.0000E+00    0.0000E+00
        111     0.0000E+00    0.0000E+00    1.000         0.0000E+00
        112     0.0000E+00    0.0000E+00    0.0000E+00    0.0000E+00
        113     0.0000E+00    0.0000E+00    0.0000E+00    0.0000E+00
        114     0.0000E+00    0.0000E+00    0.0000E+00    0.0000E+00
        115     0.0000E+00    0.0000E+00    0.0000E+00    0.0000E+00
        116     0.0000E+00    0.0000E+00    0.0000E+00    0.0000E+00
        117     0.0000E+00    0.0000E+00    0.0000E+00    0.0000E+00
        118     0.0000E+00    0.0000E+00    0.0000E+00    0.0000E+00
        119     0.0000E+00    0.0000E+00    0.0000E+00    0.0000E+00
        120     0.0000E+00    0.0000E+00    0.0000E+00    0.0000E+00
        121     0.0000E+00    0.0000E+00    0.0000E+00    0.0000E+00
```

EXAMPLE 3: MACRO EXECUTION LOG

```
flow demo6 - thermally-driven buoyant flow

    macro instructions

    macro statement       variable 1        variable 2
           tang         0.00000E+00       0.00000E+00
           loop            3.0000          0.00000E+00
           form         0.00000E+00       0.00000E+00
           solv         0.00000E+00       0.00000E+00
           next         0.00000E+00       0.00000E+00
           disp         0.00000E+00       0.00000E+00
           end          0.00000E+00       0.00000E+00
```

```
**macro instruction loop completed. time used this step =  8.23 total time = 49
.59
**macro instruction   1 initiated** tang        v1 =    0.0000E+00 , v2 =    0
.0000E+00


**macro instruction tang completed. time used this step = 42.68 total time = 92
.27
**macro instruction   2 initiated** loop        v1 =     3.000     , v2 =
6.000


**macro instruction loop completed. time used this step =  0.28 total time = 92
.55
**macro instruction   3 initiated** form        v1 =    0.0000E+00 , v2 =    0
.0000E+00
     force convergence test
          rnmax =     3.0822          rm   =    3.0822          tol   =    0.1000
0E-08


**macro instruction form completed. time used this step = 21.86 total time =114
.41
**macro instruction   4 initiated** solv        v1 =    0.0000E+00 , v2 =    0
.0000E+00


**macro instruction solv completed. time used this step = 61.29 total time =175
.70
**macro instruction   5 initiated** next        v1 =     1.000     , v2 =
3.000


**macro instruction loop completed. time used this step =  0.22 total time =175
.92
**macro instruction   3 initiated** form        v1 =    0.0000E+00 , v2 =    0
.0000E+00
     force convergence test
          rnmax =     3.0822          rm   =    0.48298E-01    tol   =    0.1000
0E-08


**macro instruction form completed. time used this step = 23.13 total time =199
.05
**macro instruction   4 initiated** solv        v1 =    0.0000E+00 , v2 =    0
.0000E+00
```

EXAMPLE 3: MACRO EXECUTION LOG

```
**macro instruction solv completed. time used this step =  6.48 total time =205
.53
**macro instruction   5 initiated**  next        v1 =     2.000     , v2 =
3.000


**macro instruction loop completed. time used this step =  0.27 total time =205
.80
**macro instruction   3 initiated**  form        v1 =    0.0000E+00 , v2 =    0
.0000E+00
      force convergence test
            rnmax =      3.0822         rm    =    0.14639E-02    tol   =    0.1000
0E-08


**macro instruction form completed. time used this step = 23.35 total time =229
.15
**macro instruction   4 initiated**  solv        v1 =    0.0000E+00 , v2 =    0
.0000E+00


**macro instruction solv completed. time used this step =  6.48 total time =235
.63
**macro instruction   5 initiated**  next        v1 =     3.000     , v2 =
3.000


**macro instruction next completed. time used this step =  0.27 total time =235
.90
**macro instruction   6 initiated**  disp        v1 =    0.0000E+00 , v2 =    0
.0000E+00

      time   0.00000E+00

      proportional load   1.0000
```

EXAMPLE 3: NODAL DISPLACEMENT LIST

flow demo6 - thermally-driven buoyant flow

nodal displacements   time 0.00000E+00

node	1 coord	2 coord	1 displ	2 displ	3 displ	4 displ
1	0.0000E+00	1.000	0.0000E+00	0.0000E+00	1.000	0.0000E+00
2	0.1000	1.000	0.0000E+00	0.0000E+00	0.9002	-0.3273E-09
3	0.2000	1.000	0.0000E+00	0.0000E+00	0.8004	-0.6521E-09
4	0.3000	1.000	0.0000E+00	0.0000E+00	0.7006	-0.9633E-09
5	0.4000	1.000	0.0000E+00	0.0000E+00	0.6007	-0.1252E-08
6	0.5000	1.000	0.0000E+00	0.0000E+00	0.5008	-0.1511E-08
7	0.6000	1.000	0.0000E+00	0.0000E+00	0.4007	-0.1741E-08
8	0.7000	1.000	0.0000E+00	0.0000E+00	0.3007	-0.1942E-08
9	0.8000	1.000	0.0000E+00	0.0000E+00	0.2005	-0.2118E-08
10	0.9000	1.000	0.0000E+00	0.0000E+00	0.1003	-0.2266E-08
11	1.000	1.000	0.0000E+00	0.0000E+00	0.0000E+00	-0.2345E-08
12	0.0000E+00	0.9000	0.0000E+00	0.0000E+00	1.000	0.0000E+00
13	0.1000	0.9000	0.4771E-03	0.4771E-03	0.9002	-0.2385E-04
14	0.2000	0.9000	0.1507E-02	0.5533E-03	0.8004	-0.7537E-04
15	0.3000	0.9000	0.2486E-02	0.4248E-03	0.7006	-0.1243E-03
16	0.4000	0.9000	0.3135E-02	0.2251E-03	0.6007	-0.1568E-03
17	0.5000	0.9000	0.3361E-02	0.7899E-06	0.5008	-0.1681E-03
18	0.6000	0.9000	0.3138E-02	-0.2239E-03	0.4008	-0.1569E-03
19	0.7000	0.9000	0.2490E-02	-0.4245E-03	0.3007	-0.1245E-03
20	0.8000	0.9000	0.1511E-02	-0.5542E-03	0.2005	-0.7556E-04
21	0.9000	0.9000	0.4785E-03	-0.4785E-03	0.1003	-0.2393E-04
22	1.000	0.9000	0.0000E+00	0.0000E+00	0.0000E+00	-0.2422E-08
23	0.0000E+00	0.8000	0.0000E+00	0.0000E+00	1.000	0.0000E+00
24	0.1000	0.8000	0.5530E-03	0.1507E-02	0.9003	-0.7536E-04
25	0.2000	0.8000	0.1752E-02	0.1753E-02	0.8005	-0.2384E-03
26	0.3000	0.8000	0.2925E-02	0.1376E-02	0.7007	-0.3948E-03
27	0.4000	0.8000	0.3737E-02	0.7351E-03	0.6008	-0.5004E-03
28	0.5000	0.8000	0.4023E-02	0.2443E-05	0.5009	-0.5373E-03
29	0.6000	0.8000	0.3740E-02	-0.7313E-03	0.4009	-0.5008E-03
30	0.7000	0.8000	0.2930E-02	-0.1375E-02	0.3008	-0.3955E-03
31	0.8000	0.8000	0.1756E-02	-0.1756E-02	0.2006	-0.2389E-03
32	0.9000	0.8000	0.5544E-03	-0.1512E-02	0.1003	-0.7558E-04
33	1.000	0.8000	0.0000E+00	0.0000E+00	0.0000E+00	-0.2542E-08
34	0.0000E+00	0.7000	0.0000E+00	0.0000E+00	1.000	0.0000E+00
35	0.1000	0.7000	0.4239E-03	0.2484E-02	0.9003	-0.1242E-03
36	0.2000	0.7000	0.1374E-02	0.2925E-02	0.8006	-0.3947E-03
37	0.3000	0.7000	0.2330E-02	0.2333E-02	0.7008	-0.6576E-03
38	0.4000	0.7000	0.2998E-02	0.1258E-02	0.6010	-0.8372E-03
39	0.5000	0.7000	0.3237E-02	0.3943E-05	0.5011	-0.9003E-03
40	0.6000	0.7000	0.3000E-02	-0.1252E-02	0.4011	-0.8379E-03
41	0.7000	0.7000	0.2333E-02	-0.2332E-02	0.3010	-0.6587E-03
42	0.8000	0.7000	0.1377E-02	-0.2930E-02	0.2008	-0.3956E-03
43	0.9000	0.7000	0.4251E-03	-0.2491E-02	0.1004	-0.1246E-03
44	1.000	0.7000	0.0000E+00	0.0000E+00	0.0000E+00	-0.2622E-08
45	0.0000E+00	0.6000	0.0000E+00	0.0000E+00	1.000	0.0000E+00
46	0.1000	0.6000	0.2237E-03	0.3132E-02	0.9004	-0.1566E-03
47	0.2000	0.6000	0.7308E-03	0.3735E-02	0.8007	-0.4999E-03
48	0.3000	0.6000	0.1252E-02	0.3000E-02	0.7010	-0.8367E-03
49	0.4000	0.6000	0.1620E-02	0.1628E-02	0.6013	-0.1068E-02
50	0.5000	0.6000	0.1751E-02	0.4984E-05	0.5015	-0.1150E-02

EXAMPLE 3: NODAL DISPLACEMENT LIST

flow demo6 - thermally-driven buoyant flow

nodal displacements   time 0.00000E+00

node	1 coord	2 coord	1 displ	2 displ	3 displ	4 displ
51	0.6000	0.6000	0.1622E-02	-0.1620E-02	0.4015	-0.1069E-02
52	0.7000	0.6000	0.1254E-02	-0.2999E-02	0.3014	-0.8380E-03
53	0.8000	0.6000	0.7329E-03	-0.3740E-02	0.2011	-0.5011E-03
54	0.9000	0.6000	0.2246E-03	-0.3141E-02	0.1006	-0.1570E-03
55	1.000	0.6000	0.0000E+00	0.0000E+00	0.0000E+00	-0.2610E-08
56	0.0000E+00	0.5000	0.0000E+00	0.0000E+00	1.000	0.0000E+00
57	0.1000	0.5000	-0.7953E-06	0.3355E-02	0.9004	-0.1677E-03
58	0.2000	0.5000	-0.2442E-05	0.4017E-02	0.8008	-0.5363E-03
59	0.3000	0.5000	-0.4044E-05	0.3237E-02	0.7012	-0.8991E-03
60	0.4000	0.5000	-0.5238E-05	0.1758E-02	0.6016	-0.1149E-02
61	0.5000	0.5000	-0.5844E-05	0.5227E-05	0.5019	-0.1237E-02
62	0.6000	0.5000	-0.5434E-05	-0.1750E-02	0.4020	-0.1150E-02
63	0.7000	0.5000	-0.3914E-05	-0.3235E-02	0.3020	-0.9005E-03
64	0.8000	0.5000	-0.1858E-05	-0.4023E-02	0.2016	-0.5376E-03
65	0.9000	0.5000	-0.3737E-06	-0.3365E-02	0.1009	-0.1682E-03
66	1.000	0.5000	0.0000E+00	0.0000E+00	0.0000E+00	-0.2475E-08
67	0.0000E+00	0.4000	0.0000E+00	0.0000E+00	1.000	0.0000E+00
68	0.1000	0.4000	-0.2249E-03	0.3129E-02	0.9005	-0.1564E-03
69	0.2000	0.4000	-0.7346E-03	0.3732E-02	0.8010	-0.4995E-03
70	0.3000	0.4000	-0.1258E-02	0.2998E-02	0.7015	-0.8359E-03
71	0.4000	0.4000	-0.1628E-02	0.1626E-02	0.6020	-0.1067E-02
72	0.5000	0.4000	-0.1761E-02	0.4478E-05	0.5024	-0.1149E-02
73	0.6000	0.4000	-0.1631E-02	-0.161BE-02	0.4028	-0.1068E-02
74	0.7000	0.4000	-0.1261E-02	-0.2995E-02	0.3029	-0.8373E-03
75	0.8000	0.4000	-0.7363E-03	-0.3736E-02	0.2024	-0.5007E-03
76	0.9000	0.4000	-0.2254E-03	-0.3139E-02	0.1014	-0.1570E-03
77	1.000	0.4000	0.0000E+00	0.0000E+00	0.0000E+00	-0.2203E-08
78	0.0000E+00	0.3000	0.0000E+00	0.0000E+00	1.000	0.0000E+00
79	0.1000	0.3000	-0.4242E-03	0.24B0E-02	0.9006	-0.1240E-03
80	0.2000	0.3000	-0.1375E-02	0.2921E-02	0.8012	-0.3940E-03
81	0.3000	0.3000	-0.2331E-02	0.2329E-02	0.701B	-0.6565E-03
82	0.4000	0.3000	-0.3000E-02	0.1255E-02	0.6024	-0.8357E-03
83	0.5000	0.3000	-0.3240E-02	0.2824E-05	0.5031	-0.8986E-03
84	0.6000	0.3000	-0.3004E-02	-0.1251E-02	0.4037	-0.8363E-03
85	0.7000	0.3000	-0.2337E-02	-0.2327E-02	0.3043	-0.6574E-03
86	0.8000	0.3000	-0.13B0E-02	-0.2923E-02	0.2040	-0.3949E-03
87	0.9000	0.3000	-0.4257E-03	-0.24B8E-02	0.1026	-0.1244E-03
88	1.000	0.3000	0.0000E+00	0.0000E+00	0.0000E+00	-0.1795E-08
89	0.0000E+00	0.2000	0.0000E+00	0.0000E+00	1.000	0.0000E+00
90	0.1000	0.2000	-0.5521E-03	0.1503E-02	0.9006	-0.7517E-04
91	0.2000	0.2000	-0.1750E-02	0.1749E-02	0.8013	-0.7378E-03
92	0.3000	0.2000	-0.2921E-02	0.1373E-02	0.7020	-0.3939E-03
93	0.4000	0.2000	-0.3731E-02	0.7326E-03	0.6028	-0.4992E-03
94	0.5000	0.2000	-0.4016E-02	0.1017E-05	0.5037	-0.5359E-03
95	0.6000	0.2000	-0.3734E-02	-0.7307E-03	0.4048	-0.4994E-03
96	0.7000	0.2000	-0.2926E-02	-0.1372E-02	0.3060	-0.3942E-03
97	0.8000	0.2000	-0.1756E-02	-0.1749E-02	0.2073	-0.2382E-03
98	0.9000	0.2000	-0.5554E-03	-0.1507E-02	0.1053	-0.7534E-04
99	1.000	0.2000	0.0000E+00	0.0000E+00	0.0000E+00	-0.1269E-08
100	0.0000E+00	0.1000	0.0000E+00	0.0000E+00	1.000	0.0000E+00

flow demo6 - thermally-driven buoyant flow

nodal displacements   time 0.00000E+00

node	1 coord	2 coord	1 displ	2 displ	3 displ	4 displ
101	0.1000	0.1000	-0.4756E-03	0.4756E-03	0.9007	-0.2378E-04
102	0.2000	0.1000	-0.1503E-02	0.5519E-03	0.8014	-0.7515E-04
103	0.3000	0.1000	-0.2479E-02	0.4236E-03	0.7022	-0.1239E-03
104	0.4000	0.1000	-0.3126E-02	0.2241E-03	0.6031	-0.1563E-03
105	0.5000	0.1000	-0.3350E-02	0.1405E-06	0.5042	-0.1675E-03
106	0.6000	0.1000	-0.3127E-02	-0.2240E-03	0.4056	-0.1563E-03
107	0.7000	0.1000	-0.2479E-02	-0.4235E-03	0.3077	-0.1240E-03
108	0.8000	0.1000	-0.1503E-02	-0.5521E-03	0.2107	-0.7517E-04
109	0.9000	0.1000	-0.4757E-03	-0.4757E-03	0.1181	-0.2378E-04
110	1.000	0.1000	0.0000E+00	0.0000E+00	0.0000E+00	-0.6574E-09
111	0.0000E+00	0.0000E+00	0.0000E+00	0.0000E+00	1.000	0.0000E+00
112	0.1000	0.0000E+00	0.0000E+00	0.0000E+00	0.9007	0.0000E+00
113	0.2000	0.0000E+00	0.0000E+00	0.0000E+00	0.8014	0.0000E+00
114	0.3000	0.0000E+00	0.0000E+00	0.0000E+00	0.7023	0.0000E+00
115	0.4000	0.0000E+00	0.0000E+00	0.0000E+00	0.6032	0.0000E+00
116	0.5000	0.0000E+00	0.0000E+00	0.0000E+00	0.5044	0.0000E+00
117	0.6000	0.0000E+00	0.0000E+00	0.0000E+00	0.4060	0.0000E+00
118	0.7000	0.0000E+00	0.0000E+00	0.0000E+00	0.3083	0.0000E+00
119	0.8000	0.0000E+00	0.0000E+00	0.0000E+00	0.2126	0.0000E+00
120	0.9000	0.0000E+00	0.0000E+00	0.0000E+00	0.1199	0.0000E+00
121	1.000	0.0000E+00	0.0000E+00	0.0000E+00	0.8903E-01	0.0000E+00

**macro instruction disp completed. time used this step = 12.19 total time =247.33

8
Closing

8.1 SUMMARY

In conclusion, we have seen several examples of the applications of finite element analysis on microcomputers. Most of these examples have been introductory in nature, as was intended, and concentrated on the traditional areas of finite element analysis. But, finite element analysis on microcomputers is not limited to such basic problem types. Indeed, finite element analysis on microcomputers may encompass any application or field of endeavor. Recent literature (24,25) has described techniques which help analyze the magnetic flux around irregular shapes and aid in modeling implants for the human eye.

8.2 COMMERCIALLY AVAILABLE FINITE ELEMENT SOFTWARE

This section describes the software used in this chapter for the sample problems in the previous chapters.

8.2.1 COSMOS/M

COSMOS/M is an FEA package by Structural Research and Analysis Corporation. This model runs on the IBM XT/AT and compatibles with 640K, at least one floppy drive, an 8087 math co-processor, MS-DOS 2.xx or higher, and a hard disk (10-20 Mb). Many of the more popular graphics cards are supported. The analysis package is capable of performing linear static, linear dynamic, nonlinear, and heat transfer analysis on one-, two-, and three-dimensional structural and thermal models. The program allows the calculation of structural buckling load calculations. In addition to the above capabilities, postdynamic analysis includes modal analysis, response spectrum analysis, and random vibration analysis. Certain types of field problems (electrical and magnetic) may also be solved. COSMOS/M has extensive mesh generation features for lines, areas, and volumes.

A very thorough manual is included that covers all aspects of model utilization. All commands are discussed in detail and examples are provided for a thorough understanding of each command. At least one-third of the manual is devoted to example and verification problems. These problems along with details on surface intersection guidelines provide one of the more complete user-manuals available.

COSMOS/M is user-friendly and utilizes a module entitled "MODSTAR" to control the flow of program operations. MODSTAR is a menu-driven interactive system with full functionality through a large number of submenus. These submenus originate from the main menu. MODSTAR is the preprocessor portion of this FEA program. Preprocessors were discussed in Chapter 3. The user will find COSMOS/M offers substantial help screens throughout the entire program run. Another particularly useful feature is the ability to interface with AutoCad and personal designer CAD systems. These two programs may be used in place of MODSTAR in the preprocessing phase of the overall analysis.

The main modules of the COSMOS/M package are as follows:

MODSTAR

MODSTAR is the preprocessing module of the COSMOS/M package that controls the overall program operations.

STAR

STAR is the static analysis module that computes the node deformations.

DSTAR

DSTAR is the module used for frequency and node shape analysis and for buckling load calculations.

MICROTAP

MICROTAP performs the heat transfer analysis and writes the temperature file used for temperature contour plots.

PLOTSTAR

PLOTSTAR is the graphics module used with both the preprocessing and postprocessing graphics. Such items as the ability of allowing multiple windows makes this feature an outstanding addition to the overall package.

PDSTAR

PDSTAR is used after the frequencies and mode shapes have been calculated by the DSTAR module. This is the module that allows post-dynamic analysis such as modal, response spectrum, shock, random vibration, and others.

NSTAR

NSTAR is used for nonlinear problems.

COSMOS/M accepts input either from the terminal in the interactive mode or from a file in the batch mode. For problems that are too large to be run on the microcomputer, COSMOS/M can be used as a preprocessor for mainframe codes.

The program is designed to provide a comprehensive set of elements to perform both the structural and thermal analysis. Table 8.1 lists the elements and if they are used in either the structural analysis or thermal analysis or both.

Complete element descriptions are given the manual provided to the user. The COSMOS/M package is hardware copy protected. User support is excellent.

8.2.2 PCTRAN Plus

PCTRAN Plus is an interactive finite element program capable of performing linear and nonlinear static and dynamic analyses, steady state and transient thermal and

Table 8.1 Finite Elements Used in COSMOS/M

Element	Structural	Thermal
Two-dimensional spar/truss	*	*
Two-dimensional elastic beam	*	*
Three-dimensional elastic beam	*	*
Three-dimensional spar/truss	*	*
Elastic straight pipe	*	*
Elastic curved pipe	*	*
Two-dimensional plane stress, plane strain, body of revolution	*	*
Triangular thick shell	*	*
Quadrilateral thick shell	*	*
Three-dimensional 8-node solid	*	*
Triangular thin shell	*	*
Quadrilateral thin shell	*	*
Composite quadrilateral plate and shell	*	
Boundary element	*	
General mass element	*	
Radiation link		*
Two-dimensional thermal bar		*
Three-dimensional thermal bar		*
Convection link		*
Three-dimensional electrical link		*
Hydraulic conductance		*

fluid flow analyses. This model was developed by Brooks Scientific, Inc. of Cambridge, Massachusetts. The model runs on the IBM PC/XT/AT and is compatible with at least 640K, an 8087 (80287) math co-processor, one floppy drive, a hard disk, and MS-DOS 2.xx or higher. The user has the option of running the individual modules that comprise the FEA package or by running the program via menus. PCTRAN Plus supports a mouse and will interface with AutoCad and CadKey. The manual is well written and presents all of the necessary information for the novice finite element user and the advanced finite element user. As with all finite element programs, the problem solution is divided into three phases: (1) preprocessing, (2) analysis, and (3) postprocessing.

All data preparation is conducted in the PCTRAN Plus preprocessor. The preprocessor module contains twelve modules as listed that represent the different aspects of the model building process:

1. NODE
2. ELEMENT
3. LOAD
4. PLOT
5. JOB CONTROL
6. ELEMENT TYPE
7. REAL
8. MATERIAL
9. COORDINATE SYSTEM
10. MISCELLANEOUS
11. MESH GENERATION
12. CAD INTERFACE

PCTRAN Plus contains many elements to ease the analysis process. These are given in Table 8.2.

After the preprocessing phase of PCTRAN Plus, the analysis phase is executed. The results of the analysis include stresses, deflections, temperatures, and vibration modes. Again, this phase is divided into four modules. These four modules are STATICS, VIBRATION, THERMAL, and FLUIDS.

These modules contain submodules to calculate the stresses, deflections, temperatures, pressures, velocities, and natural frequencies. PCTRAN Plus uses both in-core and out-of-core solvers depending upon the size of the model.

PCTRAN Plus's postprocessor is used to interpret and document the results of each analysis. The postprocessor is composed of two modules. A PLOT module displays the results of the deflection, stress, natural freuqncies, and temperatures. The LIST module organizes the output into an understandable presentation of facts.

For problems that are not able to be solved with the PC, PCTRAN Plus provides the means to create the files necessary to interface with such programs as ANSYS.

Table 8.2 PCTRAN Plus Element Library

Three-dimensional 2-Node Truss
Three-dimensional 2-Node Beam
Three-dimensional 2-Node Spring
Three-dimensional 4-Node Plate
Three-dimensional 8-Node Solid
Two-dimensional 4-Node Solid
Two-dimensional 2-Node Truss
Two-dimensional 2-Node Beam
Two-dimensional 4-Node Solid (2nd Version)
1-DOF Mass
2-DOF Mass
3-DOF Mass
Two-dimensional 3-Node Solid
Two-dimensional 3-Node Axisymmetric
Two-dimensional 4-Node Axisymmetric
One-dimensional 2-Node Thermal Truss
Two-dimensional 2-Node Thermal Truss
Three-dimensional 2-Node Thermal Truss
Three-dimensional 4-Node Thermal Plate
Three-dimensional 8-Node Thermal Solid
Two-dimensional 4-Node Thermal Plate
Two-dimensional 4-Node Fluid Element
Convective Boundary Element

PCTRAN Plus is the only commercially available FEA code reviewed that contains a fluid finite element. An example of the use of this element is presented in Chapter 7.

8.2.3 IMAGES3D

IMAGES3D V 1.4A, by Celestial Software, Inc., is a highly interactive, user-friendly finite element program. IMAGES3D will perform static stress and linear dynamic analysis. The user also has the capability for seismic response analysis. IMAGES3D also contains an AISC database for commonly used beam cross sections. Program requirements are an IBM PC/XT/AT or compatible computer, an 8087 or 80287 math co-processor, at least 640 Kb of RAM, a hard disk, monitor and CGA or EGA, and DOS 3.xx or greater. An outstanding feature of this program

Table 8.3   Elements Available in IMAGES3D

Two- and three-dimensional truss elements
Two- and three-dimensional beam elements
Membrane plates
Three-dimensional solids
Axi-solids
Spring elements

---

is that the user is able to watch the model being built while entering data. Other FEA programs require the user to run another program or another section of the program to view the process. This version does not have the thermal capabilities of the other programs; nor does the program have the capability for nonlinear analysis. The program does have the quickest stiffness matrix assembly that has been encountered.

The elements available for analysis are listed in Table 8.3.

IMAGES3D has an excellent menu system that leads the user through all phases of the preprocessing, analysis, and postprocessing.

Preprocessing

The preprocessing section of the program is divided into the geometry development, geometry check, and bandwidth optimization section and the creation of the model loads and model weights. The user can not only specify point loads but also can specify pressure and temperature loadings on the elements. The pressure loading feature is helpful in a variety of situations. Again, the graphics that accompany the model development during this phase are very helpful and can highlight modeling errors immediately.

Analysis

The analysis section consists of the creation of the model weights and the assembly of the stiffness matrix. The V 1.4A has an extremely fast matrix assembly routine.

Postprocessing

The postprocessing section of IMAGES3D gives complete element stress and strain energy output and the output can be formatted in a variety of ways. The graphical output is limited to displaced or deformed structures and the vibration or response of the particular model in question. The program does not have any stress contouring or color shading.

8.2.4 SUPERSAP

SUPERSAP, based on the SAP IV program, by Algor Interactive Systems, requires an IBM PC/XT or /AT or compatible with MS/PC-DOS 2.xx or higher, 640 Kb of RAM, a math co-processor, a hard disk (10-20 Mb), and a graphics card and monitor. This program will also run with the PRIME and DEC VAX/MICROVAX computers. SUPERSAP supports several graphics boards including the Tecmar, Hercules, Conographic, Persyst, and IBM and compatible CGA and EGA versions. Input devices are either the keyboard or such devices as the mouse or digitizer tablet. Drivers are included for most popular plotters (e.g. H-P) and printers (e.g. Epson).

SUPERSAP is by far the largest program from a storage viewpoint in these reviews. The complete system loaded onto a hard drive will take in excess of 7 Mb. With most analyses and if disk space is at a premium, the better way to run SUPERSAP is to load only those portions needed for the analysis.

As with all the programs, SUPERSAP is divided in three phases that include preprocessing, analysis, and postprocessing.

Before discussing the features of SUPERSAP, Table 8.4 lists the elements available in the package for analysis.

These element types coupled with the ability to apply point and distributed loads, temperatures, and pressures in addition to specific orthotropic and isotropic material behavior provides the user with a very complete PC-based finite element package.

Table 8.4 SUPERSAP Elements

Truss elements (three-dimensional, two-dimensional)

Beam elements (three-dimensional, two-dimensional)

Three-dimensional plane stress, plane strain (quadrilateral, triangular)

Axisymmetric solids

Three-dimensional solid (8 node, 21 node)

Three-dimensional plate/shell

Boundary elements (rigid support, elastic support, temperature)

Stiffness matrix elements

Pipe elements

Rigid elements

Convective boundary elements

Radiative boundary elements

Three-dimensional, two-dimensional thermal conductivity elements

---

Preprocessing

The preprocessing portion of the package is very thorough in helping the user to build the model for analysis. An outstanding feature of this program is the variety of mesh generators included. These are listed below with a brief explanation:

MSHGEN. A general purpose program for the generation of two-dimensional finite element models (axisymmetric, plane stress, plane strain).

PLTGEN. A general purpose system for three-dimensional mesh generation using the quadrilateral plate and shell elements.

PROGEN. Generates a model of a given profile along a given path using the thin shell or plate elements.

RADGEN. A special purpose cylindrical generation program to create finite element meshes for containers or other shells of revolution using the thin plate or shell elements.

WARPGEN. Generates a model by merging one profile into another by sweeping along two constraining paths using the thin shell/plate elements. The two profiles are linearly averaged along the paths allowing a smooth surface generation.

RADGENBR. A special purpose cylindrical generation program to generate finite element meshes using the 8-node three-dimensional solid brick element.

LAYERGEN. A general purpose three-dimensional mesh generator for solid elements. This is used when the model can be constructed from a number of two-dimensional layers.

THREEGEN. A general purpose three-dimensional mesh generator using the three-dimensional solid brick elements.

In addition, SUPERSAP has a special program SUBSTRUC. SUBSTRUC is used to join two models together to make a combined model from the two parts.

To interface with the popular CAD programs, ALGOR has included drivers that allow conversion from the CAD packages to SUPERSAP and vice versa. In the review of all programs, this package actually worked as described.

Animation as well as AISC Steel Manual Beam data and extensive edit facilities are provided. A time history analysis for constant frequency, constant amplitude sinusoidal excitation, and for response spectrum analysis is provided. The manual included with the program is one of the best with respect to program introduction, detailed explanations and examples. A separate problem verification manual is included. A program by Algor called SUPERDRAWII is a terrific CAD type preprocessor for SUPERSAP. It is easy to use and it works.

Analysis

SUPERSAP analysis involves large files. To optimize performance, the programs are supplied in binary form along with programs to load run-only versions. The options given are listed below:

Stress analysis
Dynamic modal analysis
Dynamic modal analysis with time history
Dynamic modal analysis with response spectrum
Direct integration
Frequency response
Buckling analysis
Only dynamic restart
Dynamic with geometric stiffness
Steady state heat transfer
Transient heat transfer

The execution of the program with its run-time options may be handled automatically by issuing one command. This allows the user to obtain just the desired response.

Postprocessing

The output options from the SUPERSAP program are many and varied. All available stress outputs are available for each element type. These are discussed in detail in the manual. The strong area in the postprocessing section is the color shading of the stress contours. This is a desirable feature in any finite element program. In addition, the program has output options for plotters and graphics printers.

8.2.5 MSC/Pal2

MSC/Pal 2 Version 2, developed by the MacNeal-Schwendler Corporation, is an interactive finite element analysis program. The hardware requirements are an IBM PC/XT/AT or compatible with 640K RAM, one floppy disk drive, an 8087 math co-processor, at least a 10Mb hard drive, MS-DOS (PC-DOS) 2.1x or higher. MSC/Pal 2 is a structural package. MSC/cal contains the thermal analysis capabilities. MSC/cal performs linear and nonlinear steady state and transient analyses. MSC/Pal 2 and MSC/cal contain a good selection of elements to be used in an analysis. Table 8.5 lists the elements available.

The MSC/Pal 2 system is composed of several programs with each program having several suboptions. Many of the programs generate data for use in subsequent programs much as do the other FEA programs. The program options for MSC/Pal 2 are listed below.

1. PAL2. Generates the mathematical model.
2. STAT2. Performs the static analysis that includes nodal displacement and element force and stress results.
3. DYNA2. Performs the dynamic analysis: natural frequencies and mode shapes; transient response to time-varying forces, displacements, and accelerations; and response for frequency dependent forces, displacements, and accelerations.

Table 8.5 Elements Available with MSC/Pal 2

Beam type 0 - constant cross-section beam for mass only
Beam type 1 - constant cross-section beam element
Beam type 2 - variable cross-section rectangular beam element
Beam type 3 - circular tube beam
Curve beam - curved beam element of type 1, 2, 3 above
Damping element - three-dimensional viscous damping element
Link element - rigid link
Offset - connector element to define attach points
Quad plate - three-dimensional quadrilateral plate element
Spring - single-axis spring element
Stiffness - three-dimensional stiffness element
Triangular plate - three-dimensional triangular plate element

Elements used in MSC/cal

Bar
Perfect conductor
Planar
Plate
Axisymmetric
Quadrilateral
Triangular

4. VIEW2. Creates undeformed and deformed structural plots, animation, and contour plots from static and dynamic results.
5. XYPLOT2. Creates X-Y screen plots of dynamic response as well as static stress distribution displays.
6. ADCAP2. Performs the conversion to MSC/NASTRAN format, model data set expansion, and printing system equations.
7. PALPREP2. Functions as an interactive preprocessor for simple models.

The MSC/Pal 2 program comes with an excellent manual consisting of a user's section and a reference section. This manual goes into a greater discussion of finite element analysis than other manuals reviewed. An application manual is provided with many examples. These examples cover every model definition command and are very useful for the first time or novice user.

The program is software copy protected.

MSC/cal requires the same hardware as MSC/Pal 2. Models with up to 1000 nodes can be analyzed. Steady state nonlinear heat transfer associated with temperature dependent material properties and boundary conditions is performed in an iterative manner using the automated restart files. Transient nonlinear analysis requires a piecewise-linear representation of material properties and boundary conditions. This solution proceeds by applying the linear transient solution scheme over each selected time segment utilizing the average values of the boundary conditions and material properties for each time frame. The subsequent time segment then uses the temperature state of the last solution frame as its initial condition. Boundary condition options include temperature, flux, convection, discrete nodal power, and linearized radiation specifications. Material property data includes conductivity, density, specific heat, and volumetric internal heat generation. MSC/cal graphical postprocessing capabilities include isothermal color contour plots and X-Y transient plots of specified nodal temperatures versus time.

8.2.6 SAP86

SAP86 Release 3.3, by Number Cruncher Microsystems, Inc., is a finite element analysis program derived from the SAP IV program. SAP IV is a mainframe code developed originally at the Earthquake Engineering Research Center of the University of California. SAP86 is a structural analysis program only. Both two-dimensional and three-dimensional static linear and dynamic linear problems are treated. Hardware requirements are typical. SAP86 requires an IBM PC/XT/AT with one 5-1/4 floppy drive, a 10-20 Mb hard drive, an 8087 or 80287 math co-processor, 640K RAM, and MS-DOS (PC-DOS) 2.1 or higher.

SAP86 Version 3.3 has an overlay architecture and provides to the user the elements listed in Table 8.6. The maximum problem size is 2000 nodes with the simplest two-dimensional elements and minimum material differences, load sets, and geometric data sets.

The SAP86 program is controlled by a menu-driven shell that minimizes direct DOS interaction. The program is structured in modules similar to that of the other FEA programs. The SAP86 modules are listed below:

1. PreSAP. Prepares input data for the analysis module.
2. Analysis. Runs the static or dynamic analysis.
3. PlotSAP. Prepares the data generated in the analysis phase for graphical output.

SAP86 can produce plotting files for direct use in three graphics postprocessors. These are AutoCad, MicroCAD, and mTAB. The files generated by SAP86 are ASCII files that contain the nodal coordinates and element connectivity for plotting two-dimensional and three-dimensional finite element grids, deformed structures, and vibration mode shapes.

Table 8.6   Elements Available with SAP86

Three-dimensional truss

Three-dimensional beam

Two-dimensional elements
    membrane
    plane stress and plane strain
    axisymmetric
    quadrilateral or triangular

Three-dimensional solid
    8-node brick
    isotropic
    isoparametric

Thick shell and three-dimensional solid element
    variable number of nodes (8-21)
    orthotropic
    isoparametric or subparametric

Thin plate or thin shell element
    triangular or quadrilateral
    anisotropic

Boundary element

Pipe elements
    tangent element
    bend element

For novice and first time users, the manual will prove to be difficult to follow and thoroughly understand. It is advised that the SAP IV manual be studied to fully grasp the use of SAP86.

SAP86 is hardware copy protected.

8.2.7 LIBRA

LIBRA is a FEA package by InterCept Software. This model will run on an IBM PC/XT/AT or compatible with at least 512K RAM, two floppies or one floppy drive and a hard disk, DOS 2.0 or higher. In order to run the postprocessor in order to view graphically the output, a Hercules Graphics Board or compatible and a monochrome monitor is needed. This is a serious limitation. The model is capable of performing linear static analysis and thermal analysis (both steady state and transient) on one-, two-, and three-dimensional models.

There are two manuals available for the LIBRA software. One manual is for the structural analysis while the other manual is for the thermal analysis. Each manual is thorough in its coverage of the aspects of the LIBRA software. In addition, the on-screen help and instructions minimize the need for constant referring to the manual. The overall program is user-friendly and utilizes the module approach to build, analyze, and perform the postprocessing. Adequate sample problems are given to illustrate program usage. LIBRA does not have a CAD interface. LIBRA for the PC does interface with the Libra mainframe version.

The main modules of Libra are as follows:

Interactive Model Builder

The interactive model builder helps the user build a structural model in the form of a data file that will be used by the analysis module. Libra menus guide the user through a series of prompts that collect the data for the finite element analysis. This module is also used to modify previously created model files. The file output is in ASCII format and can be modified by any standard word processing program.

The problem solution menu is contained in the interactive model builder.

Table 8.7 Elements Available in LIBRA

Element	Structural	Thermal
Three-dimensional spring	*	
Two-dimensional spring	*	
One-dimensional beam		*
Three-dimensional beam	*	
Triangular (3 node)	*	*
Triangular (6 node)	*	*
Axisymmetric shell	*	*
Triangular shell (3 node)	*	*
Triangular shell (6 node)	*	*
Quadrilateral shell (4 node)	*	*
Quadrilateral shell (8 node)	*	*
Quadrilateral plane or axisymmetrical (4 node)	*	*
Quadrilateral plane or axisymmetrical (8 node)	*	*
Three-dimensional solid (8-node brick)	*	*
Three-dimensional solid (6-node wedge)	*	*
Convective boundary (1-8 nodes)		*
Radiation boundary (1-8 nodes)		*

Table 8.8   ANSYS-PC/Linear Elements

Two- and three-dimensional solids
Two- and three-dimensional shells
Two- and three-dimensional beams
Two- and three-dimensional spars
One-, two-, and three-dimensional spring
One-, two-, and three-dimensional mass
Three-dimensional stiffness or mass matrix

Post-Processor

The Libra postprocessor allows the user to view the original model and the analysis results graphically. The postprocessor is menu driven.

The program is designed to provide a comprehensive set of elements for both structural and thermal analysis. Table 8.7 lists the elements available in Libra.

Complete element descriptions are given in the manuals for both the structural and thermal elements.

LIBRA is not copy protected.

8.2.8   ANSYS

ANSYS-PC/Linear, ANSYS-PC/Thermal, and ANSYS-PC/OPT are subsets of the mainframe ANSYS finite element analysis program by Swanson Analysis Systems, Inc. These programs will run on an IBM PC/XT/AT or compatible with a math co-processor, 512K of memory, a 10 Mb to 20 Mb hard disk, and with DOS 2.1 or later. Two- and three-dimensional linear static and dynamic analysis may be performed as well as thermal analysis. ANSYS also has the option of an optimization program to further enhance this program as a design tool.

The manual is adequate for the more advanced FEA user and the user familiar with the mainframe version of ANSYS. All applicable commands are covered and

Table 8.9   ANSYS-PC/Thermal Elements

Radiation Link
Two- and three-dimensional conducting bars
Convection Link
Two- and three-dimensional conducting solids
Three-dimensional conducting shell
Three-dimensional lumped thermal mass

---

several examples are given to illustrate these commands. The concrete dam problem discussed previously is from the ANSYS manual.

The ANSYS-PC/Linear element library given in Table 8.8 contains thirteen elements from the ANSYS mainframe library. The ANSYS-PC/Thermal element library given in Table 8.9 contains eight ANSYS elements.

The element options include axisymmetric, plane stress, and plane strain.

ANSYS is not as user-friendly as the other packages reviewed. The premise is that the user is somewhat experienced with the ANSYS mainframe program. The programs feature a preprocessor and postprocessor and an unusually good graphics presentation of the output. ANSYS has an option that allows design optimization. The program uses approximations of the user defined objective, design variables, state variables, and appropriate constraints and tolerances. An automatic optimization loop capability is used to aid the analyst in making design modifications to the analysis model.

ANSYS is unlike other programs in one area. A user cannot purchase the programs. All programs are leased by the month by the user. At the end of the lease period, the program ceases to function. ANSYS-PC/Linear is hardware copy protected.

Appendix A

FEA Package	Static	Dynamic	Thermal	Fluid
PCTRAN Plus*	x	x	x	x
IMAGES3D	x	x		
MSC/Pal 2	x	x		
MSC/cal			x	
COSMOS/M*	x	x	x	
LIBRA	x		x	
SAP86	x	x		
SUPERSAP*	x	x	x	
ANSYS-PC/Linear*	x	x		
ANSYS-PC/Thermal*			x	
ANSYS-PC/OPT				

*Nonlinear capabilities

PCTRAN Plus

Brooks Scientific, Inc., 55 Wheeler Street, Cambridge, MA 02138, (617) 491-9220
Recently purchased by Swenson Analysis Systems.

Base Module:	
Preprocessors and postprocessors, and stress analysis	$1295
Add-on modules:	
Vibration	695
Heat transfer	695

Fluid flow	1095
Nonlinear	1095
Three-dimensional solid modeling (RoboSOLID)	1095
Solids–FEA interface	295
CAD–FEA interface	295
PCTRAN Plus documentation	75
Learning Finite Element Analysis (Book)	35
Demonstration diskettes	free

PCTRAN

Base Module:	
Preprocessing and postprocessing, and stress analysis	$495
Add-on modules:	
Vibration	250
Heat transfer	250

IMAGES3D
V1.4A

Celestial Software, Inc., 125 University Avenue, Berkeley, CA 94710
(415) 420-0300

Three-dimensional static analysis	$ 595
Three-dimensional static and dynamic analysis	995
Three-dimensional static, dynamic, and seismic analysis	1390
Three-dimensional static, dynamic, seismic, and any one (1) translator (to Ansys, AutoCAD, etc.)	1995
AISC code check	195
User manual	45
Technical manual	45
6-month user support	free

COSMOS/M
V1.51

Structural Research and Analysis Corporation, 1661 Lincoln Blvd., Suite 100,
(213) 452-2158

COSMOS/M	$1995
Heat transfer option	1500
Nonlinear option	1500

SAP86

Number Cruncher Microsystems, Inc., 1455 Hayes Street, San Francisco, CA 94117, (415) 922-9635

SAP86 Release 3	$1495

SUPERSAP

Algor Engineering, Inc., Essex House, Pittsburgh, PA 15206, (412) 661-2100

SUPERSAP	$ 995
Heat transfer option	795
SUPERSAP and heat transfer option	1395

LIBRA

InterCept Software, A Div. of S. Levy, Inc., 3425 South Bascom Avenue, Campbell, CA 95008

Structural analysis	$995
Thermal analysis	$795

MSC/Pal 2 VERSION 2

The MacNeal-Schwendler Corporation, 815 Colorado Boulevard, Los Angeles, CA 90041-1777, (213) 258-9111

MSC/Pal 2 (structural)	$1995
MSC/cal (thermal)	1495
MSC/AutoFEM (AutoCAD interface)	295
MSC/CASE	895
MSC/mate (matrix equation solver)	495

ANSYS

Swanson Analysis Systems, Inc., Johnson Rd. P.O. Box 65, Houston, PA

ANSYS-PC/Linear	Lease @ $300/mo.
ANSYS-PC/Thermal	Lease @ $300/mo.
ANSYS-PC/OPT	Lease @ $ 75/mo.

Appendix B
Pre- and Postprocessors

mTAB

Structural Analysis, Inc., 1701 Directors Blvd., Suite 360, Austin, TX 78744
(512) 444-0555

mTAB (PRE-)	$1295
mTAB (POST-)	595

Translators for SAP86, SAP IV, SAP V, MSC/NASTRAN, MSC/Pal 2, ANSYS, and STARDYNE.

Bibliography

1. Clough, R. W., "The Finite Element Method in Plane Stress Analysis," *Proceedings of 2nd ASCE Conference on Electronic Computation*, Pittsburgh, PA., September 8 and 9, 1960.
2. Courant, R., "Variational Methods for the Solutions of Problems of Equilibrium and Vibrations," *Bull. Am. Math. Soc.*, Vol. 49, 1943.
3. Morse, P. M., and Feshback, H., *Methods of Theoretical Physics*, McGraw-Hill Book Co., New York, 1953.
4. Greenstadt, J., "On the Reduction of Continuous Problems to Discrete Form," *IBM L. Res. Dev.*, Vol. 3, 1959.
5. Turner, M. J., Clough, R. W., Martin, H. C., and Topp, L. C., "Stiffness and Deflection Analysis of Complex Structures," *J. Aeronautical Sci.*, Vol. 23, No. 9, 1956.
6. Melosh, R. J., "Basis for the Derivation of Matrices for the Direct Stiffness Method," *AIAA J.*, Vol. 1, 1963.
7. Jones, R. E., "A Generalization of the Direct-Stiffness Method of Structural Analysis," *AIAA J.*, Vol. 2, 1964.
8. McLay, R. W., "Completeness and Convergence Properties of Finite Element Displacement Functions—A General Treatment," *AIAA 5th Aerospace Science Meeting*, New York, 1967.
9. Johnson, M. W., and McLay, R. W., "Convergence of the Finite Element Method in the Theory of Elasticity," *J. Appl. Mech.*, Vol. 35, No. 2, June, 1968.
10. Tong, P., and Pian, T. H. H., "The Convergence of the Finite Element Method in Solving Linear Elastic Problems," *Int. J. Solids Struct.*, Vol. 3, 1967.

11. Zienkiewicz, O. C., and Cheung, Y. K., "Finite Elements in the Solution of Field Problems," *Engineer*, Vol. 220, 1965.
12. Girault, V., and Raviart, P. A., "Finite Element Approximation of the Navier-Stokes Equations," *Lecture Notes in Mathematics*, Springer-Verlag, New York, 1979.
13. Teman, R., *Theoretical Studies on the Finite Element Method Applied To The Navier-Stokes Equations*, North Holland, Amsterdam, 1977.
14. Fortin, M., "Old and New Finite Elements for Incopressible Flow," *Int. J. Numerical Methods Fluids*, 1981.
15. Griffith, D. F., "An Approximating Divergence-free 9-node Velocity Element for Incompressible Flows," *Int. J. Numerical Methods Fluids*
16. Thomasset, F., *Implementation of Finite Element Methods for Navier-Stokes Equations*, Springer-Verlag, New York, 1981.
17. Heywood, J. G., and Rannacher, R., "Finite Element Approximations of the Non-Stationary Navier-Stokes Problem, Part I: Regularity of Solutions and Second Order Spatial Discretisations," *SIAM J. Num. Analysis*, 1982.
18. Cullen, M. J. P., "Analysis and Experiments with some Low Order Finite Element Schemes for the Navier-Stokes Equations," *J. Computational Phys.*, 1982.
19. Chung, T. J., *Finite Element Analysis in Fluid Dynamics*, McGraw-Hill, New York, 1978.
20. Gallagher, R. H., *Finite Element Analysis Fundamentals*, Prentice-Hall, Inc., Englewood Cliffs, New Jersey.
21. Segerlind, Larry J., *Applied Finite Element Analysis*, John Wiley & Sons, New York, 1976.
22. Bear, F. P., and Johnson, Jr., E. R., *Vector Mechanics for Engineers, Statics and Dynamics*, McGraw-Hill Book Co., New York, 1962.

Index